Electronic Business in Developing Countries:

Opportunities and Challenges

Sherif Kamel
The American University in Cairo, Egypt

Acquisitions Editor:	Renée Davies
Development Editor:	Kristin Roth
Senior Managing Editor:	Amanda Appicello
Managing Editor:	Jennifer Neidig
Copy Editor:	Bernie Kieklak
Typesetter:	Cindy Consonery
Cover Design:	Lisa Tosheff
Printed at:	Yurchak Printing Inc.

Published in the United States of America by
Idea Group Publishing (an imprint of Idea Group Inc.)
701 E. Chocolate Avenue
Hershey PA 17033
Tel: 717-533-8845
Fax: 717-533-8661
E-mail: cust@idea-group.com
Web site: http://www.idea-group.com

and in the United Kingdom by
Idea Group Publishing (an imprint of Idea Group Inc.)
3 Henrietta Street
Covent Garden
London WC2E 8LU
Tel: 44 20 7240 0856
Fax: 44 20 7379 3313
Web site: http://www.eurospan.co.uk

Library of Congress Cataloging-in-Publication Data

Electronic business in developing countries : opportunities and challenges / Sherif Kamel, editor.
p. cm.
Summary: "This books tackles issues of e-business with a vision to the future on how to bridge these gaps and close down the barriers between the different corners of the world"--Provided by publisher.
Includes bibliographical references and index.
ISBN 1-59140-354-5 (hardcover) -- ISBN 1-59140-355-3 (soft cover) -- ISBN 1-59140-356-1 (ebook)
1. Electronic commerce--Developing countries. 2. Electronic commerce--Developing countries--Case studies. I. Kamel, Sherif.
HF5548.325.D44E55 2006
381'.142'091724--dc22

2005004531

British Cataloguing in Publication Data
A Cataloguing in Publication record for this book is available from the British Library.

All work contributed to this book is new, previously-unpublished material. Each chapter is assigned to at least three expert reviewers and is subject to a blind, peer review by these reviewers. The views expressed in this book are those of the authors, but not necessarily of the publisher.

Electronic Business in Developing Countries: Opportunities and Challenges

Table of Contents

Preface

During the last two decades, information and communication technology have empowered societies in different and diversified ways with their continuously emerging tools and techniques. Implications of these technologies are affecting different sectors including; education, health, governance, trade, tourism, and entertainment amongst other sectors. Moreover, social and economic relationships are dramatically changing giving societies, businesses, governments and other economic establishments the platform to increase and diversify productivity as well as contribute to business and economic development and growth. Information and communication technology have led over the last decades the transformation process at the individual, organizational and societal level. However, for information and communication technology to succeed and continuously remain effective in the future, it must change and adapt to global and local environmental elements.

The platforms, vehicles and tools of information and communication technologies are numerous and are changing all the time. Newly emerging tools are being introduced and gradually the collectivity of such tools are having implications on most sectors of the economy and for many of the different aspects of our daily life such as the way we do business, conduct research, communicate with people, get entertained and study. More is envisioned in the near future where state-of-the-art information and communication technology is perceived to dominate in the 21st century. One of the growing platforms that were enabled by the Internet and the World Wide Web evolution has been the establishment of the marketspace through the new momentum of doing business in the digital economy or "electronic business."

Understanding how much electronic business and its different business models will have implications on development and growth and how will it introduce change to the society at large is an issue that is extensively under study especially that some of the research reports indicate that by the end of 2005 more than 25% of global trading will be conducted through the digital economy. This is increasingly becoming feasible with the growth in the volume of Internet users that has reached by September 2004 around 750 million users representing the Internet community worldwide. With these figures in place, it is important to address the issue of the digital divide and how wide is such a problem growing and what are the implications associated with the digital divide (mainly negative) that are contributing to widening the gap with the developing world. At the end of 2003, the developing world accounted for 32% of the community of Internet users. However, this percentage is expected to increase in the years to come. Moreover, even if reports and research activities indicate that electronic readiness in the developing world is lower than in the developed countries, with the massive investments in information and communication technology, the developing countries are gradually becoming increasingly integrated into the digital economy.

Policies, regulations, liberalization of services, government-private sector partnerships, "infostructure development" and infrastructure deployment are among the important steps information and communication technology adopters in developing nations are taking seriously in their quest to close the digital gap with the developed world and becoming more integrated into the information society. However, it is important to note that for some nations in the developing world, classical problems still prevail including: the need for more economic development, low literacy rates, cultural resistance, low income levels, lack of comprehensive and structured information and communication technology infrastructure and the need for more comprehensive and effective awareness campaigns on the usefulness and added values and competitive advantages information and communication technology can bring to the community.

The global economy is becoming an information and communication technology-based economy. Distance and time barriers are fading away and the world is becoming an integrated community of buyers and sellers that interact within an internetworked economy that includes a series of internetworked enterprises. Such global market is witnessing a phase of transition from a tangible-oriented marketplace to an information-based marketspace. This transition promises to lower transaction costs, save time, and remove distance barriers. All these elements are contributing to the general digital sense of the world as being one global digital village where people can communicate anytime and anywhere. The structure of the organization is being re-engineered to fit in the new global perspective of the marketplace.

Throughout the developing world, many efforts have been taken to attempt to minimize the digital divide within the developing world and also between the developed and the developing nations. This included the formulation of national information and communication technology plans and programs as well as medium- and long-term strategies that can help keep pace with the developments taking place in the developed world and at the same transfer the knowledge and technology infrastructure innovations to different corners of the world. These policies include: awareness raising, infrastructure development and deployment, communication deregulation and liberalization, investment in human resources through education and training, and the partnership and involvement of the private sector with the government in the development process. Many such developments and directions are geared by the continuous diffusion of information and communication technology throughout the world, coupled with the massive introduction of electronic business as a new model and an innovation in conducting business that is affecting many aspects of life including different products and services and creating numerous opportunities for collaboration and growth unlimited by time or distance or varying cultures.

Electronic business represents a new evolution of trading, offering a borderless global marketspace. For developing countries, electronic business represents a tremendous challenge, but also a great opportunity for growth and development. Business on the Internet is booming and many retailers, brokers and suppliers around the world are actually changing habits in the way they do business by introducing new methods and creating new opportunities through the marketspace. Electronic business is creating a virtual market with innovative virtual businesses and trading communities where firms are outsourcing many of their services and operations, distribution services are becoming online, and buyers, sellers and intermediaries frequently called infomediaries in fields such as computing and the automobile industry are actively integrating their supply chains, thus creating interactive value-added services to businesses and consumers. There is no doubt that electronic business and the Internet represent the opportunity for developing nations to leap forward to more economic development and growth, where the creation of added value will be driven in the 21st century by information, knowledge, and the adoption, diffusion and adaptation of information and communication technology.

Electronic business is estimated to have a value in terms of goods and services traded online around the world to being for both business2business and business2consumer 6.8 trillion US dollars in December 2004 with only around 1% to be estimated for developing nations. However, development and growth indicate that the growth of electronic business in developing countries and the ability of enterprises to benefit from the efficiency gains in their production and distribution processes will be largely dependent on their adoption of business2business practices and not business2consumer. It is through such an electronic business mode that most of the development will take place because they translate into improved competitiveness for enterprises and higher levels of productivity, and hence incomes for the economy as a whole. However, the role to be played by electronic business models such as business2consumer, government2business as well as platforms such as electronic learning and electronic government will also contribute to electronic business on a more macro level.

The changes brought about by the Internet in the global economy will have implications for the competitiveness of the enterprises of developing countries. While some of the factors affecting the evolution of global electronic business including information and communication technology do not respond directly to the national policies of developing countries, it is important to note that governments, business players and other stakeholders will have to play a role in the formulation and implementation of national strategies to ensure that the new opportunities for creating, transforming, applying and exchanging information and value are used to improve the productivity of developing economies and their enterprises.

The effects of electronic business on the organization of the global economy continue to spread and reshape the context in which enterprises, including those from developing countries, must compete in local, regional and international markets. These developments and changes relate to the way business is being done, the business processes and the effective response to customer needs and requirements. Today, the leading region in electronic business in developing nations is the Asia-Pacific region with 50 million new Internet users a year which is faster than any other region in the world including regions from the developed world in terms of Internet community development and growth. It is expected that more regions in the developing world will be following the same direction with the massive investments taking place in that domain supported by government plans and private-sector support.

Born global, electronic business encompasses a wide spectrum of activities, some well established, most of them very new, and all are blended through a hybrid model of traditional and unconventional components. Driven by the Internet evolution and the World Wide Web, electronic business is dramatically expanding and undergoing radical changes engendering a wide array of innovative businesses, markets and trading communities creating new functions and new revenue streams. Therefore, developing nations around the world should always focus on the strategies that can yield positive payoffs in terms of investment in information and communication technology. The focus should also be on the policies that will lead to positive economic and productivity return in the long term. In that respect, when it comes to electronic strategies and the diffusion of electronic business with its different components such as electronic commerce, electronic learning, electronic government, electronic payment and electronic banking amongst other components of the digital economy, policies followed by the governments of developing nations should attempt to provide an environment in which information and communication technology should realize their full potential. This includes the problems and challenges of awareness, infrastructure development, information access, regulations, human resources and capacities development, local content creation depending on the languages used and the cultures prevailing in the local environment.

In the long term, Internet and electronic business will generate social and economic returns to different countries irrespective of whether they are developed or developing. But those that will invest in adaptable information and communication technology that can provide added values to the community will be felt across the different levels in the society. It is perceived that a well thought business model and the adaptation to the local conditions and the needs of the community represent some of the initial steps that could be taken for the development and implementation of a successful electronic business environment that could have effective implications on the economic and societal development and growth in the context of developing nations.

This book intends to provide a comprehensive survey of developing countries in terms of its information and communication technology platforms that enable the introduction and diffusion of electronic business among its community. The book will address the issues related to electronic business in a number of developing countries including vision, strategies, policies, government and private sector roles, plans, business models, achievements, success and failure cases, and challenges and opportunities amongst other issues. There are many potentials that could be realized at the business and socio-economic development front and this book will try to demonstrate that and present the lessons learned to set supporting guidelines for future implementations combining theoretical foundations with real practical cases.

Organization of the Book

This book covers through its 20 different chapters a variety of issues and concepts that relate to the changing nature of doing business online. A brief description of each of the chapters follows.

Chapter I titled "Electronic Business in Developing Countries: The Digitalization of Bad Practices?" addresses the Internet and other advanced digital information technologies and their effective role between developed and developing nations and the discrepancy between the two environments in terms of infrastructure readiness. The chapter discusses the effect of the Internet in general and electronic business in particular with respect to the digital divide through the development of a framework that defines four types of information technologies that shows why the Internet does not provide the same development thrust to all the countries that use it and identifies the differences prevailing. Moreover, the chapter provides an understanding of the difference between information systems that only process information and those that transform or distribute physical goods or control work and how it is frequently believed that the globalization of information networks such as the Internet does not reduce but increases the digital divide — an issue that has raised so many conflicting arguments since the establishment and diffusion of the Internet and the World Wide Web.

Chapter II titled "Role of Culture in Electronic Business Diffusion in Developing Countries" opens the discussion on the different cultural factors and their varying implications on the diffusion of electronic business while studying the different challenges faced and opportunities presenting themselves. In developing countries, culture is considered invaluable and adaptation to the local needs and conditions represents a key element in the successful implementation of electronic business. A number of cases are demonstrated to validate such argument, to assess the levels of technological usage as well as to provide a list of lessons learned to work as a model for possible and similar future implementations in similar contexts and environments of development.

Chapter III titled "International E-Commerce: Language, Cultural, Legal, and Infrastructure Issues, Challenges and Solutions" addresses the issues of language, cultural, legal, and infrastructure as barriers and deterrents to global electronic business through the formulation of a framework with an objective to help organizations meet the challenge of providing online customers with a premium quality experience, regardless of location, language, business practices, and culture, while addressing regulatory issues. The chapter covers a number of the primary issues that represent the major challenges and barriers for proper electronic commerce diffusion around the world and how they could be transformed into platforms for opportunities for these countries to capitalize upon and benefit from the advantages presented.

Chapter IV titled "E-Commerce Infrastructure and Economic Impacts in Developing Countries: Case of India" demonstrates how discrepancies and huge disparities are prevailing in the adoption of electronic commerce across developed and developing countries with examples from the Unites States, the Asia-Pacific region and Africa. The focus of the chapter is the debut of electronic commerce in India in 1999 with the case of electronic selling of merchandise such as musical CDs and books standing currently at a volume of business2consumer estimated to be about 460 million US dollars. This chapter attempts to address the issue of electronic commerce research in developing countries with a focus on: (a) what are the infrastructure enablers of electronic business and their status in a developing country like India?; (b) what is the relationship between electronic business and socioeconomic variables that affect the development and sustainability of electronic business in a developing country like India? The chapter suggests a causal model that highlights factors affecting electronic commerce adoption while exploring issues such as infrastructure enablers, socioeconomic factors relevant in developing countries and the impact of electronic commerce on the nature of markets.

Chapter V titled "E-Marketplace Adoption Success Factors: Challenges and Opportunities for a Small Developing Country" addresses the use of the Internet and electronic commerce in the late nineties and how it caused the disappearance of intermediaries between buyer and seller and with the advances in the development of electronic commerce technologies, new types of intermediaries were established offering new value-adding services. The new way of doing business is described as electronic marketplaces representing a new business model using innovative and emerging information and communication technology. Doing business in the E-Marketplace enables sellers to enter new markets, to find new buyers, and to increase sales. Electronic marketplace services support the exchange of large amounts of data regarding supply and demand between buyers and sellers, and the implementation of the business transactions. This chapter explores the opportunities and challenges facing electronic marketplaces and the need to develop an infrastructure that can enable the adoption, diffusion and adaptation of electronic business.

Chapter VI titled "E-Commerce Activity, Opportunities, and Strategies in Latin America" analyzes international differences in electronic business activities and strategies with a focus on the region of Latin America and between Latin and non-Latin markets both from a macro and micro perspectives. The chapter focuses on three cases of electronic business evolution comparing cases in Brazil, Chile, and Mexico with an identification of the business models and strategies that are being implemented with an outlook for the future and how electronic business can develop and grow and benefit the economies of the developing world.

Chapter VII titled "Assessing Job Seekers' Acceptance of Online Recruitment in Egypt" defines the implications of the development in information and communication technology on Internet adoption growth rate with a focus on Web technology and addressing the case of Egypt. The chapter focuses on the evolution of the online recruitment industry that started in Egypt in 1998 and then spread across the community with many businesses addressing this profession. The model proved to be an efficient way to provide better interaction and information exchange between job seekers and employers. The chapter presents a model using the technology acceptance model discussing the main constructs affecting job seekers in embracing emerging information and communication technology. Moreover, the chapter includes appropriate business strategies stemming from the cultural, social and economic variables of a country like Egypt, a developing nation that could be set in order to boost the diffusion and efficiency of online recruitment business.

Chapter VIII titled "Evolution of Telecommunications and Mobile Communications in India: A Synthesis in the Transition from Electronic to Mobile Business" addresses the electronic business evolution and the efforts played by many countries and how they transformed their operations to cater for the platform needs of the digital economy. The chapter also addresses the migration to mobile business and the growing potentials made available for developing nations. The chapter focuses on the case of India with a demonstration of the transition of its domestic economy into the digital economy and the possibilities and regulations of permitting foreign direct investment, and the deregulation of the telecommunications industry opening up to private competition. There is also the coverage of the role of the government. The chapter aims at capturing the transition of India into the digital world, by examining the key influencers such as the evolution of telecommunications and mobile communications and the effect of information technology on government initiatives.

Chapter IX titled "Critical Success Factors for E-Brokerage: An Exploratory Study in the Brazilian Market" investigates the critical factors for the success of stock brokerage processes via the World Wide Web, using financial portals on the Brazilian Internet, from the perspective of the investor. The chapter presents the framework of the online stock trading system to be able to compare the traditional form of stock brokerage with the one introduced and made possible by the Internet. Issues such as intermediation and disintermediation as part of the overall process are also covered in the chapter.

Chapter X titled "E-Readiness and Successful E-Commerce Diffusion in Developing Countries: Results from a Cluster Analysis" addresses the continuing growth of electronic commerce and the prospects in the years to come and the strategies formulated that can help diffuse electronic business. The chapter addresses the issue of the digital divide and the gap between developed and developing nations mainly due to lack of electronic readiness. Therefore, the chapter attempts to explore the relationship between electronic readiness and successful electronic commerce diffusion. The objectives of the chapter are to: (a) present a conceptual framework of e-readiness and e-commerce success; (b) identify the constructs and variables to operationalize the framework; (c) report on the e-readiness and e-commerce success assessments of a sample of business organizations from South Africa; (d) explore the effect of e-readiness on the successful diffusion of e-commerce; and, (e) identify the theoretical and practical implications and lessons for businesses in other developing countries.

Chapter XI titled "An Overview of E-Commerce Security and Critical Issues for Developing Countries" addresses electronic commerce issues such as infrastructure, lack of awareness, network payment and secure transaction services as challenges to developing countries. The chapter focuses on the need for a secure infrastructure which makes possible the electronic exchange of financial transactions as a necessary prerequisite.

Chapter XII titled "Venturing the Unexplored: E-Readiness Assessment of Small and Medium Enterprises in Egypt" focuses on the issue of electronic readiness. Some of the issues covered are combinations of indicators such as e-connectivity, human capital, business climate, leadership and others. Few studies have attempted to evaluate e-readiness from a micro perspective. This chapter attempts to adopt a micro approach to assess e-readiness of SMEs in an Arab country. The objective of the research is to assess the e-readiness of SMEs in the textile, specifically garment, industry in Egypt, and in particular, evaluate their preparedness for electronic commerce. The research is exploratory in nature, and represents a first step towards more extensive research to assess the preparedness of different sectors in Egypt for embracing and internalizing ICTs.

Chapter XIII titled "Offshore Outsourcing: An E-Commerce Reality (Opportunity for Developing Countries)" focuses on investments and promoting foreign direct investment in developing countries. This chapter discusses the solutions enabled through information technology through offshore outsourcing. Today, the exchange of information in real time has made possible for companies to widen their horizons and look for independent suppliers in different nations. This chapter focuses on India as a country which has managed to exploit the opportunities enabled via offshore outsourcing.

Chapter XIV titled "Adoption and Implementation Obstacles of E-Banking Services: An Empirical Investigation of the Omani Banking Industry" explores the potential impeding factors that could inhibit the wide adoption and use of electronic commerce applications in the Omani banking industry. Top management perspectives and attitudes toward electronic commerce adoption and use of internal information systems/information technology competencies and success are some of the focal points in the chapter. Other factors such as power relationships, such as conflict between managers that arise during the process of information systems adoption, and lack of investment in electronic commerce applications are also covered. The findings are vital for electronic commerce strategies development and implementation in developing nations.

Chapter XV titled "Open Sourcing E-Learning for Developing Countries" addresses one of the critical success factors for developing countries in advancing their intellectual capacity. The Internet offers countries information access to the entire world, offering many educational resources that can effectively develop a global knowledge base. This chapter realizes that much of the Internet content is oriented towards the needs and interests of developed countries. Therefore, the chapter proposes an approach for developing countries to pool efforts to create a digital commons of electronic learning resources that are appropriate and relevant to their specific needs.

Chapter XVI titled "Strategic and Operational Values of E-Commerce Investments in Jordanian SMEs" aims to articulate lessons learned from a quantitative study of Internet technology use in Jordan. The chapter addresses the issues related to electronic commerce as a new way of doing business for SMEs. However, it also focuses on the challenges faced, including the technological infrastructure, legislation and organizational infrastructures ready for the digital economy.

Chapter XVII titled "E-Commerce Infrastructure in Developing Countries" and how it caters for the potentials of developing nations is the focus of this chapter. The preparation and deployment of the building blocks is invaluable for the success of electronic commerce diffusion and adoption. Many other elements are important such as the financial and legal framework. This chapter highlights the technologies required to establish electronic commerce and also provides an overview of the present situation of infrastructure development for electronic commerce in developing countries.

Chapter XVIII titled "An E-Commerce Longitudinal Case Study from Ukraine" identifies key value-added e-business applications and focuses on success factors that clearly support small enterprise goals. Performance factors are presented that show the impact of electronic business applications on the organization's bottom line. The chapter explores some lessons learned from a case in the Ukraine with investigation of a number of issues including telecommunications, energy, hardware, software, and the availability of technical skill. The chapter covers a number of electronic business issues such as: (a) difficulty of the user interface, (b) lack of a planning process for e-business applications, (c) development and testing concerns, (d) finding the Web site and a lack of branding, and (e) lack of evidence to support a formal budgeting process.

Chapter XIX titled "E-Business in India: Early Evidence from Indian Manufacturing Industry" explains how the manufacturing and especially automobile industry has changed from traditional brick and mortar business to click and brick e-business within the digital economy.

Chapter XX titled "Challenges and Opportunities for Information Brokers in Brazil: A Study of Informational Needs of Southern-Brazilian Enterprises When Expanding Their Businesses" covers a study that was developed as a part of a program of research into information intermediation via the Internet, focusing on the informational needs of Southern-Brazilian enterprises when expanding their businesses. The research was conducted with the support of SEBRAE-RS. This chapter presents information needs and identifies the challenges and opportunities for information brokers in Brazil.

Acknowledgments

This book is the compilation of different invaluable efforts of so many parties including but not limited to the authors, the reviewers, the editor and the publisher. I would like to seize this opportunity and acknowledge the help of all those who were involved in the different stages of the production of this book, without whose support and assistance this effort could not have been satisfactorily completed on time and with such high quality. The different stages and preparation elements a book of such magnitude goes through require a tremendous and continuous cooperation and assistance by everyone involved for a long duration of time that extends to over a year. Therefore, I would like to express my sincere gratitude to all the contributors of this book who provided through their submitted manuscripts a comprehensive overview of electronic business in developing countries that was coupled with an in-depth analysis of a number of selected cases and projects. This is an important topic that is interesting for researchers and academics worldwide and the authors' contributions through the different chapters will undoubtedly enrich the literature in that domain. I would like to specifically acknowledge the authors for sharing their valuable experiences and accumulated knowledge. Moreover, I would like to acknowledge those selected authors who in addition to contributing to chapters to the book have contributed as reviewers of manuscripts and played an important role in a rigorous double-blind review process. Thanks go to all those who provided constructive and comprehensive reviews throughout the different stages of the review process. In closing, I wish to thank my wife and children for their love, support, and encouragement throughout the different stages of producing this book.

Sherif Kamel, PhD

Cairo, Egypt

May 2005

Chapter I

Electronic Business in Developing Countries: The Digitalization of Bad Practices?

Carlos Ferran, Penn State University Great Valley, USA

Ricardo Salim, Cautus Networks Corporation, USA

Abstract

This chapter uses information theory to study the effect of the Internet and e-business over the digital divide. It develops a framework that defines four types of information technologies and Internets based on four dimensions of information: physical, syntactic, semantic, and pragmatic. Technologies like e-mail that only make use of the first three hardly interact with the industrial infrastructure and superstructure. Nonetheless, technologies like e-commerce that require the pragmatic dimension, which is related to matter and energy (products and services) need such structures. For example, e-commerce requires a trustworthy transportation and payment infrastructure. Unfortunately, developing countries are lacking them. This differential capacity to use the pragmatic dimension of AdvIT/IS (pragmatic fragility) explains why developing countries are not able to skip industrialization and jump into the information era. This pragmatic fragility increases the existing digital divide since implementing e-business in developing countries ends up being the digitalization of bad business practices.

Introduction

Many believe that the Internet and other advanced digital information technologies (AdvIT/IS) are going to be the great equalizer between developed and developing countries. Unfortunately, that seems to be far from truth. The Internet and other AdvIT/IS provide less benefit to developing countries than to developed ones. This holds true even when network access is similar in both countries. We observe that the technological infrastructure is easy to put in place. However, the sole importing of AdvIT/IS is not enough to close that breach. A set of infrastructural and superstructural deficiencies hinder the successful accomplishment of the goals that are normally achieved in developed countries when using the same technologies.

In this chapter we discuss the effect of the Internet in general and electronic business (e-business) in particular over the digital divide (between developed and developing countries). We draw upon information theory to develop a framework that defines four types of information technologies and Internets based on four dimensions of information: (a) physical: the physical support of information, (b) syntactic: the formal aspect of information, (c) semantic: the meaning or relationship between two informational items, and (d) pragmatic: the relationship between information and action (energy or matter exchanges). A comprehensive discussion of the dimensions of information technology is included in the section "Using Information Theory to Classify the Internet."

This framework improves our understanding of why the Internet does not provide the same development thrust to all the countries that use the Internet. We find that the digital divide occurs specifically in the pragmatic dimension of information. Successful implementations of AdvIT/IS in developing countries occur only in projects that require just the first three dimensions of information. Implementations of communication networks and databases that already have a proper physical, syntactic, and semantic definition are usually successful in developed and developing countries. The difference in success comes in the pragmatic dimension of such systems, in other words, in the interaction of AdvIT/IS with the existing (or non-existing) infrastructure and superstructure. We refer to the differential capacity to make use of the pragmatic dimension of AdvIT/IS as "pragmatic fragility" and discuss it depth in the "Pragmatic Fragility" subsection of this chapter.

Fragility in the Pragmatic dimension of AdvIT/IS projects causes the digitalization of bad practices. Such practices are not easily changed, particularly by individual companies, because they are rooted in the country-wide infrastructure and superstructure. When individual institutions implement advanced information technologies they are forced to adapt them to those bad practices and, therefore, they end up computerizing bad business practices. This discussion is made in depth in the "Digitalization of Bad Practices as a Result of Pragmatic Fragility" section of this chapter.

As a result of the above, implementing e-business in developing countries is conducive to the digitalization of bad practices. Therefore, e-business and AdvIT/IS implementations not only reflect but increase the digital divide (as stated earlier) as they computerize bad business practices which in turn consolidate and empower even more of those bad business practices. This is further discussed in the section "Digitalization of Bad Practices as a Result of Pragmatic Fragility."

In a nutshell, e-business implementations generally reflect the digital divide. Successful implementations of AdvIT/IS in developing countries occur in projects that encompass just the first three dimensions of information. Fragility in the Pragmatic dimension of AdvIT/IS projects causes the digitalization of bad practices. Therefore, implementing e-business in developing countries tends to force the digitalization of bad practices.

At the end of this chapter, the reader will understand the difference between information systems that only process information and those that are also connected to systems that process matter and energy (i.e., transform or distribute physical goods or control work). This knowledge will help managers understand why a given AdvIT/IS that works under certain conditions, like those present in developed countries, does not work under different conditions, as those characteristic of developing countries. Furthermore, they will be able to distinguish those parts — or dimensions — of the system that are more sensitive to those conditions and, therefore, focus their efforts on them when making investment decisions. Moreover, the reader will be able to use this knowledge to understand why the sole globalization of information networks like the Internet does not reduce, but actually increases the digital divide.

Background

The E-Business Divide

E-Business

It is common to use the terms e-business and e-commerce as synonyms. However, there are differences between the two: e-business is broader than e-commerce. E-commerce is the exchange, procurement, and distribution of products, services, and/or payments between two or more economic entities via computers or other electronic means (Pitre, 2000). This definition is consistent with the first definition of e-commerce given by Kalakota and Whinston (1997). However, it excludes intra-organizational applications that do not interface with external entities. In contrast, e-business includes the strategies, tactics, practices, activities, and methodologies that companies apply to use information technologies to improve their business practices (Abu-Musa, 2004; Pinero, 2001). E-business practices are re-engineered internal and external business processes that take advantage of AdvIT/IS (Pinero, 2001).

E-business also refers to automating business practices using AdvIT/IS. This automation improves business practices by reducing the time it takes people to carry them out and by reducing the errors that are associated with human intervention and discretion. E-business does not imply e-commerce. However, e-commerce does imply e-business.

Companies may use IT/IS only to improve their internal processes, but for them to interact electronically with other parties, they need to first have e-business capabilities in place.

While many believe that e-business is just technology, technology is only a small part of what is involved. According to Pitre (2000), 70% of e-business is the adoption of new models and corporate culture and only 30% technology. Successful e-business implementations requires a holistic approach that goes far beyond simple technology and a process that is different depending on the maturity of the company (Ash & Burn, 2003). E-business poses new and unique challenges at all levels of the organization (Kohli, Sherer, & Baron, 2003).

Developing Countries and the Digital Divide

Historical Background

Agricultural, industrial, and information revolutions are historic productivity landmarks. Tribes and societies can be classified according to whether or not they have reached each of those milestones. We still find a few tribes that never mastered agriculture and tribes dedicated merely to hunting and gathering. Some of these tribes are found deep in the Amazon. Their productivity barely covers the basic alimentary needs of the individual and those of their young children. On the other hand, the productivity of agricultural tribes covers the needs of more children, for a longer period of time, and that of a few relatives that cannot work due to old age or physical impediments. Furthermore, in some cases, it even extends to cover the necessities of another class of people or caste that, instead of doing physical labor in the field, does intellectual work. Intellectual work (observations, research, planning, and leadership) brings about higher levels of overall productivity.

Industrialized societies increased their overall productivity by several orders of magnitude in comparison to agricultural ones (Slesser, 1978). Societies that are now undergoing the information revolution are advancing at an even higher pace. Unfortunately, with those great leaps accomplished by a few societies, the split between them and those many others that have not yet accomplished the prior stages is becoming larger and larger. A breach that is not only limited to productivity but also to the forms of social organization (Beniger, 1986, 1990).

Industrial Infrastructure and Superstructure

Nations that developed their industry also developed an infrastructure and superstructure that could handle their speed and volume. Their infrastructure encompassed the proliferation of large and novel production centers (industrial zones with highly mechanized factories powered by combustion engines or remotely generated electricity) and efficient distribution systems (extended road networks, extensive railway systems, far-reaching power lines, comprehensive aerial and maritime routes, telegraph, telephone, telex). Their superstructure included innovative and specialized legal systems (laws that properly regulated industry and commerce), judicial systems (courts, prosecutors, police, and other institutions capable of enforcing the new laws), financial systems and public policy (banks oriented to support small and large industrial projects, incentives for entrepreneurship), entrepreneurial R&D and education systems (development of specialized management techniques, technical and higher education oriented towards the immediate and long-term needs of the industry).

Competitive industry was not able to flourish in nations that did not develop such infrastructure and superstructure. Nonetheless, we can find some scattered islands of industries immersed in otherwise agricultural, mining, or mono-productive economies. Unfortunately, such industries are rarely competitive under global standards.

The postindustrial economy or service economy made obsolete some of the prevailing industrial practices. Quantity over quality is no longer acceptable, mass production of identical products is not as effective and we now refer to mass production of customized products or mass customization (Gilmore & Pine, 2000; Pine, 1999; Stuart & Eichenberger, 1968). However, the new economy did not dismiss the industrial infrastructure and superstructure. In fact, quite the opposite happened. They provide a solid ground to support the information era. Information systems do not replace the industrial processes (that make intensive use of matter and energy), but optimize them.

Countries that lack an adequate industrial infrastructure and superstructure, in other words developing countries, are not able to skip the industrialization phase and jump into the information era and the service economy to stand in the same ground as the developed nations. They enter this new era with a handicap. They keep on having a disadvantage that preserves (and even increases) the existing divide.

The Internet Boom

Today's historians are continuously perplexed by the fact that year after year — and in some cases, in shorter periods of time — informatization standards are realized and soon become obsolete. New informatization goals are set and carried out yearly by the top industrialized countries while, at the same time, the rest do not seem to even get started. A decade ago, the Internet was a technology reserved for the experts. Today, the experts are lost inside the great masses of children, elders, and professionals who use the Internet daily. During the 90s the World Wide Web (WWW) interconnected the world (or at least a major part of it). In 2000, the last country still missing, Eritrea, joined the rest of the world in hyperspace (Jensen, 2002). Since then, large parts of the population of industrialized nations and the elite from not–so-developed countries have been linked together. They are connected to chat, to exchange information, to teach, and to give orders to one another.

The speculative boom of the dot-coms ended less than four years ago, but it seems as it was long ago when anybody could have an idea to build a dot-com and tens of venture capitalists would line up at their door to finance the idea. Today, venture capitalists ask for much more than attractive Web pages filled with information. Now the minimum standards include the use of practical services, e-business, extensive databases, useful applications with ample accessibility, and several payment modalities (like renting, pay-as-you-go, and pay per transaction). Moreover, most of today's winners are a combination of dot-coms with brick-and-mortar companies where the Internet is only one more vehicle for the company to connect with the customer. And the latest investment boom is going towards Application Service Providers, Internet applications to control or monitor geographically dispersed vehicles, machinery, packages, or employees (e.g., Sharoff, 2000). As a result, if we examine the Internet boom in more detail, we will find that it has not been a single boom but a sequence of booms where each boom creates greater expectations, generates larger investments, and has a more profound effect on society than the previous one.

Developing countries have been passive spectators of the development of the Internet, or in Internet terms, lurkers. Developing countries have hardly contributed to the technology on which the Internet is based. But what is even more serious, their contribution in useful information, in sites, and in economic processes (e-business) is almost null. And unfortunately, those same activities are what make the Internet an important factor of productivity. At the extreme, their main contribution has been in the number of users. But when those numbers are compared with the number of users from industrialized nations, they look very small (Ali, 1999) even though their total population is far larger than that of the industrialized nations. In addition, the Internet is often ignored or even mocked in developing countries. When it is embraced, it is in a peculiar mode of a consumer of imported "fashions" from the large world centers of production (e.g., Wilson, 2001).

Furthermore, some countries are avoiding the Internet because their leaders see it as a tool of colonization used by developed nations to indoctrinate and put a curb on the development of the poor ones. Unfortunately, the foregoing observations contradict the hope expressed by many (i.e., Nevin, 2002) who expect information technologies — and the Internet in particular — to reduce the breach between the rich and the poor nations (which long ago lost every hope of catching up by the process of industrialization because they arrived too late). And they arrived too late because long before they arrived, the industrialized nations had already raised the barriers of entry to levels that became too high for the newcomers.

Communication technologies are believed to lower barriers of entry because what a decade ago used to take years, now takes only months or weeks and costs a small fraction of what they used to cost (Peha, 1999). Still, the communication infrastructure, the part of the information revolution that is easily implemented, is only a part of what is needed to successfully participate in the information revolution (Lane, 1997). Working and accessible payment and logistics systems are also required. Business and industries are not self contained. They cannot function without the appropriate intermediaries and ancillary services to support it (Sowinski, 2001).

The Digital Divide

The Digital Divide is the breach or division between those societies that are involved in the accelerated process of informatization and those that are left behind in inferior stages of productivity (Anonymous, 2001a). This divide is replacing the current division between the so-called industrialized and non-industrialized countries that had in turn replaced the distinction between developed and underdeveloped countries.

Unfortunately, countries tend to stay on the same side of the breach no matter the term that is used. On one side we find the rich countries and on the other the poor ones.

However, the real problem is that the breach has been increasing through time. In 1820 the distance was 3 to 1, in 1913 11 to 1, in 1950 35 to 1, in 1973 44 to 1, and in 1992 72 to 1 (UNDP, 1999). There was a time when power was gained by having natural resources. Afterwards, that changed to having the means to exploit those natural resources. Still later, those instruments evolved and the requirement was to have industrialized means of production. In today's information society, the most important assets are no longer physical nor energetic assets, not even the industrialized means of production. Today's most valued assets are informational. Those who have the information, the patents, the expertise, the knowledge, and the means to process and distribute them are the ones that hold the power (Liikanen, 2000).

The use and implementation of technology does not work equally well in all conditions (Markus & Robey, 1988). We need to consider many factors. The effect of a given technology in each company and in each country will be slightly different. Management intent and organizational culture also have an effect on the outcome (Markus & Robey, 1988). Furthermore, management behavior and organizational culture are highly dependent on internal and external factors and are continuously evolving.

Developing countries make up the majority of the world population and land extension. They have a large diversity of religions, cultures, ethnic backgrounds, races, colors, weathers, political structures, and natural resources. However, they also have similar problems that include a deficient industrial infrastructure and superstructure, a lack of understanding of the role of scientific and technical information in development, and economic, administrative, cultural, educational, technological, and other barriers to adequate information flow (East, 1983).

The Internet allows people from countries around the world to communicate better and more often. However, while this effect is observed in developed and developing countries it is also more pronounced and profound in developing countries where public information such as television broadcasting, telephone services, educational institutions, and public libraries are not as efficient and readily available as in developed countries. In developing countries this infrastructure is seriously deficient and people do not have the same capability to gather information, coordinate with each other, and solve their problems. Therefore, the capability of the Internet to efficiently communicate people and disseminate information becomes far more important for developing than developed nations (Okoli & Mbarika, 2003). On the other hand the benefits that AdvIT/IS offer to developing countries are capped because their basic problems remain unsolved. While electronic publishing and delivery may be useful for developed countries, the required investment and volume of use does not match the current capabilities of developing countries (East, 1983). However, not even the best e-publishing and delivery technologies can be used if the user does not have the needed education. Basic education and literacy are prerequisites to benefit from them (Sowinski, 2001).

Developed countries used AdvIT/IS during the 60s and 70s to implement automation, during the 70s and 80s to implement informatization, and during the 90s to empower their employees (Bellamy & Taylor, 1997). Each stage provided a unique and added advantage. Newly created companies achieve all of them at the same time and reap the combined advantage. On the other hand, developing countries approach automatization and informatization the same way as developed countries did 20 years ago: assuming that it will reduce headcount (not improve the productivity of the existing employees) and eliminate (not increase) the need for the operator to understand the process (Zuboff, 1988). However, that is not applicable to the current wave of AdvIT/IS. They are developed to empower and increase the productivity of current workers, requiring higher levels of knowledge on the part of the person being empowered. Furthermore, studies indicate that developing countries tend to have a skilled human resource deficit that prevents them from fully profiting from these technologies (Chin, 2004).

Imposing modern information technology may polarize developmental activities and may retard instead of encourage the growth of indigenous resources. Many databases and practices available in the industrialized world are not appropriate nor adequate for the needs of developing countries (East, 1983). Furthermore, free markets are just an idea — not a reality — in most developing countries. The telecommunication sector is generally a state-owned monopoly that uses antiquated technologies and provides very poor services (Mbarika, Musa, Byrd & McMullen, 2002). Moreover, when those monopolies go to the private sector, they continue being monopolies with little incentive to invest on improving service. Furthermore, developing countries do not encourage entrepreneurship and institute many forms of bribes (Mbarika et al., 2002).

The Dimensions of Information

Theoretical Framework / Information Theory

Many authors have developed frameworks to classify information per se. In general there are three concepts or types of information: (1) Information as form, format, structure, or syntax. In other words, information that is independent of all meaning. This is called syntactical information. (2) Information as a symbol, a representation, a map, a reference, a description, news or data about people, things, or actions. This is called semantic information. And (3) information as a signal, stimulus, or activation code to a behavior in an organism or system. This is called pragmatic information (Morris, 1971; Mosterín, 1987; Nauta, 1972). In grammar, pragmatic information corresponds to an imperative sentence, an order, an instruction like, "open sesame" or "love thy neighbor." Semantic information corresponds to a declarative sentence like, "Ali Baba entered the cave" or "That wall is blue." Syntactic information corresponds to the drawings used to write the sentence (calligraphy) or to the sounds that are used to pronounce the sentence (phonetics). We understand the calligraphy as a set of lines and the phonetic as a set of sounds, both of which lack meaning by themselves. They are like symbols that belong to an unknown or non-existent language. From the point of view of communication or of the relation between sender and receiver, pragmatic information is a message that activates in the receiver a process or behavior expected by the sender. Semantic information is a dataset that is copied or somehow linked from the knowledgebase or database of the sender to that of the receiver. Syntactic information is the format used to transmit a message or dataset from a sender to a receiver. It could be taken as the waveband used in the case of an analog transmission or a set of empty boxes to be filled with ASCII (American Standard Code for Information Interchange) characters in the case of digital communications. An illustrative example of the three types of information is a scribble written on a piece of paper. From the syntactic point of view, it is no more than that: a scribble. From the semantic point of view, it may be the symbol that represents a specific person. And from the pragmatic point of view it could be the signature used to authorize or activate an action, for example, a payment. Another example is a bank check. By itself, an empty check is a form that has very little meaning and therefore almost completely made up of only syntactic information. That same check gains meaning when filled up with a beneficiary, an amount, a bank name, an account number, and a valid signature. That check now means that a specific amount of money will be given by the bank to the beneficiary and debited from the given account based on the authorization given by the signer, all of which is semantic information. Moreover, that same check can be used to produce an action like providing a service or exchanging cash or merchandise, and, in such a case, the check is then pragmatic information.

Another illustrative and current example of the types of information is found in the genetic code. Until recently the great work of deciphering the human genome has consisted mainly in identifying repetitive molecular sequences without understanding their meaning. In other words, we have identified the sequences but not the parts of the anatomy or the human behavior that is encoded into them. Therefore, all we have learned is syntactic genetic information. It is expected that in the next five to 20 years, we will progressively determine the meaning of each of those sequences. In information terms, we will slowly gain genetic semantic information. Nevertheless, we will only have pragmatic information when we start using genetic engineering to manipulate the genetic code to cure diseases, correct deficiencies, improve organs, or stimulate behavior. But we are referring to the deciphered genetic code. The genetic code in itself is by definition pragmatic information since the beginnings of life in the planet. That is because since then DNA has been able to activate biochemical processes that create life. However, it is pragmatic information that we cannot control.

There is also a material dimension of information, which is made up of its physical support. All information requires some form of physical support or media. Even when information travels in the form of a wave it does so between a sender and a recipient which are both physical and without which it would not be information but just radiation or noise. The physical support for a bank check is its paper and for genetic information are the DNA macromolecules.

In synthesis, information in general is a form inscribed in a material support (like a drawing in a stone) or in an energetic support (like a wave in an electromagnetic field). That information is called semantic information if it reminds the observer or receiver of an object or behavior (which can be a machine or living organism). It is called pragmatic information if it activates a given behavior in the observer or receiver and it is called syntactic information if it does neither of the above.

Using Information Theory to Classify the Internet

The Material Internet and the Syntactic Internet

The Internet has a material dimension which is made up of the physical support of the information that flows through it. This dimension comprises the physical computers, the networks, the cables, etc. The Internet also has a syntactic dimension, a dimension that consists of the structures that organizes and gives shape to the information flux. A dimension composed of operating systems like Windows and Unix, protocols like TCP, HTTP, and FTP, formats like HTML and XML, etc.

The Semantic Internet

The Internet is filled with semantic information. Common examples are advertisements (marketing information), news, encyclopedias, catalogs, and many different pieces of data. An Internet user can take information from a Web page or an e-mail and copy it into its own brain (read it, learn it) or he could copy it into a secondary storage facility like the hard drive of his PC or print it and file it in a folder. This information is generally found as a sequence of symbols (letters, icons, sounds, etc.) and when read or interpreted, it increases the knowledge of the individual. Generally this knowledge refers to objects that are of interest to that individual. It may be marketing descriptions of products that are for sale, newsworthy descriptions of people or current events, encyclopedic descriptions of almost anything under the sky, numerical or graphical descriptions of the financial behavior of a specific stock, descriptions of the climate, etc. This is primarily descriptive information. Descriptive information on the Internet may stimulate a user to carry out a future act based on that knowledge. But until that act takes place, the individual is only consuming information. He is neither using the objects described, nor executing the behaviors learned. The *semantic Internet* adds meaning to symbols and increases the knowledge of individuals, but it does not cause actions to take place.

The Pragmatic Internet

Pragmatic information is also common over the Internet. A purchase order that includes a valid credit card number is not just increasing the knowledge base of an individual or of a database, but it is activating a comprehensive delivery system of the product ordered. It may be even initiating a whole production process (of the item being ordered or of an item that will replace the shipped one in the warehouse). It is also initiating a process of fund transfer from one account (that of the purchaser) to another (that of the seller). A marketing advertisement may be limited to providing product information to the potential buyer, but it may also be activating a purchase behavior that converts the potential buyer into an actual customer. That advertisement may neglect to transmit much information about the product as long as it directs the potential buyer to the right place at the right time and with the right intentions (that of purchasing the product or service advertised). Such an advertisement is a successful piece of pragmatic information. A typical example of the latter is an erotic image that guides the user's attention to a place (or a Web page) where the product is located or advertised. The erotic image has little or no information regarding the product, but it serves its purpose by leading the potential buyer towards the product.

While the *semantic Internet* adds meaning to symbols, the *pragmatic Internet* does not provide information about objects. The *pragmatic Internet* activates (or allows activation of) processes as well as controls them. In other words, it makes practical use of already existing (and described) objects and processes. At its heart, while the *semantic*

Internet is information that interacts with information. The pragmatic Internet is information that interacts with matter and energy, activating or controlling the processes that transform or transport matter from a form or starting point to another one. The article Infoenergetic Systems (Salim & Ferran-Urdaneta, 1997) defines in depth this relation between information, energy, and the matter moved by the latter:

We define an Infoenergetic System as a system that responds to information with physical work, particularly those systems using digital or biological brains... The information that constitutes input to the infoenergetic systems is by definition pragmatic information... We can now say that an infoenergetic system is a system that is capable of amplifying an energetically minimal but informationally large input to mechanical energy levels, or work (i.e., levels that are manifested by macroscopic changes in the matter).

E-Business is Pragmatic Internet

The Internet is no longer in the semantic stage (or the stage of publishing descriptive information). The current importance of the Internet is based on the Internet as a medium to transmit activation or pragmatic information. What makes the Internet what it is today is the commerce that takes place through its veins. The common name that we currently use for the *pragmatic Internet* is e-business. Of course the traffic of semantic information is still higher than that of pragmatic information, but R&D (as well as other gigantic financial investments) are mostly directed towards the *pragmatic Internet* (Iyer, Taube & Raquet, 2002).

E-business is an infoenergetic system based on the *pragmatic Internet*. In e-business, digital information flows through electronic networks where purchase orders and other informational instruments (or "forms") activate productive and distribution processes that incorporate large amounts of physical matter and energy. It also controls some of those processes by accelerating or stopping them accordingly to the informational feedback that flows through the networks. Moreover, it also guides the inventory and liquidity levels by regulating the different exchanges within and between organizations.

Other Forms of Pragmatic Internet

Even though the *pragmatic Internet* is primarily e-business, as we just said, it is important to recognize that e-mail (including all non-commercial communications) can be considered, in some cases, to be pragmatic information (Ferran & Salim, 2004). A non-commercial e-mail can activate a pre-determined artistic or political behavior in the receiver (Becker, 2001; Meeks, 1997). In such a case, those e-mails exchanges are part of the *pragmatic Internet*. Those e-mails are often found in developing countries. Furthermore, when those communications are done by using instant messaging, their effect is even faster and therefore their pragmatic value is even higher.

Once the capacity to induce behavior is known, it only takes a good businessperson to exploit it. Even the anti-consumism fashion of the 60s was commercially exploited (i.e., blue jeans). Microsoft® acquired Hotmail™ for more than 200 million dollars (Wingfield, 1998). All they got was a very large base of free e-mail users and the knowledge and the technology required to run such a site. Today, they are recouping their investment from the advertisement revenue obtained from the publicity trailers added to all the e-mails sent through their system. Once more, this type of pragmatic use has very little effect on reducing the digital divide.

Another occurrence of the *pragmatic Internet* that is not part of e-commerce is virtual communities. The effects of virtual communities are many. They have changed working and commuting behaviors. They have stimulated a multi-disciplinary approach to problem solving. They have brought together people with similar interests who would have never met one another due to geographical constraints. They have helped in the creation of political movements (Meeks, 1997; Moon & Yang, 2003). Last year South Korea presented a very interesting use of the *pragmatic Internet* and virtual communities. As in many other countries, a large part of presidential campaign financing is done through corporate donations. However, in this case, the wining candidate, Mr. Roh, was able to raise about one billion dollars from over 180,000 individual donors by using an Internet-based campaign (Anonymous, 2003; Moon & Yang, 2003). However, while some virtual communities involve work (Robb, 2001), most of them have a more social purpose (Leung, 2001).

The *pragmatic Internet* has also allowed people who live in underdeveloped countries to participate in projects that otherwise would be out of their reach. This also helps in bringing worldwide expertise to projects in underdeveloped countries. But once more, this capacity can only be exploited if there are: (1) organizations willing to hire personnel under such conditions and (2) institutions respected in those countries that would help enforce the contracts entered therein. Thus, this segment of the *pragmatic Internet* still needs those institutions that are not generally found in developing countries.

While many of the above uses of the *pragmatic Internet* prosper and multiply in underdeveloped countries, their importance is much lower than that of e-business. That difference in the effect is similar to that of the large distribution and retail enterprises to the traditional political and artistic movements. The former engendered today's capitalistic societies while the latter inspired and originated large social movements whose capacity to promote development was dependent upon the creation of viable economic structures.

The Dimensions of the Divide

Using Information Theory to Classify the Divide

The digital divide is not as wide in all the dimensions of information. The computer, networks, cables, backbones, and other elements of the physical dimension of information are or can be (in the short term) the same for all countries independent of their stage of development. "The largest ISP, which developed from one company's network service, supports dedicated wireless links and provides virtual private network services for many businesses in Haiti. In effect, this ISP bypasses the government-owned national telephone network. In addition, access to the electromagnetic spectrum is not regulated, which means that any wireless device can be added anywhere. This considerably speeds up the growth of networks and, consequently, of e-commerce offshoots" (Travica, 2002). And let's not forget that Haiti is the poorest country in North America and one of the poorest countries in the world.

The operating systems (i.e., Windows, Unix), programming languages (i.e., COBOL, C, Java), protocols (i.e., PCP/IP, FTP, HTTP), the basic software (i.e., word processor, spreadsheet, DBMS, statistical system) are not very different from one country to another. Most of them are designed and programmed in developed nations – or at the request and specifications of those nations in the case of the now common outsourcing of programming — to conform to a given standard under which most of the world's AdvIT/IS operates. Moreover, in the case that a developing nation was able to invent a new technology that was accepted as a standard around the globe this would not increase but at the most lessen even more the syntactical digital divide.

Something similar occurs with the informational contents and their semantic relations in the different countries. For example, Web page formats (syntactic dimension) are the same and their contents (semantic dimension) are very similar. Product and service descriptions over the Web are very similar in developed and developing countries. The content of official and commercial Web pages in developed and developing countries equally talk about the excellence of their products and services. However, it is in practice (pragmatic dimension) where we find the big differences. It is in the practical or pragmatic use of AdvIT/IS that developed and developing countries are very different.

We can then say that there are different dimensions to the digital divide. The physical, syntactic, and semantic dimensions of the digital divide equate to the few existing differences in physical support, formal systems and semantic relations, between developed and developing nations. On the other hand, the pragmatic dimension of the digital divide is characterized by the pragmatic fragility that we will now describe.

Pragmatic Fragility of IT in Developing Countries

Pragmatic Fragility

As previously discussed, developing, underdeveloped, and poor countries are characterized by their deficient service and industrial infrastructure. They also tend to have weak currencies. And in a society where the main object of fiduciary value — in other words the local currency — is subject of continuous devaluation, the rest of the assets whose value are based on faith in the value of that currency — like the *pragmatic Internet* — will be subject to losses. We can observe that the countries with relatively stable currency (US dollar, Euro, Yen, etc.) are where the Internet is primarily pragmatic and intensively used. We can also observe that in countries with unstable currencies, where the currency is devaluated in respect to others (mainly developing countries), the *pragmatic Internet* is weak or nonexistent even though such countries may have and use a very extensive *semantic Internet* (Ferran & Salim, 2004).

It is generally accepted that the process of becoming a developed and industrialized country is slow and full of obstacles. On the other hand, recent experiences have shown that implementing the *semantic Internet* does not require long periods of time and most of the obstacles are easily overcome. Many underdeveloped countries have been able to build the *semantic Internet* to levels comparable to those of industrialized nations (Guillen & Suarez, 2001). We have seen how current technologies allow the creation of a country's post-industrial communication infrastructure from practically nonexistence to very sophisticated levels in a matter of a few years and at costs that can be afforded by even the poorest nations (e.g., Peha, 1999). And when we add to this the low prices of information technology, we find that the *semantic Internet* is now serving not only the rich but also very poor sectors of the world population (Dutta & Roy, 2001).

Previously we discussed how the *pragmatic Internet* requires a highly developed industrial and service infrastructure. We also discussed how underdeveloped countries lack such infrastructure (even though they may have a post-industrial communication infrastructure). We then have to conclude that while underdeveloped countries may make extensive use of the *semantic Internet* — for example 95% of the Internet use in Kuwait is limited to chatting (Wheeler, 1998) — they will not be able to make much use of the *pragmatic Internet*.

In the United States the payment of e-commerce transactions is supported by a sophisticated culture of credit card use while in developing nations the culture has not assimilated credit and credit cards properly, bringing e-commerce to almost a standstill (Dadashzadeh, 2002). Bank customers in the UAE have rejected Internet banking because of security concerns but also because they do not know how to use the service and their reluctance to change the way they currently deal with banks (El-Adly, 2003). In Malaysia bank customers had trouble accessing the Internet and affording the cost of Internet banking but also many were not even aware of their existence nor were they willing to change the way they used to deal with banks (Sohail & Shanmugham, 2003).

Research has shown that the diffusion of the Internet, e-business, and AdvIT/IS in general works differently in developing and in developed nations (Dutta, 1997; Mbarika, Jensen & Meso, 2002; Montealegre, 2001; Travica, 2002; Wolcott, 2001). Furthermore, to effectively implement in developing countries the e-business technologies that were created in developed nations we first need to have a better understanding of the context and environment of developing nations (Okoli & Mbarika, 2003). The local context must be taken into consideration (Bada et al., 2004). A recent study by Tigre and Dedrick (2004) provides some corroboration that in Brazil local development and the use of local paradigms provides for better e-business implementations. However, Brazil is an atypical developing country because the international integration of their manufacturing industry is relatively low and it has historically been internally oriented. Moreover, many AdvIS/IT projects in developing nations have failed because they were centered on the technology and not the people nor the processes (Rao, 2003).

The importance of AdvIT/IS is starting to be recognized in developing countries. However, most of their AdvIT/IS is under-utilized. Utilization of those system is highly dependent on the human factor and unfortunately that is one of the biggest weaknesses of developing countries (Anandarajan, Igbaria & Anakwe, 2002).

The *pragmatic Internet* as a medium of exchange has a fiduciary value similar to paper money (bills) or a signed check. It has no intrinsic value. The value of a bill resides in faith, in the belief of people that the assets or services given in exchange for it will be equal in value to the ones that they will receive later, when they in turn exchange it with someone else. It is not a gratuitous faith. It is a faith supported by the existence of social institutions that are materially and morally prepared to upkeep its value. Institutions like banks, courts, and the police force.

For this reason, the value of the *pragmatic Internet* is similar to that of the other objects of fiduciary value in a given society. The *pragmatic Internet* requires the prior existence of people, organizations, or systems ready to obey an order or activation signal that is transmissible over a telecommunication network. A purchase order with a valid credit card number cannot accomplish its objective if the receiver cannot deliver the requested product under the conditions specified (for example in a given time frame) or if there are no credible financial institutions capable of transferring the funds issued in payment to the account of the seller.

In order for e-commerce to operate, it requires a set of structural pillars. These structural pillars include financial conditions (credit card and other forms of e-payment), a legal framework (e.g., consumer protection legislation, e-contracts), a fair, effective, and prompt judicial system, telecommunication networks, and transportation networks (Markus & Soh, 2002).

In that respect, the public service of the judicial system is critical. The legal framework governing e-business within a country enhances or inhibits the development of e-commerce (Sowinski, 2001). A weak, forgiving, or inefficient judicial system does not offer the trust required by individuals or corporations to open themselves to new or innovative types of businesses. There is very little inherent trust between parties transacting over the Internet because the delivery of funds generally takes place at a different moment than the delivery of the goods. Parties transacting over the Internet require some way of assuring the enforcement of agreements. The judicial system needs to provide just and timely solutions to such disagreement. According to Tigre and Dedrick (2004) the absence of a suitable legal and regulatory framework has been an important barrier to the adoption of e-commerce in Brazil. The Internet is reducing the time it takes to do many things but if the judicial system is going to operate at a totally different pace, users will not agree to transact over the Internet (for a more comprehensive discussion of this point see, for example, Cowcher, 2001).

The *pragmatic Internet* requires trust, particularly when there is a lack of practical enforcement mechanisms that impose some correspondence between what is advertised over the 'Net and what the user finally receives at the end of the transaction. Arms-length transactions do not need trust as long as one can see the end of the arm, but as the arm gets longer and we cannot see the end product or service that we will receive in exchange for our money, trust becomes an important variable. A good example of this problem is illustrated in a "Letter to the Editor" found by the authors in a third world country well known to both authors:

Letter to the Editor: I was not able to sit at the game! In January 6, 2002 I bought over the Internet from XYZCo.com four tickets to the baseball game between Navegantes and Cardenales that was to take place on January 11 at the University Stadium. I let them pick the seats because the site does not allow me to select the seats since they argue that they will select the best places available. I trusted them, but on the day of the game, when we arrived at the stadium we learned (the hard way) that two of the assigned seats did not exist. I had nobody to complain to and had no other course of action but leave with a bitter sensation...

A letter that was later followed by another one where that individual claimed that after several weeks of sending complaints by e-mail he finally received a reply where he was notified that in not less than 90 days they would refund the purchase price, a refund that would be issued in a local currency that a month later had already devaluated by more than 30% (Ricardo & Ferran-Urdaneta, 2002).

This deficient implementation of the *pragmatic Internet* and of IT/IS in general is what we call *pragmatic fragility*. And, it is this *pragmatic fragility* that we observe in the underdeveloped countries that is increasing the technological distance or digital divide between developed and underdeveloped countries.

Exceptions to Pragmatic Fragility

There is a case to be made that developing countries could make intensive use of the *pragmatic Internet,* even more intensive than that of developed countries. That is when developing countries opt to undertake a set of corrective measures — primarily economic — that have been proven and continue to be proven effective by many countries (like the case of the Asian-pacific and Chile, Spain and other European countries in clear social and economic ascent, and some regions of India). Those measures are not perfect and tend to be very painful. They may require custom adjustments along the way. Nevertheless, they are considered effective by the international community and this introduces an element of trustworthiness that provides confidence to locals and foreigners. Countries obtain credibility when they truly embrace such measures. Once credibility is attained, the value of their currency and other fiduciary assets begin to stabilize. And the trust on the Internet is, as we said earlier, among those assets — a trust that is observed through the increase the use of the *pragmatic Internet.* At that point, those countries should advance on the road to development as fast as possible and build over the same foundations of those who arrived there first. Furthermore, their use of technology will be more intensive than that of the developed countries because they will be used as tools of accelerated leveling (Foster, 1986).

Singapore and Spain are good examples of this intensive use of information and communication technologies. For Singapore these technologies are at the same time part of their production infrastructure and a product to be exported (Kam, 1998; Lenatti, 1998; Man, 1998; Wong, 1998). In Spain they helped to raise in an accelerated manner their service industry to the level of far more industrialized European countries (like England and Germany) while at the same time they skipped many of the painful steps that those first-comers had to follow in their industrialization process (Foster, 1986; Jalava & Pohjola, 2002; Wong, 2002).

While the *pragmatic Internet* will certainly help in the accelerated development of those countries that opt for the road that has been proven effective and is accepted as such by the international community, it will not induce underdeveloped countries to opt for a given road. As a matter of fact, most underdeveloped countries have not developed because they have not opted for those clear and generally accepted development roads. And it is those countries that will continue to make very little use of the *pragmatic Internet*. At the most, they will be able to make use of the *pragmatic Internet* for marketing purposes, but this will still be restricted by the limited trust that users of those countries have of the Internet.

The Reach of Pragmatic Fragility

The foregoing discussions bring us to conclude that users from places where the public and private industrial and service infrastructure has not reached the information revolution will hardly be able to have access to the *pragmatic Internet* (e.g., Anonymous, 2001b; Mueller, 1999). Internet users in those regions will not only be unable to activate processes taking place in their own region but they would not be able to activate most processes that would take place in regions that do have such infrastructure. That is because while pragmatic information can freely flow from one region to the other, those users would not have the means and methods to control them or enforce them. For example, an international purchase order may bump into custom barriers, deficiencies in the transportation infrastructure, lack of international treaties that would assure compliance by the parties involved, etc. (see for example D'Amico, 2001).

In conclusion, the information that flows through the Internet in developing countries is primarily syntactic and semantic. And in contrast to developing countries, this information is rarely pragmatic. Therefore, information processing in developing countries is more like a conversation where the flux in both directions is limited to words. It is not a process whose end-result is a decision that is later translated into actions, cash flow, exchange of merchandise, or services rendered. In other words, it is a conversation that is not translated into business.

Fragility in the Pragmatic Dimension and the Digitalization of Bad Practices

The Automation of Bad Business Practices in Developing Countries

Technological development has meant process automation from the first agricultural machines to the latest advances in robotics. Process automation is a logical step to take when the process prior to automation produces some benefit and the automation in itself increases that benefit in an amount larger to that invested in the automation process.

It has been standard practice to apply new technological breakthroughs to automate or accelerate business practices that are working properly. And, in itself, this is considered a good business practice, a business practice that tends to impress and signal others that the company is doing a good job. However, the acceleration and automation of business processes is not always indicated. Some processes require some re-engineering before automation and others (that were already wrong) require a comprehensive change. Unfortunately, there are many managers that are so easily impressed and dazzled by the simple application of technology that they do not realize that the process beneath it does not work. They do not realize that there is nothing to gain and sometimes a lot to lose from accelerating or automating a bad business practice. Furthermore, it could be counterproductive to accelerate or restrict human intervention in a practice that has problems to start with. It is not uncommon for managers in less developed countries to go for the show of applying technological breakthroughs without first analyzing the quality of the practices that are being automated.

Developing countries tend to import best practices in the same manner as they import physical merchandise. However, unlike tangible assets, the intangible nature of these assets makes it very difficult to timely appreciate the potential lack of adaptation or applicability. Furthermore, the effect of the infrastructural and superstructural holes previously mentioned compound the problem and make it the most frequent reason for inadequately adopting foreign practices.

Digitalization is the process of automating and accelerating business practices by using advanced IT/IS. Developed countries have digitalized many manual and mechanical practices as well as practices that have a large intellectual or informational content. Among these practices we find many manufacturing, optimization, operational, administrative, and managerial tasks. In fact, we tend to categorize the level of development of a country by looking at the level of effectiveness, efficiency, and sophistication of these same processes.

Digitalization also takes place in less developed countries. However, we frequently observe that the digitalization of existing practices does not improve their efficiency nor their effectiveness but quite the contrary; it increases its already existing deficiencies. This in turn exacerbates the problem and hides it even deeper.

Neither is it useful to automate best practices of developed countries that are not previously adapted nor applicable to the different condition prevalent in developing countries. In effect, these unadapted and automated best practices become bad practices. In both cases we are creating an innovative bad practice, that of *digital malpractice.*

One potential solution is for developing countries to develop their own scientific, technological, and e-business capabilities. Unfortunately, when this is done they are usually modeled after industrialized nations and often do not match local needs (Raisinghani, 2003; Sangster, 1979). Therefore, the important factor is not the place where the initial technology or capabilities is developed but that the technology is either modeled or later adapted to the local customs, infrastructure, and needs without idealizing the models and paradigms of the developed nations.

Long before the birth of the Internet, software engineers, management consultants, and other professionals, warned their clients in developing countries about this potential malpractice. They foresaw that acquiring and installing electronic information systems alone was not enough to improve business practices. However, that did not stop them from making such unsuitable implementations. They badly wanted or needed such systems but were not able or willing to afford the additional expense of analyzing and redesigning their business processes. More recently, with the height of the Internet and its infinite possibilities, particularly e-business, the effect of digitalization of bad practices transcends the scope of single organizations or government agencies to the entire business web of a country making the digitalization of bad practices a social problem.

Propensity of Less Developed Countries Toward Digitalization of Malpractices

By definition less developed countries do not count with the same industrial infrastructure and superstructure that exists in developed countries. This is the cause for at least two problems related to the digitalization of business practices: (1) In developing countries bad practices find larger infrastructural and superstructural holes to sift-through, increasing their propensity to incur in digitalization of bad practices. This occurs at both the organizational level (microeconomic) as well as at the national level (macroeconomic). This even occurs inside systems that are initially considered safe and comprised of good practices. (2) Some of these infrastructural and superstructural deficiencies hinder any quantitative analysis that analysts may want to do to compare developed and less-developed countries. This also affects managers since they do not count with the needed information to identify and correct the bad practices.

In their study of the Banking Industry in Oman, Khalfan & Alshawaf (2004) showed an example of superstructural problems typical of developing nations. Regulatory authorities failed to formulate the needed substantive guidelines. While freedom is good, too much freedom leads to anarchy which does not help anyone. In some cases the market is capable of providing the needed guidelines but real markets do not yet exist in developing countries.

There are many commonalities in the bad practices present in developing countries around the world. They cannot be easily pinpointed, particularly by those who are themselves involved in the bad practice. To help managers identify bad practices we will disaggregate individual cases and summarize them in factors that help in the detection of bad practices. Once the bad practices have been identified, managers can later avoid their digitalization, particularly in e-business projects. The factors we found are:

Lack of Benchmarks

The first factor is a complete lack of benchmarks. Organizations in less-developed countries argue that they cannot be compared with similar organizations located in developed countries. Some of their arguments are valid, mainly those based on the lack of benchmarks for activities in their environment. This is compounded by the fact that many of the economic indexes of their environment are conflictive and many times untrustworthy. However, many of the arguments are nothing else than excuses to explain notorious deficiencies.

The lack of benchmarks allows the organization to claim achievements that cannot be compared to that of other organizations, triumphs which in some cases are not such but, since they cannot be compared, they cannot be disputed either. However, those triumphs are not only claimed at the organizational level. Mid-level managers also make those claims to top-level management who cannot compare or argue that the results presented were mediocre instead of outstanding. Thus, neither stockholders nor top-level managers who are not familiar with IT implementations can critically evaluate the quality, efficiency, or effectiveness of the implementation.

Low Management Accountability

The second factor is the low management accountability typical of less-developed countries and of family owned organizations. Accountability would at a minimum assure that the established standards (however low they may be) would have to be satisfied. Underdeveloped countries have very little public participation in most companies. Many companies (even the very large ones) are family-owned or family-controlled and nepotism is quite common (Okoli & Mbarika, 2003). Very few companies are listed in the stock exchanges. Therefore, most companies are free from public or government control. Very few companies undergo a third party audit of their financial statements. In fact, most of the few companies that go through the trouble of auditing their statements still keep the statements and audits private. Not even the local offices of multi-national companies have to publish their sales or projection figures, much less their profit margins. The few indexes and figures attributed to those companies are nothing more than gossip, speculations, or assumptions made by specialized media trying to make sense of the economic environment. And no matter how professional those publications try to be, it has to be taken into consideration that most of the advertisement they publish comes from the same companies that they are trying to evaluate.

Triple Concealment

The third factor is a triple concealment of the deficiency. It is triple because it involves the technology user (company acquiring the technology), the technology supplier, and the auditing (or consulting) firm. They all tend to conceal any deficiencies in an IT/IS implementation. The technology user hides the deficiencies because it does not want to associate its image (towards stockholders or the public in general) with a deficiency in IT/IS right in the middle of a very sensitive topic as modernization. The technology supplier does not want to point out any deficiencies with the same technology that it is trying to keep selling nor to add to its list of dissatisfied customers. Finally the auditing firm, while it has the duty of reporting the deficiencies, it is also very common (in less developed countries) that they already had a stake in the project since most of them are initially recommended by the auditing firm. It is only in the rare cases that the auditing firm had no involvement in the selection and implementation of the technology that they report deficiencies. However, in most cases they end-up involved in the "repair" of those deficiencies and thus lose their independence.

Triple concealment is not exclusive of developing countries (e.g., Enron and Parmalat). However, they are far more frequent in less developed countries. Furthermore, given the lack of benchmarks and low management accountability, it becomes very difficult if not impossible to discover the triple concealment. The study of IS/IT in less developed countries is constrained by various forms of peculiar cultural, social, and political practices. While the technology used is the same, its use in real life is intermixed with those practices.

One of the main problems of developing nations is corruption. It not only affects the implementation of AdvIS/IT but all aspects of the economy. Most of the governmental attempts to promote free markets or large investments in AdvIS/IT fail due to corrupt mismanagement of the funds (Okoli & Mbarika, 2003). Simple things like obtaining a land phone line for a home in a residential area may take years and always requires greasing a few hands along the way. In many cases the salaries of police officers are so low that bribes become a standard operating procedure that is already factored into many budgets. Therefore, all the parties involved in corruption make an extra effort to make sure that all information and statistics in these regards are kept secret.

Examples of Digitalization of Malpractices in Underdeveloped Countries

Due to the lack of reliable and comprehensive statistics, case analysis is not only a good methodology but also the only one available for a comprehensive theoretical discussion. Case studies also provide easy-to-use advice and the essential background information needed to appreciate the challenges of implementing AdvIT/IT in developing nations (Dadashzadeh, 2002).

Transportation Services

An example of this occurs in systems for secured transportation services. When safe routes also include frequent obstructions, a practice of route deviation without activating exception procedures may become institutionalized. Such practice could also appear due to negligent personnel that are kept on the payroll because of overly protective national labor policies or inefficiencies in the justice system, both of which are very common in developing countries. As an illustration of this problem, we have the systems for armored transportation which rely (as they do in developed countries) on a timely police response. If the system is implemented without prior determination of such response times (using reliable statistics) it then becomes an unsecured practice, in other words, a bad practice (Salim & Ferran, 2004)

Document Storing

A typical and very common practice is that of requiring and storing of several physical copies of the same document. This is contrary to basic database theory which recommends the reduction of data redundancy to a minimum. In developing countries it is very common to find the co-existence of physical and electronic databases for the same data.

This would be acceptable if the physical database was used for legal backup of the electronic data. Unfortunately, we normally observe that the physical database is the primary source and the electronic database is an additional redundancy used as a contemporary ornament (Salim & Ferran, 2004).

National Voting System

From the early 60's until the late 90's Venezuela had several presidential elections. During that time it was common to hear "Acta mata votos," which can be translated as "Minutes (record, summary) kill votes." That phrase was used to express the claims of many that the process to write the minutes allowed, with the consent of some of the official witnesses, the alteration of the results. Instead of the minutes reflecting the real vote count, they reflected the wishes and agreements of the parties represented in the official witness pool of the voting center. And since what was later summarized as the results of the election were the respective minutes of each voting center and not the votes themselves, the minutes actually killed the votes. In a recent election the authorities decided to computerize it to eliminate that lack of reliability (among other reasons). In this new system instead of the official witnesses counting the physical votes and writing a minute, the minutes were automatically produced by the machine. Official witnesses only controlled who was allowed to vote in the machine but did not see or counted votes. Election authorities advertised that such a system would no longer allow the old manipulation of the minutes that used to favor those who were now in the opposition. However, the opposition was never allowed to do a timely and satisfactory audit of the computer algorithms and therefore the new claims of fraud in these elections argue that this time it was the computer algorithm (and thus the printed minutes) that did not produce the real vote count but one already pre-programmed in it. We now have a digital "Minutes kill votes" or "Computer kills votes." Independently of the veracity of either claim, the truth is that both processes or practices generate the same type of doubts regarding their credibility. The digitalization of the process did not solve the problem but compounded it.

Digitalization of Bad Practices as a Result of Pragmatic Fragility

Digitalization of bad practices is probably the clearest expression of *pragmatic fragility*. In it we can observe all of its factors and effects doubled. They are doubled because the *pragmatic fragility* is present in both the digitalization process and the bad practice in itself.

We find that in the majority of the IT/IS implementations in less developed countries the syntactic and semantic dimensions of information are accomplished. Nonetheless, most of them lack the pragmatic dimension. And in those few cases where the pragmatic dimension is accomplished it is largely underestimated because in truth that was not what the stakeholders expected. Furthermore, we foresee that this underestimation on the part of the stakeholders of the rare pragmatic accomplishment is going to affect it negatively in the midterm. This will probably occur through the lack of maintenance or by steering the next project toward what the stakeholders involved really value (Salim & Ferran, 2004). As a result, the pragmatic *digital divide* is not reduced. Neither a deficient automation of good practices nor a successful automation of malpractices will reduce the breach.

We also observe that the *pragmatic fragility* is consistent with the infrastructural deficiencies easily found in less-developed countries, deficiencies that other authors point to as the root for many of their problems (Simon, 2004) and of their bad business practices.

There is also a more subtle way in which *pragmatic fragility* brings about digitalization of malpractices. There is some agreement — in both developed and underdeveloped countries — that information systems, like mechanical systems, require at one point or another some form of maintenance. However, the suppliers of parts and maintenance for mechanical systems have separated from the original manufacturers (generating a much larger secondary market accessible in most countries) while the ones for information systems are still closely tied to the original developers. In developed countries most of the IS maintenance is provided by the original developers while in developing countries the maintenance is either not provided at all or it is provided by independent consultants that have very little knowledge of the internal workings of the system (Salim Koussa & Ferran-Urdaneta, 2001).

Every automobile owner knows and counts on the existence of independent car repair shops along most routes. Companies that own large and expensive machinery assume that even after the original manufacturer stops providing parts and services they will still be able to obtain them from third parties. Unfortunately, IS maintenance is still tied to the original developers, developers that tend to have high prices and place their service support centers close to their own offices which tend to not be located in underdeveloped countries.

Systems (mechanical or informational) that are not properly maintained tend to malfunction. The more the system is used the higher the probability that a malfunction is going to occur (Salim Koussa & Ferran-Urdaneta, 2001). The *pragmatic fragility* present in underdeveloped countries hinders the appropriate maintenance of IT/IS, which in itself is a bad business practice.

Electronic Business as Digitalization of Bad Practices

The implementation of e-business in developing countries is generally done under the assumptions of either the technological or the managerial imperative (Markus & Robey, 1988). Industrial infrastructure and superstructure as well as other emergent factors are generally underscored. However, the success or failure of an AdvIT/IS implementation depends on making the required organizational and technical changes (Dologite, Mockler, Bai & Viszhanyo, 2004). Furthermore, we have observed that the strength of *pragmatic fragility* and in turn of digitalization of malpractices is proportional to the amount of interaction that the IT/IS applications have with systems that process physical matter (merchandise) and energy (or work), in other words, with the more traditional industrial systems. By the same token, e-business is the typical case of IT applications interacting with traditional industrial systems since the information embedded in it are mainly purchase orders that end-up creating cash flows, moving merchandise, and/or activating services. Therefore, the digitalization of bad practices is more intense in the implementation of e-business in less-developed countries.

Some IT applications help or even force the identification and correction of bad practices before or during the automation process. They help and promote either simple redesign processes or complex re-engineering. Informatization requires a holistic approach that takes into account skills, roles, and organizational structures. A comprehensive learning environment needs to develop during a successful AdvIT/IS implementation (Zuboff, 1988). Changes in management and organizational structure are needed for an AdvIT/IS implementation to succeed (Morton, 1991). Partial change efforts or technology-limited initiatives are unlikely to succeed (Zuboff, 1988). Implementing AdvIT/IS enables fundamental changes in the way that an organization works. It changes the way production is done, it helps coordinating the work of the different units, and it empowers both workers and management (Morton, 1991). However, these applications are not what we generally refer to as e-business. The objective of e-business is not that of correcting practices but of automating them. It is possible that the process of automating the commercial interactions between organizations facilitates the identification of bad practices in both the external and internal processes and therefore helps in correcting and redesigning them. Furthermore, it is possible that in order to implement e-business, organizations may be forced to automate and digitalize their back office. However, the above conditions are not necessarily present or required in many e-business projects and they are certainly not their objective. Thus, it is possible that in the process of implementing e-business the organization also automates bad business practices.

As an example, let us assume: (1) in a manual process certain data is essential and (2) that the organization incurs in the bad practice of leaving them blank or filling them with dummy data. When such a process is automated we could identify the bad practice and redesign the process to avoid such troubles. Or we could just automate the current process and keep omitting a verification mechanism. The prior examples of digitalization of malpractices are a testimony that the latter is the more frequent case in underdeveloped countries. Therefore, it is more common to find that in developing countries the implementation of e-business ends-up in digitalization of malpractices than in better, more efficient, and more effective business practices.

E-business implementation is a higher level or upper stage type of informatization process. In order to implement it one needs to have successfully developed an internal platform that can properly interface with the e-business platform. Implementing higher level phases that redesign the business network before the internal platform is a sure formula for failure (Venkatraman, 1991, 1994; Zaheer & Venkatraman, 1994). E-business is not only automation of processes but an informatization of the organization that brings about the transformation of the business processes. A successful implementation of e-business requires all three of them as well as good leadership, strategic vision and ongoing organizational empowerment (Morton, 1991).

Solutions and Recommendations

While ideal, it is also utopian to expect a substantial short-term improvement in the industrial infrastructure and superstructure of less-developed countries. However, there are some steps that may be taken by governments (macro level) and companies (micro level) to stop e-business from being a mere digitalization of malpractices. These steps would contribute to lessen the digital divide and allow less-developed countries to make better use of the pragmatic dimension of IT/IS.

E-commerce provides a unique set of challenges for developing nations. To overcome those challenges, avoid islands of information and repeated failures in the public and private sectors have to make an orchestrated nation-wide effort. Such combination is the only one that will provide the critical mass that e-commerce requires (Kamel & Hussein, 2002). Companies cannot implement e-business in a vacuum (Rao, 2003). However, they need to look for solutions that do not wait until government solves all the industrial infrastructure and superstructure problems but at the same time are not too limited, unidimensional, or out of context (see Li & Chang, 2004).

The most important changes come from the private and not the public sector. They depend on the established agents that offer and demand IT/IS goods and services. They are the chambers and professional associations that bring together IT/IS developers, manufacturers, providers, and supporters as well as the chambers that group IT/IS users. These chambers and associations tend to make up for the deficiencies in the public justice system by offering mediation and arbitration services. Nonetheless, public and private IT/IS mediators and arbiters need special training and knowledge to efficiently and effectively take care of IT/IS disputes. Fortunately, such training may be provided by the same IT/IS professional associations.

Professional IT/IS associations have to establish local benchmarks and adapt international certification standards to the local requirements and capabilities. In some cases the adaptation requires the establishment of less stringent standards. However, a less stringent standard that is followed is far better than a stricter one that cannot be followed. Using strict standards that are not followed but falsified through a process of triple concealment only helps to increase the digital divide. On the other hand, a slow but continuous elevation of the standards that is confidently certified at each step reduces the digital divide. In fact, developed countries did not jump directly into strict standards but progressively increased the stringency of their standards.

IT/IS professional associations as well as chambers that group IT/IS users should found and finance an institution to create realistic standards, develop a plan that gradually elevates those standards, and certify that those standards are being met. Academic institution may also help. However, not much should be expected of them since academic institutions in underdeveloped countries have many problems of their own, particularly economic ones. Funding for the benchmarking and certification institution must be provided as a statutory percentage permanently included in the chambers' and professional association's yearly budget. The amounts invested by the members of those chambers and associations will be less than what most of them currently spend in artificial, unreal, or "forged" certification.

Applying realistic standards is especially important for e-business. A purchasing company in a developing nation cannot realistically require a supplier to have the latest and generally over-dimensioned platforms, systems, and procedures for e-business that are standard in developed countries. If they were to realistically apply such requirements, the universe of suppliers would shrink to a point were there could be none available, or just a few multi-national companies with local operations that would have monopolistic power over the quality and prices of the goods and services. They could also try to provide the appearance of stringency while looking the other way when shown "forged" certification by smaller local suppliers. However, this would hardly help them and would certainly contribute to an increase in the digitalization of malpractices.

A cheaper, easier, and more efficient option is for companies to require from their suppliers to have local and realistic certifications, certifications that would certainly be more modest but at the end more effective. This way the whole e-business chain would digitalize modest but good practices instead of the current ambitious digitalization of malpractices.

Another approach that does not wait for the big change from government but neither proposes a solution that is enclosed in an out-of-context crystal ball was made by Sangster (1979). He proposes the creation of a Foundation for International Technological Cooperation that would help localize and remodel research and development (R&D) resources towards the real needs of developing countries. He argues that developed nations cannot drop crumbs of technology from 10,000 feet in the air and expect that it would land on its feet. Research and development for developing countries should be modeled according to their paradigms and not to the paradigms of the developed nations.

Future Trends

AdvIT/IS solutions are developed to work under a given set of conditions that is dependent on many internal and external factors. Generally, these factors are not well described. The outcome of implementing one technology in a given organization is always different to the outcome of implementing that same technology in a different organization (Markus & Robey, 1988). The further apart those organizations are in their history and resources the further apart those outcomes will be (Wernerfelt, 1984, 1995). Cultural and environmental factors shape how people and companies make use of AdvIT/IS (Wheeler, 2001). The history, culture, environment, and resources available to companies in less-developed countries are very different than those of companies in developed countries. Therefore the outcome of implementing the same technologies will also be very different unless additional steps are taken to modify and adapt the technology to the new conditions.

Companies in developed countries have many commonalities among themselves. Commonalities based on factors present in developed countries and absent in less-developed ones. Factors that require or assume the existence of resources that may be unique to developed countries. Therefore, many of the best practices that have been packed inside technology solutions do not work directly when implemented in underdeveloped countries. Furthermore, many times neither is the knowledge acquired in developed countries useful for implementations in less-developed countries.

The developing countries will not catch the already developed nations without the use of AdvIT/IS. However, the simple use of AdvIT/IS will not be enough to accomplish it either. In parallel, developed nations cannot expect to successfully impose a global apartheid to separate themselves from the misery present in developing countries. For many (Calhoun, Price & Timmer, 2002; Carr, 2002; Hershberg & Moore, 2002; Levy, 2004; Verrill, 2002) the September 11 episode is living proof of that. Incursions from the poor countries into the richer ones (violent or not) are unavoidable. Neither the Rio Grande nor a river of border patrols stop the constant immigrations from Mexico and Central America into the Unites States. In Europe the Mediterranean Sea is not able to stop similar migrations. Even the turbulent Caribbean Sea and the treacherous Cuban prisons are challenged by those living in misery trying for a chance for wealth (Bates, 2001; Griffin & Khan, 1978; Sen, 1981). Therefore, developed nations (for their own benefit) cannot limit themselves to dropping technologies from 10,000 feet in the air. They need to help in customizing them to the conditions existing in the third world (Gore & Figueiredo, 1997; McCullock, Winters & Cirera, 2001; Vreeland, 2003).

Richer countries have adopted a free political (democracy) and economic (capitalism) market. Many experiments with other models have already taken place resulting in expensive failures from both the social and the economic point of view. During the past decades the political left and right parties of the most successful countries have democratically alternated in power as two phases of capitalism: the wealth-generation phase or phase that benefits capital (low taxes, less government interventions, etc.) and the wealth-distribution phase or phase that benefits labor and reduces exclusion (higher taxes, subsidies to promote health, housing, and education, more government interventions, etc.). There may be other formulas that provide even greater welfare. However — up to this date — the proven and currently prevailing formula is that of the democratically alternating phases of the market economy (Lefort, 1994; Riker, 1982; Rothschild & Russell, 1986; Schmitter & Karl, 1991; Weber, 1930).

Democracy and capitalism make up the environment that is assumed and required by the existing AdvIT/IS. Therefore, the countries that want to import such AdvIT/IS as a means to get over their lack of development will need to start by taking as much advantage as possible by adapting them to their current conditions (weak democracy and/or capitalism). This way — as we argued earlier — they will reap the benefits of the semantic dimension and exceptionally some of the benefits of the pragmatic dimension. Then, supported on these adapted AdvIT/IS, advance toward the comprehensive implementation of the required and assumed environment where those technologies are known to provide the maximum benefit: developed democracy and capitalism (and not their mere mockups of democracy without capitalism or capitalism without democracy).

Alternatively, we could develop AdvIT/ISs that do not require nor assume the existence of democracy and capitalism to overcome the problems of underdeveloped countries. However, we have yet to find a way.

Conclusion

This chapter discussed how certain macroeconomic and social variables mediate the effect of the Internet on national development hindering the closing of the breach of the digital divide. It also developed a framework based on information theory to distinguish between the *semantic Internet* and the *pragmatic Internet*. Among other benefits, this framework improves our understanding of why the Internet does not provide the same development thrust to all the countries that implement it. Furthermore, it explains why e-business is not exempt from the digital divide nor does it help to reduce the breach. Underdeveloped countries primarily use the *semantic Internet* while developed countries make use of both. Underdeveloped countries make a very limited use of the pragmatic dimension of the Internet and IT/IS in general, which is what we call *pragmatic fragility*. We observe this *pragmatic fragility* in the unadapted digitalization of best practices taken from developed countries and in the blind automation of current (and deficient) manual practices. At first sight, any form of automation and digitalization seems to help lessen the digital divide. However, when analyzed in depth, they show a set of deficiencies that we call *digitalization of malpractices*.

References

Abu-Musa, A. A. (2004). Auditing e-business: New challenges for external auditors. *Journal of American Academy of Business, Cambridge, 4*(1/2), 28.

Ali, A. J. (1999). Digital divide: A challenge that must be faced. *Advances in Competitiveness Research, 7*(1), 1-4.

Anandarajan, M., Igbaria, M., & Anakwe, U. P. (2002). IT acceptance in a less-developed country: A motivational factor perspective. *International Journal of Information Management, 22*(1), 47.

Anonymous. (2001a). The 'digital divide'. *Far Eastern Economic Review, 164*(14), 6.

Anonymous. (2001b). OECD and developing countries join Dot.force to tackle digital divide. *Organization for Economic Cooperation and Development. The OECD Observer, 225*, 48.

Anonymous. (2003, January 10). Korea's Internet election. *Social design.*

Ash, C. G., & Burn, J. M. (2003). A strategic framework for the management of ERP enabled e-business change. *European Journal of Operational Research, 146*(2), 374.

Bada, A. O., Ikem, F., Omojokun, E., Eyob, E., Adekoya, A., & Quaye, A. (2004). *Globalization and the Nigerian banking industry: Dynamics of context/process interaction.* Paper presented at the Americas Conference on Information Systems (AMCIS 2004), New York.

Bates, R. H. H. (2001). *Prosperity and violence.* W. W. Norton.

Becker, T. (2001). Rating the impact of new technologies on democracy. *Association for Computing Machinery. Communications of the ACM, 44*(1), 39-49.

Bellamy, C., & Taylor, J. A. (1997). *Governing in the information age*. Philadelphia: Open University Press.

Beniger, J. R. (1986). *The control revolution: Technological and economic origins of the information society*. Cambridge, MA: Harvard University Press.

Beniger, J. R. (1990). Conceptualizing information technology as organization, and vice versa. In J. Fulk & C. W. Steinfeld (Eds.), *Organizations and communication technology*. Newbury Park, CA: Sage Publications.

Calhoun, C., Price, P., & Timmer, A. (Eds.). (2002). *Understanding September 11*. New York: The New Press.

Carr, C. (2002). *The lessons of terror*. Random House.

Chin, P. O. (2004). An examination of factors that affect the management of information technology in organizations. Paper presented at the *Americas Conference on Information Systems (AMCIS 2004)*, New York.

Cowcher, R. (2001). E-trust. *The British Journal of Administrative Management, 28*, 22-23.

D'Amico, E. (2001). Global e-commerce. *Chemical Week, 163*(36), 24-29.

Dadashzadeh, M. (2002). *Information Technology Management in Developing Countries*.

Dologite, D. G., Mockler, R., Bai, Q., & Viszhanyo, P. F. (2004). A manager's critical role as IS change agent: A case of packaged software implementation in China. Paper presented at the *Americas Conference on Information Systems (AMCIS 2004)*, New York.

Dutta, A. (1997). The physical infrastructure for electronic commerce in developing nations: Historical trends and the impact of privatization. *International Journal of Electronic Commerce, 2*(1), 61-83.

Dutta, A., & Roy, R. (2001). The mechanics of Internet diffusion in India: Lessons for developing countries. Paper presented at the *Twenty-Second International Conference on Information Systems*, New Orleans, LA.

East, H. (1983). Information technology and the problems of less developed countries. *Information Society, 2*(1), 53.

El-Adly, M. I. (2003). Internet banking usage by bank consumers in the UAE: Exploratory study. *Arab Journal of Administrative Sciences, 10*(2), 141-169.

Ferran, C., & Salim, R. (2004). The Internet and the digital divide. *Asian Information-Science-Life, 2*(1).

Foster, R. N. (1986). *Innovation: the attacker's advantage*. New York: Summit Books.

Gilmore, J. H., & Pine, B. J. (2000). *Markets of one: creating customer-unique value through mass customization*. Boston: Harvard Business School.

Gore, C., & Figueiredo, J. B. (Eds.). (1997). *Social Exclusion and Anti-Poverty Policy*. Geneva: International Institute of Labour Studies.

Griffin, K., & Khan, A. R. (1978). Poverty in the world: Ugly facts and fancy models. *World Development, 6*(3), 295-304.

Guillen, M. F., & Suarez, S. L. (2001). Developing the Internet: Entrepreneurship and public policy in Ireland, Singapore, Argentina, and Spain. *Telecommunications Policy, 25*(5), 349-371.

Hershberg, E., & Moore, K. (Eds.). (2002). *Critical Views of Sept. 11: Analyses from Around the World.* New York: The New Press.

Iyer, L. S., Taube, L., & Raquet, J. (2002). Global E-commerce: Rationale, digital divide, and strategies to bridge the divide. *Journal of Global Information Technology Management, 5*(1), 43-69.

Jalava, J., & Pohjola, M. (2002). Economic growth in the New Economy: Evidence from advanced economies. *Information Economics and Policy, 14*(2), 189-210.

Jensen, M. (2002). *The African Internet - A Status Report.* Port St. Johns: The Association for Progressive Communications.

Kalakota, R., & Whinston, A. B. (1997). *Electronic commerce: a manager's guide.* Reading, MA: Addison-Wesley.

Kam, W. P. (1998). Technology acquisition pattern of manufacturing firms in Singapore. *Singapore Management Review, 20*(1), 43-64.

Khalfan, A. M., & Alshawaf, A. (2004). Adoption and implementation problems of e-banking: A study of the managerial perspective of the banking industry in Oman. *Journal of Global Information Technology Management, 7*(1), 47.

Kohli, R., Sherer, S. A., & Baron, A. (2003). Editorial—IT investment payoff in e-Business environments: Research issues. *Information Systems Frontiers, 5*(3), 239.

Lane, C. (1997). Island of disenchantment. *The New Republic, 217*(13), 17-22.

Lefort, C. (1994). *The political forms of modern society: Bureaucracy, democracy, totalitariansim* (J. B. Thompson, ed.). MIT Press.

Lenatti, C. (1998). Bandwidth battle: Singapore vs. Malaysia. *Upside, 10*(8), 46-50.

Leung, L. (2001). College student motives for chatting on ICQ. *New Media and Society, 3*(4), 483-500.

Levy, B.H. (2004). *War, Evil, and the End of History.* Melville House Publishing.

Li, P. P., & Chang, S. T. (2004). A Holistic Framework of E-Business Strategy: The Case of Haier in China. *Journal of Global Information Management, 12*(2), 44.

Liikanen, E. (2000). Europe & eBusiness. *Presidents & Prime Ministers, 9*(4), 10-12.

Man, S. N. W. (1998). High speed connection between vBNS and SingAREN in Singapore. *Computer Networks & ISDN Systems, 30*(1-7), 723-726.

Markus, L. M., & Robey, D. (1988). Information technology and organizational change: Causal structure in theory and research. *Management Science, 34*(5), 583-597.

Markus, L. M., & Soh, C. (2002). Structural influences on global e-commerce activity. *Journal of Global Information Management, 10*(1), 5-12.

Mbarika, V. A. W., Jensen, M., & Meso, P. (2002). Cyberspace across sub-Saharan Africa. *Communications of the ACM, 45*(12), 17-21.

Mbarika, V. A. W., Musa, P. F., Byrd, T. A., & McMullen, P. (2002). Teledensity growth constraints and strategies for Africa's LDCs: "Viagra" prescriptions or sustainable development strategy? *Journal of Global Information Technology Management, 5*(1), 25-43.

McCullock, N., Winters, L. A., & Cirera, X. (2001). *Trade liberalization and poverty: A handbook*. London: Centre for Economic Policy Research.

Meeks, B. N. (1997). Better democracy through technology. *Association for Computing Machinery. Communications of the ACM, 40*(2), 75-78.

Montealegre, R. (2001). Four visions of e-commerce in Latin America in the year 2010. *Thunderbird International Business Review, 43*(6), 717-735.

Moon, J. Y., & Yang, S. (2003). The Internet as an agent of political change: The case of Rohsamo in the South Korean presidential campaign of 2002. Paper presented at the *24th International Conference on Information Systems*.

Morris, C. (1971). *Writings on the general theory of signs*. Hague: Mouton.

Mosterín, J. (1987). Conceptos de información. Paper presented at the *Actas del II Congreso de Lenguajes Naturales y Lenguajes Formales*, Barcelona.

Mueller, M. (1999). Emerging Internet infrastructures worldwide. *Association for Computing Machinery. Communications of the ACM, 42*(6), 28-30.

Nauta, D. (1972). *The Meaning of Information*. The Hague: Moton.

Nevin, T. (2002). Africa's new challenge: Crossing the digital divide. *African Business* (277), 24-26.

Okoli, C., & Mbarika, V. A. W. (2003). A framework for assessing e-commerce in Sub-Saharan Africa. *Journal of Global Information Technology Management, 6*(3), 44-66.

Peha, J. M. (1999). Lessons from Haiti's Internet development. *Association for Computing Machinery. Communications of the ACM, 42*(6), 67-72.

Pine, B. J. (1999). *Mass customization: the new frontier in business competition*. Boston: Harvard Business School.

Pinero, E. (2001). eBusiness or eCommerce—"need-to-know" information for economic developers. *Economic Development Review, 17*(3), 26.

Pitre, B. (2000). The difference between e-commerce and e-business. *Business Advisor*.

Raisinghani, M. S. (2003). Key perspectives on the global e-readiness of Websites: A reality check. *Journal of Global Information Technology Management, 6*(2).

Rao, M. (2003). Checklist for national e-readiness. *International Trade Forum*, (3), 10-12.

Riker, W. H. (1982). *Liberalism against populism: A confrontation between the theory of democracy and the theory of social choice*. San Francisco: W.H. Freeman.

Robb, D. (2001, July). Ready or not...instant messaging has arrived as a financial planning tool. *Journal of Financial Planning*, 12-14.

Rothschild, J., & Russell, R. (1986). Alternatives to bureaucracy: Democratic participation in the economy. *Annual Review of Sociology, 12*, 307-328.

Salim K. R., & Ferran-Urdaneta, C. (1997). Infoenergetic systems. Paper presented at the *World Multiconference on Systemics, Cybernetics and Informatics (SCI '97/ISAS '97)*, Caracas, Venezuela.

Salim K. R., & Ferran-Urdaneta, C. (2002). The Internet and the Digital Divide. Paper presented at the *First International Conference on Information and Management Sciences (IMS2002)*, Xi'An, China.

Salim, K. R., & Ferran-Urdaneta, C. (2001). Neither a perpetuum mobile nor a perfect software: Sincerity in the relationship between the manufacturer and the client with respect to software defects. Paper presented at the *Fifth World Multi-Conference on Systemics, Cybernetics and Informatics (SCI'01/ISAS'01)*, Orlando, FL.

Salim, R., & Ferran, C. (2004). Pragmatic Fragility of Information Technology in Latin America: Case studies. Paper presented at the *Americas Conference on Information Systems (AMCIS 2004)*, New York.

Sangster, R. C. (1979). R&D for developing countries: The role of industrial R&D. *Research Management, 22*(3), 34.

Schmitter, P. C., & Karl, T. L. (1991). What democracy is...and is not. *Journal of Democracy, 2*, 75-88.

Scott Morton, M. S. (1991). *The corporation of the 1990s*. New York: Oxford University Press.

Sen, A. K. (1981). *Poverty and famines: An essay on entitlement and deprivation*. New York: Clarendon Press.

Sharoff, R. (2000). Dot-com gamblers hit the jackpot. *Marketing News, 34*(23), 19.

Simon, S. J. (2004). Critical Success Factors for Electronic Services: Challenges for Developing Countries. *Journal of Global Information Technology Management, 7*(2), 31-53.

Slesser, M. (1978). *Energy in the economy*. New York: St. Martin's Press.

Sohail, M. S., & Shanmugham, B. (2003). E-banking and customer preferences in Malaysia: An empirical investigation. *Information Sciences, 150*(3-4), 207-217.

Sowinski, L. L. (2001). Which countries are best positioned in the e-business race? *World Trade, 14*(9), 32-33.

Stuart, D. R., & Eichenberger, F. (1968). *The mass production of unique items*. Raleigh: Design Research Laboratory of the School of Design North Carolina State University.

Tigre, P. B., & Dedrick, J. (2004). E-commerce in Brazil: Local Adaptation of a Global Technology. *Electronic Markets, 14*(1), 36.

Travica, B. (2002). Diffusion of electronic commerce in developing countries: The case of Costa Rica. *Journal of Global Information Technology Management, 5*(1), 4-25.

UNDP. (1999). *Human Development Report*. New York: Oxford University Press.

Venkatraman, N. (1991). IT-induced business reconfiguration. In M. S. Scott Morton (Ed.), *The corporation of the 1990s* (pp. 122-158). New York: Oxford University Press.

Venkatraman, N. (1994). IT-enabled business transformation: From automation to business scope redefinition. *Sloan Management Review, 35*(2), 73-86.

Verrill, C. (2002). Terrorism and energy: Bush's 2020 vision? *Palo Alto Weekly Online Edition.* Retrieved January 2, 2002, from *http://www.paloaltoonline.com/weekly/index2.shtml*

Vreeland, J. R. (2003). *IMF and economic development.* Cambridge University Press.

Weber, M. (1930). *The Protestant ethic and the spirit of Capitalism* (T. Parson, Trans. 1930 ed.). New York: Charles Scribner.

Wernerfelt, B. (1984). A resource-based view of the firm. *Strategic Management Journal, 5*, 171-180.

Wernerfelt, B. (1995). The resource-based view of the firm: Ten years after. *Strategic Management Journal, 16*, 171-174.

Wheeler, D. L. (1998). Global Culture or Culture Clash: New Information Technologies in the Islamic World - A View from Kuwait. *Communication Research, 25*(4), 359-376.

Wheeler, D. L. (2001). The Internet and Public Culture in Kuwait. *Gazette, 63*(2), 187-201.

Wilson, D. (2001). Somewhere over the rainbow. *Communications International, 28*(3), 56-60.

Wingfield, N. (1998, January 2). Microsoft to buy e-mail start-up in stock deal. *Wall Street Journal.*

Wolcott, P. (2001). A Framework for Assessing the Global Diffusion of the Internet. *Journal of the Association for Information Systems, 2*(6).

Wong, P.K. (1998). Leveraging the global information revolution for economic development: Singapore's evolving information industry strategy. *Information Systems Research, 9*(4), 323-341.

Wong, P.K. (2002). ICT production and diffusion in Asia: Digital dividends or digital divide? *Information Economics and Policy, 14*(2), 167-187.

Zaheer, A., & Venkatraman, N. (1994). Determinants of Electronic Integration in the Insurance Industry: An Empirical Test. *Management Science, 40*(5), 594-566.

Zuboff, S. (1988). *In the age of the smart machine: The future of work and power.* Basic Books.

Chapter II

Role of Culture in Electronic Business Diffusion in Developing Countries

Marwa M. Hafez, The American University in Cairo, Egypt

Abstract

This chapter includes discussions pertaining to the role of culture in influencing electronic business diffusion in developing countries. In this chapter, the author discusses specific cultural factors and their influences on the individual components required for Internet rollout and use in developing countries. Cases from developing countries are also presented to illustrate the effects culture produces on the levels of electronic business technology usage. The discussions of the lessons learned from those cases along with the theoretical foundations presented throughout the chapter, culminate in the author's provision of recommendations to the reader, deemed necessary for effectively increasing the diffusion of electronic business in developing countries as well as for reaping the potential benefits generated from its use.

Introduction

The importance of the role of culture as a major factor influencing electronic business diffusion in developing countries has been acknowledged for quite some time. Only recently, however, have researchers begun to investigate the specific cultural elements propagating technological change and electronic business diffusion (Volken, 2002). As such, this chapter provides discussions of the cultural specificities involved in moderating the rate of diffusion of electronic business in developing countries.

The chapter starts by providing a definition of developing countries. It then proceeds to provide a general introduction to the role of culture in technological diffusion. The rollout of the Internet in developing countries in general is then discussed, before moving on to investigate specific electronic business technologies and challenges facing their diffusion, culture being one such major challenge. The specific cultural elements which play a role in influencing electronic business diffusion are also discussed, with cases from developing countries presented so as to illustrate the effects culture produces on the levels of electronic business technology usage. Through combining the lessons learned from the cases discussed and the theoretical foundations presented throughout the chapter, the reader is finally provided with supporting guidelines useful for future electronic business implementation in developing countries.

Background

Introducing Developing Countries

What are *developing countries*? Before discussing the role of culture, an answer to this question provides an adequate first step to achieving the stated objectives of this chapter. According to the World Bank (1996), the main criterion for classifying countries and distinguishing different stages of economic development is Gross National Product (GNP) per capita. In this respect, countries are classified into three categories, namely, low-, middle-, and high-income countries, with the middle-income countries subdivided into lower-middle and upper-middle income groups. It is the low-income and middle-income countries that are generally referred to as *developing countries*.

The term *developing*, however, does not imply that the other countries have reached a preferred or final stage of development. It must be noted that the classification of countries by income does not accurately reflect the level of Internet or technology diffusion within a country, especially since high-income countries experience different levels of technology diffusion. The Internet and technology developments in the US, Canada and Australia, for instance, are far more advanced than in Kuwait or Qatar despite the fact that all of these countries belong to the same income group. The classification of countries by GNP per capita, however, provides a sufficient starting point for the definition of developing countries as deemed relevant to the objectives of this chapter.

The Role of Culture

Defined as such, it is safe to assert that developing countries are in fact usually perceived as being problematic hosts of information and communication technologies. This is primarily due to the fact that this group of countries generally lacks the economic resources and indigenous techno-scientific capabilities to develop and deploy modern information systems infrastructures. It also tends not to make the best use of opportunities of technology transfer, one of the factors largely hindering this transfer rate being again culture! Thus, a complete picture of the diffusion process of electronic business technology in the developing world cannot be created without an understanding of the role of culture in this process.

A review of the literature which has been published on the role of culture in electronic business diffusion over the past ten years confirms that culture remains to be cited by researchers as being related to the rate of national technology diffusion. The works of researchers such as Volken (2002), and Kumar, Maheshwari and Kumar (2004), support this assertion. It has also been acknowledged by researchers of cross-country comparisons that local culture is related to differences in information technology (IT) adoption amongst countries. This is since culture influences the perception of the extent to which IT is favourable for economic growth purposes (Gibbs, Kraemer & Dedrick, 2003).

Upon comparing advanced and developing countries, studies also demonstrate differing levels of IT adoption as well as differing business management concerns regarding IT adoption. It is observed that advanced countries tend to be more concerned with strategic issues regarding IT adoption such as the use of information systems for competitive advantage (Swan & Newell, 1995). In contrast, developing countries tend to have more operational concerns such as the availability of technical skills for information system development. In comparison with their advanced counterparts, developing countries also tend to have a relatively low appreciation for the potential and benefits of management information systems. This in turn naturally influences their rate of use of electronic business as well as its diffusion amongst their business community. Hence, it may be claimed that as long as developing countries have basic problems regarding their acquisition and operation of technology, with management and strategic issues neither being raised nor perceived as being of relevance, the hindrance of electronic business diffusion will continue to be the natural outcome (Cole & O'Keefe, 2000).

Issues, Controversies, Problems: Specific Cultural Influences on Internet Rollout in Developing Countries

Conceptual Framework for Electronic Business Diffusion in Developing Countries

A survey of the literature on Internet rollout in developing countries reveals five categories of components necessary for Internet rollout and use (Goodman et al., 1994). Researchers such as Lange (1995), Rogers (1995), and Youssef (2001) imply these to be: (1) national and organisational needs; (2) technology; (3) people and skills; (4) capital

Figure 1. Electronic business diffusion in developing countries

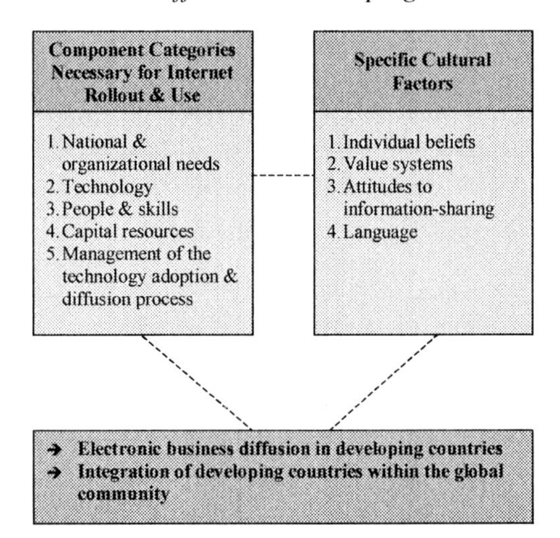

resources; and (5) management of the technology adoption and diffusion process. Youssef (2001), for instance, has claimed technological change to have proceeded slowly in developing countries due to poor research facilities, business cultures which do not support research in information technology (IT), and traditions which serve as obstacles to the quest for new knowledge. With these component categories being subject to cultural influences, this allows culture to play a significant role in shaping Internet rollout as evidenced by the experiences of a number of developing countries. This influential role played by culture is next illustrated by the author through discussions of cases of developing countries pertaining to each of the component categories necessary for Internet rollout and use. These discussions are based on the conceptual framework (Figure 1) which was developed by the author to conceptualise the roles of specific cultural factors in hindering or promoting electronic business diffusion through their influences on these five Internet rollout component categories.

Component Categories Necessary for Internet Rollout and Use

This section includes discussion of the five categories necessary for Internet rollout and use. First of all, concerning *national and organisational needs*, it must be noted that Internet use in a country should not just be applied for the sake of technology itself. Rather, it should be implemented with the vision of improving *the social as well as the economic conditions in both the country and its organisations*. Thus, the objective should be the improvement of both country as well as organisational performance and competitive advantage. *Adequate infrastructure* is also a requirement for Internet rollout as it facilitates Internet adoption at both the individual as well as the organisational levels. In fact, poor infrastructure presents serious barriers to Internet rollout in many developing countries, with connecting in rural areas being a major common challenge. Very few countries experience minor problems in connecting to remote and rural areas, such as China and Russia. In this respect, many researchers such as Rogers (1995) suggest the reliance on wireless technologies for solving this challenge as was successfully implemented in the case of Indonesia.

In fact, developments in *data communications technology* (particularly wireless technology) makes it possible to bypass poor infrastructure and gain access to faster and more reliable Internet connection. Besides, the use of these technological options is cheaper than using the old infrastructure and ensures equipment and protocol compatibility with newer services that the Internet might have to offer (Nelson, 1997).

People and capital resources are also essential influences in propagating Internet use. This is since employee training, as recommended by researchers such as McClure, Bertot and Beachboard (1995), becomes an important strategic investment for implementing Internet-based information systems in organisations. As for capital resources, the actual amounts are observed to differ from country to country depending on factors such as the quality of the available telecommunications infrastructure, government policies, and levels of competition in the Internet service provision market. Based on 1996 figures, for instance, a dial-up SLIP/PPP connection to the Internet cost US$16-$20 for unlimited use in Australia in comparison to US$90 in Mongolia (Goodman et al., 1994). Nowadays, however, free Internet access is not uncommon, facilitating Internet use and access to new knowledge for users. Nevertheless, ease of access from any location within the country (especially from remote and rural areas) remains a problem in most developing countries due to poor infrastructure.

Finally, the actual *management of technology adoption and diffusion* must be highlighted as an important factor for propagating Internet rollout and use. This is since, as evidenced by the experiences of developing countries such as Egypt and other countries in Africa (Mutume, 1997) and in South East Asia, Internet diffusion emerges as passing through two stages. This is since, although there is no doubt that the Internet is a technology-driven phenomenon, as noted by authors such as Lange (1995), for widespread use of this technology, it must meet: (1) business requirements and then also (2) serve as a strategic tool for creating future opportunities for organisations. For businesses to increasingly adopt such technology, however, adequate user support is required, as well as attendance to user needs and easy access. In this respect, Zamnet, a Zambian ISP, serves as a good example.

Specific Cultural Factors

Discussions so far have referred to culture in general without mentioning the specific cultural factors which tend to play a dual role in: (1) promoting or hindering electronic business diffusion; and (2) influencing the integration of developing countries' communities within the global community. These factors, as advocated in the works of researchers such as Al-Abdul-Gader (1999), Inglehart and Baker (2000), and Volken (2002), include: (1) individual beliefs; (2) value systems; (3) attitudes to information-sharing; and (4) language. The influences of these cultural factors become evident when their roles in specific country cases are examined. Discussion of such cases follows.

Looking at the case of India for instance, *language* emerges as an important factor having influenced business practices. This is since the emphasis of its *education system* on the importance of learning the English language has provided English-speaking companies with a rich pool of educated English speakers. This in turn has created preference for the software business in a country like India over a country like China. China on the other hand is preferred over India when it comes to manufacturing. This is partly because the government in China has intentionally encouraged the cultural elements needed to attract manufacturing business (Spiegel, 2004).

The importance of language can also be illustrated through citing two further examples. Firstly, during the mid-1990s, language served as a significant barrier in spreading the Internet in non-English countries, and secondly, in recent years, the integration of non-English languages into communications software packages such as in Mongolia, Vietnam, and Indonesia, seems to have contributed significantly to Internet rollout. There still remains, however, a need for creating databases and value-added information services in local languages.

The role of cultural factors in influencing business practices also emerges in the case of the poorer Asian countries. This is since in those countries, Internet service providers are usually accessible in industrial and business locations whereas extensions of their services to rural areas and more remote places are often given poor attention, especially since such extensions tend to be rather costly (Ang & Loh, 1996). This is the case in numerous developing countries, causing poorer and underprivileged citizens to have limited access to new knowledge resources. In fact, underlying cultural influences play a role in aggravating this lack of extension to rural and remote areas. This is due to the diversity in literacy rates amongst the urban and rural areas and extensions of Internet services to rural areas remain to be perceived by many as not worth the costs involved (Xiaoming & Kay, 2004).

As for the influences of *attitudes to information sharing* as a cultural factor moderating technology diffusion, these can be illustrated by the period covering the years 1995 and 1996. During this period, Internet developments in developing countries were highly accelerated as more countries started connecting to the Internet, hoping this would reduce their isolation and facilitate their economic development. Despite the currently increasing adoption of the Internet in developing countries and the increasing awareness of the need for global information-sharing, user attitudes remain oriented towards obtaining rather than providing information (Ang & Loh, 1996). This can be attributed to the fact that in most third world countries, information remains to be perceived as a source of power. As a result, the existence of a networking culture within and between organisations remains lacking (Beeharry & Schneider, 1996).

The developing world's *perception of information* also serves as a key influence shaping behaviour towards technology diffusion. Information tends to be perceived as a "source of power," thus by sharing information, people of the developing world believe they will lose this power. A good example illustrating this is the case of Tunisia and its early efforts in developing information systems infrastructure in scientific research institutes. At the very times when more strategic issues required addressing, questions of network ownership and management seemed to emerge, causing disputes amongst institute directors regarding information and network control (Ben Hanada & Cracknell, 1992).

Solutions and Recommendations

From the discussions presented in this chapter, several inter-related recommendations highlight themselves. These recommendations were derived by the author as being useful for effectively increasing the diffusion of electronic business in developing countries as well as for reaping the benefits generated from its use. Figure 2 was developed by the author to summarise and illustrate these recommendations and their inter-relations.

First of all, the discussions in this chapter imply that Internet and technology diffusion in general, and electronic business diffusion in particular, can be discussed at three levels: (1) national; (2) organisational; and (3) individual. These three levels, as has emerged in the discussions presented in this chapter, can be claimed to be interdependent, with all three being required for the overall economic and technological development of a country. This is since Internet diffusion within a country tends to consist of four areas of application, namely: (1) research; (2) education; (3) commercial (business) use; and (4) use by individuals (Figure 2).

Figure 2. Increasing electronic business diffusion in developing countries

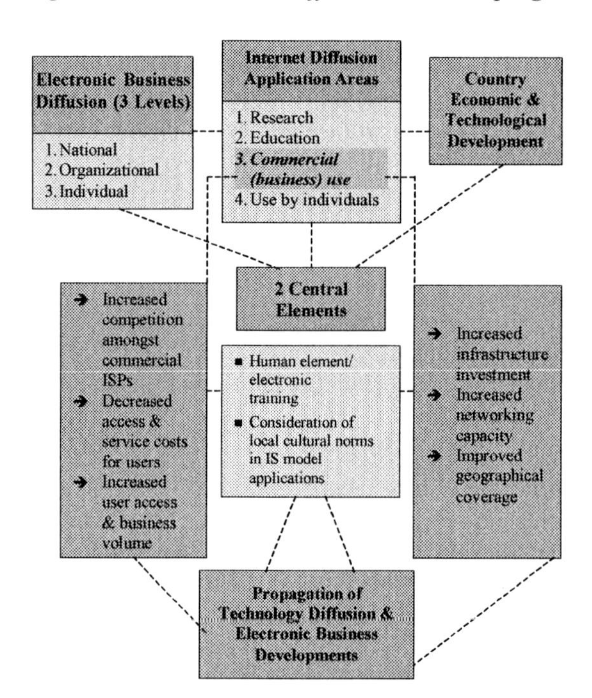

In fact, in both developing and industrialised countries, the Internet was first introduced by research institutes. In the US, for instance, the Internet was established by the NSF. As for the cases of Thailand and Morocco, it was introduced via the efforts of researchers. Kindly note, however, that although research institutes and educational organisations are important and active Internet users, Internet developments only accelerated when businesses joined the Internet community, since it is business use that seems to have not only continued but also propagated technology diffusion and electronic business developments.

Consequently, developing countries must keep this importance of the commercial use of the Internet in mind, especially since increased business use tends to lead to automatic increases in two areas, which in turn serve Internet diffusion in general and electronic business diffusion in particular, at all of the national, organisational, and individual levels. These two areas are: (1) increases in investments in infrastructure which lead to increased networking capacity, thus improving geographical coverage; and (2) increases in competition amongst commercial Internet service providers (ISPs) thus serving to decrease access and service costs for users, which in turn further increases user access and business volume (Figure 2).

Future Trends

The discussions presented in the previous section, pertaining to the important role played by culture in the electronic business diffusion process, imply a future need for directing attention to the "human element" (Figure 2) involved in this process (Efendioglu & Yip, 2004). As claimed by Kamel and Ibrahim (2003), electronic training is expected to increasingly become an essential factor for developing countries in meeting technological challenges and closing the technological gap between them and the developed world.

Volken (2002) provides further support for the importance of this "human element." This researcher concludes that even in cases where countries resemble one another in terms of material resources, differing levels of Internet diffusion are observed due to differences in human elements such as "trust," which nevertheless also indirectly affect material resources. This is since societal qualities such as "trust" are perceived to substantially reduce transaction, control and monitoring costs, thus freeing more resources for productive use. Moreover, "trust" between companies tends to expand the scope of action, allowing companies to enter into co-operative electronic business agreements and exchanges.

Discussions of the future of electronic business diffusion in the developing world, however, are not complete unless one additional issue is mentioned, namely the use of consulting services in information technology. Numerous organisations are resorting to the use of consulting services when it comes to information technology. In principle, this is not wrong. But it can be detrimental to business success in developing countries when local cultural norms are ignored and information systems (IS) models from developed countries are implemented in organisations and left to work flawlessly (Figure 2). Unfortunately, this is too often the case and thus is an important consideration for future information technology developments in the organisations of the developing world (Al-Abdul-Gader, 1999).

Despite all the challenges discussed, it can not be denied that the Internet is evidently spreading around the world. Governments and organisations in developing countries are increasingly realising the benefits and opportunities that the Internet presents for social and economic development, as well as for improvements in organisational and thus national competitiveness. Developing countries, however, must not forget that while the Internet may sometimes be a little more than a digital playground or vast shopping mall for many users in Western countries, for them it remains an essential key to economic survival.

Conclusion

In conclusion, it may be claimed that although there is no doubt that today's Internet-enabled world is driven by infrastructure, appliances, and e-services, the role of culture should never be undermined (Sheats, 2000). Technology is never a solution in itself. Delivery of computing services, their availability, and the provision of appliances suited to user needs are all essential for delivering technology with sustainable value to the developing world. But unless such delivery is consistent with cultural requirements, it will not be delivered (Inglehart and Baker, 2000).

The importance of the electronic business diffusion process in the developing world in general, and the role of culture in this process in particular, is emphasised further when viewed from a broader perspective. This is because electronic business diffusion in the developing world is in fact affecting global electronic business at large. With businesses and customers in different parts of the world constantly at differing technical maturity levels, this will continue to prohibit cross-cultural electronic business practices and dealings. Effective Internet rollout in developing countries is essential if these countries are to gain maximum benefit to be shared throughout their communities.

As such, unless cultural factors are given serious attention in developing countries and are acknowledged as being significant promoters or inhibitors to electronic business diffusion, "universal access to basic communication and information services... [will remain]... a distant dream" (Xiaoming & Kay, 2004).

References

Al-Abdul-Gader, A. H. (1999). *Managing computer based information systems in developing countries: A cultural perspective*. Hershey, PA: Idea Group Publishing.

Ang, P., & Loh, C. (1996). Internet development in Asia. *Proceedings of the INET '96 International Conference,* June 5-28, Montreal, Canada.

Beeharry, A. & Scheider, G.M. (1996). Creating a campus network culture in a newly developing economy. *Information Technology for Development, 7,* 3-16.

Ben Hanada, E. & Cracknell, M. (1992). Feasibility of a sustainable development network – Tunisia. *UN Development Programme* 30 (New York).

Cole, M. & O'Keefe, R,M. (2000, January). Conceptualising the dynamics of globalisation and culture in electronic commerce. *Journal of Global Information Technology Management, 3.*

Efendioglu, A.M. & Yip, V.F. (2004, February). Chinese culture and e-commerce: An exploratory study. *Interacting with Computers,* 16, 45-63.

Gibbs, J., Kraemer, K.L. & Dedrick, J. (2003, January). Environment and policy factors shaping global e-commerce diffusion: A cross-country comparison. *Information Society, 19,* 5-19.

Goodman, S. E., Press, L.I., Ruth, S.R. & Rutkowski, A.M. (1994). The global diffusion of the Internet: Patterns and problems. *Communications of the ACM, 37,* 27-31.

Inglehart, R. & Baker, W.E. (2000). Modernization, cultural change and the persistence of traditional values. *American Sociological Review, 65,* 19-51.

Kamel, S. & Ibrahim, M. (2003). Electronic training at the corporate level in Egypt: Applicability and effectiveness. *Industry & Higher Education, 17,* 409-417.

Kumar, U., Maheshwari, M. & Kumar, V. (2004). A framework for achieving e-business success. *Industry & Higher Education, 18,* 47-52.

Lange, L. (1995). The Internet: Where is it all going? *Information Week, 536,* 30-35.

McClure, C. R., Bertot, J.C. & Beachboard, J.C. (1995). Internet costs and cost models for public libraries. *US National Commission on Libraries and Information Science.* Washington, DC.

Mutume, G. (1997, February 25). Africa-telematics: Small service providers need to watch out. *African Network of IT Experts and Professionals.*

Nelson, R. (1997, March). Satellite is developing offline browser for Web access through e-mail. *African Network of IT Experts and Professionals.*

Rogers, E. M. (1995). *Diffusion of innovations* (4th ed.). New York: The Free Press.

Sheats, J. R. (2000). Information technology sustainable development and developing nations. *Greener Management International, 32,* 33-42.

Spiegel, R. (2004, March 3). A tale of two countries. *Electronic News (North America), 50.*

Swan, J. A. & Newell, S. (1995). The role of professional associations in technology diffusion. *Organizational Studies, 16,* 847-874.

Volken, T. (2002). Elements of trust: The cultural dimension of Internet diffusion revisited. *Electronic Journal of Sociology, 6.*

World Bank. (1996). *The World Development Report*: 188-239. Oxford University Press.

Xiaoming, H. & Kay, C.S. (2004, February). Factors affecting Internet development. *First Monday, 9.*

Youssef, S. (2001, June). Globalization and the challenge for developing countries. *World Bank*. DECRG.

Chapter III

International E-Commerce: Language, Cultural, Legal, and Infrastructure Issues, Challenges, and Solutions

Magdi N. Kamel, Naval Postgraduate School, USA

Abstract

The phenomenal worldwide growth of the Internet and the World Wide Web has made it an important vehicle for both business-to-business (B2B) and business-to-consumer (B2C) commerce. While the Web offers great opportunities for international electronic commerce by eliminating the barriers of time and space, language, cultural, legal, and infrastructure issues present major impediments to global Internet commerce. In this chapter we address language, cultural, legal, and infrastructure issues as barriers to international electronic commerce. Specifically the chapter presents a framework for identifying and addressing language, cultural, legal and infrastructure issues. The goal is to help organizations meet the challenge of providing online international visitors with a high quality experience, regardless of location, language, business practices, and culture, while complying with legal requirements.

Introduction

The phenomenal worldwide growth of the Internet and the World Wide Web has made it an important vehicle for both business-to-business (B2B) and business-to-consumer (B2C) commerce. Forrester Research predicts that in 2004, online commerce for both B2B and B2C will reach $6.8 trillion (Worldwide eCommerce Growth, 2001). International Data Corporation predicts that while only 26% of Internet commerce was conducted outside the US in 1998, the ratio is expected to be over 50% in 2004 (Glasheen, Gantz & Emberley, 2004). The opportunity is particularly great for certain countries with current low participation rates, like China, who in 2003 had about six million Internet users, but potentially can have over a billion users in the future.

While the Web offers great opportunities for international electronic commerce by eliminating the barriers of time and space, language, cultural, legal, and infrastructure issues present major impediments to global Internet commerce.

In this chapter we address language, cultural, legal, and infrastructure issues as barriers to international electronic commerce. Specifically, the chapter presents a framework for identifying and addressing language, cultural, legal and infrastructure issues. The goal is to help organizations meet the challenge of providing online international visitors with a high quality experience, regardless of location, language, business practices, and culture, while complying with legal requirements.

The chapter is organized as follows. A background section introduces the concepts of localization, internationalization, and globalization, presents the main issues of the chapter, and provides a literature review emphasizing the importance of the issues. The section entitled "Language Issues" addresses important issues of translating Web sites to other languages. These issues include what to translate, writing for translation, language tone and formality, automated translation and maintenance, translation for different dialects, text expansion from translation, translating graphics, serving the correct localized site, directing traffic to a multilingual site, and content management workflow. The section entitled "Cultural Issues" addresses considerations for developing Web sites for different locales and cultures. These issues include cultural subtleties, local standards and conventions, graphic and design elements, color, layout, and cultural attitudes toward e-commerce. Legal issues are tackled in the "Legal Issues" section and include discussion of jurisdiction, recognition of electronic contracts, consumer protection, dispute resolution, and privacy. A discussion of Infrastructure as a barrier of electronic commerce follows and includes issues of limited bandwidth, language representation technical issues, local connection costs, and support infrastructure. Finally the chapter concludes with a summary and discussion of future and emerging trends.

Background

It is estimated that about 70% of the Internet content available today is in English, yet 93% of the world are non-native English speakers, and about 65% of today's Web users are non-English speaking (Global Reach, 2004). Over the next few years, Internet use is expected to grow by 79% in Asia, 123% in Latin America, and over 2000% in Japan. As a result, by 2004, it is expected that more that two-thirds of Internet users will be outside the United States and that 65% of Web use and 52% of electronic commerce sales will involve at least one party outside the United States (Glasheen, Gantz & Emberley, 2004). Further, research have shown that customers are more likely to buy goods and services from Web sites in their native languages, even if they can speak and understand English well.

Realizing the global potential of the Internet, many companies in their efforts to reach customers in other countries begin by providing local language versions of their sites. However, translation alone is usually not adequate. Web sites need to be localized for each national audience by taking into consideration multiple elements of the local environment. *Localization* involves translating and culturally adapting software user interfaces, Web pages, visual and graphical elements, help files, documentation, and other content for one specific national audience (Schwartz, 2000).

Numerous studies in the literature stress that, in order to attract and retain more customers, it is vital to adapt the Web site to a local community. For example, Nielsen and Del Galdo (1996) emphasize that localization should encompass more that a "surface-level" adaptation, by acknowledging underlying cultural differences such as interface design preferences and the local culture's perception of usability. Evers and Day (1997) discuss the role of culture in interface acceptance. Barber and Badre (1998) suggest the existence of cultural markers, Web design elements that are prevalent in Web sites of different cultures. Further, Sheppard and Scholtz (1999) conducted pilot studies that suggest the absence or presence of cultural markers affects the user's performance. Marcus and Gould (2000) apply Hofstede's five cultural differences dimensions (1997) to global Web interface design and provide suggestions and guidelines to produce successful localized Web sites.

It is important to note that before attempting to provide local versions of their sites, through localization, companies wanting to do business in different countries need to consider first globalizing their e-business by making their products and services appealing enough to persuade people in other countries to buy them. *Globalization* addresses all of the logistical and organization issues, its supporting content, assets, and message, across cultures and markets.

To facilitate the process of localizing their Web sites, companies should consider Internationalization during the development of their main Web site. *Internationalization* involves making programs and their user interfaces more readily localizable by including support for issues such as handling non-English characters, sorting and string comparison, case conversion, parsing input, word delimiters, concatenated strings, date/time formats, measurement scale, etc. For example, the ASCII encoding scheme widely used to represent English characters uses a single byte to represent Latin characters. This scheme is not suitable for representing characters of other languages such as Japanese, Korean, and Arabic. An extended encoding scheme that uses two bytes, such as Unicode, is needed to represent characters of these languages. Internationalized software must be able to understand this type of extended encoding scheme. Additionally, a fully internationalized code should not contain hard-coded, locale-dependent content. Text strings should be stored in external files and then translated. The software must be able to recognize the coding scheme of the newly translated string and know how to process that language. Internationalization prior to localization allows companies to avoid the full development and testing cycle for each localized version. It also reduces maintenance cost since there is only one set of source code to maintain (Perinotti, 2001).

Language Issues

In this section we address some important translation and language issues, as translation activities are central to localization. These issues include what to translate, writing for translation, language tone and formality, automated translation and maintenance, translation for different dialects, text expansion from translation, translating graphics, serving the correct localized site, directing traffic to a multilingual site, and content management workflow.

What to Translate. Many companies provide a complete translation of all of the Web pages on their site. This approach however may not be feasible for others. The home page and all first-level links should be translated in all supported versions. Pages describing product information, marketing and establishing brand should also be included in the translation. The decision on which pages get to be translated should be made by the department responsible for maintaining that page. Pages devoted to local news and interests, might be created and maintained only in the local language (Schneider, 2003).

Writing for Translation. Writing for translation assists both the human translator and the translation software. Good writing practices not only improve the readability of a Web site, but they almost always make it easier to translate. In addition some practices improve the efficiency of the parsing operations used by the translation software, other practices improve the translation itself, and still other practices mitigate locale-dependent content, thus simplifying the translation and maintenance of local sites.

Practices that improve the efficiency of the parsing operation when translating from English to other target languages include the following (Harkus, 2000):

- Using an article or descriptor to clarify the part of the speech of a word. For example, using words such as "a" and "the" provide syntax cues to both translation software and human translators.
- Including relative pronouns even when they are not required. For example, "the book that he wrote" instead of "the book he wrote."
- Writing list items as complete clauses or complete sentences.

- Avoiding phrasal verbs, such as "set up" or "shut down," and using single word alternatives. Most commercial translation products are unable to analyze phrasal verbs, except for the most common ones.

Similar practices could be adapted from the list when translating from other languages. Practices that improve the translation itself include:

- Minimizing ambiguity.
- Avoiding homographs (words that are spelled alike, but have different meanings). If an English word has several meanings, it should be used consistently with one meaning and its alternate word to express other meanings. For example, if "right" is used as the opposite of "left," then the word "correct," not "right" should be used as the opposite to "wrong."
- Using words with their primary dictionary meaning.

Recommendations provided by the translation software should followed as much as possible to optimize the translation output.

Practices that minimize the impact of locale-dependent content include:

- Restricting the use of acronyms
- Avoiding abbreviations
- Using standard ISO notation
- Using month name in date to simplify interpretation
- Avoiding special characters that have varying applications (e.g., $, @, and #)
- Tagging all items that may be locale dependent (acronyms, measurements, dates, currency, etc.)

Language Tone and Formality. There are certain "levels" or "registers" of language used in particular situations in different countries that determine formality and directness. The use of humor, metaphors, and idioms varies from one country to another. It is important to understand what the norms in a target market are for particular types of texts in order to connect with potential clients and successfully market a product or service. For example, it is common in the US to see a Web site in which a registered user is greeted with the text: "Welcome back, Bob!" While it may be appropriate to address an American by his first name, a registered Japanese visitor to the above site could take offense at such a casual greeting. A more formal greeting such as, "We are honored by your return visit, Mr. Tanaka" would make the sales pitch more receptive.

Automated Translation and Maintenance. Translating and maintaining Web pages in multiple languages could be a costly endeavor. Companies should consider use of translation software for translating their Web sites. Software translation can reach speeds in excess of 400,000 words per hour compared to 400 to 600 words per hour for a human translator. Automated translation however is generally not good enough to be published without considerable human post-editing, and for key portions of the site a competent human translator is essential to capture the exact and subtle meanings of the text to be translated.

An attractive alternative to machine translation software is translation memory software, which is designed to enhance human translation effort rather than replace it. Translation memory software typically stores matching source and target language segments that were translated by a human translator in a database for future reuse. As the translation effort progresses, the translation database grows. New segments to be translated are compared with translated segments in the database, and the resulting output is reviewed and completed by a translator (Heuberger, 2001).

More importantly companies should consider software that automates the process of maintaining a Web site in multiple languages. This type of software typically tracks the portion of the site that needs translation and inserts the translations into the proper section of all sites that include that language. Using XML tags in the text of each translated Web site is a particularly good approach. Software that employs an XML approach relates each text element with a corresponding text element at the company's main Web site. When text in a page at the main Web site changes, the software sends a notification the shows which pages need to be updates and tracks the exact location of the change in every page needing updating. When the translation is completed, manually or automatically, the software automatically inserts the translated text in the correct locations.

Translations for Different Dialects. Some languages may require different translations for different dialects. For example, the Spanish used in Spain is different from the one used in Mexico, which is different from the one used in Latin America. The US spelling and meaning of certain words is different from the British Dialect. For example, the US spelling of "color gray" becomes the "colour grey" in Great Britain, and the meaning of "bonnet" changes from a type of hat in the US to the hood of a car in Great Britain. The Chinese language has two systems of writing: a traditional one used in Hong Kong and Taiwan, and a simplified one used in mainland China.

Text Expansion from Translation. Text will expand or contract when translated from one language to another. For example, when translating from English to European languages text expands by as much as 20% (Yunker, 2001). Conversely, Asian target text often contracts when translated from English. It is therefore important to allow enough space for the text to expand without impacting the overall design of the site. Particular attention should be given to providing extra white space for columns, frames and tables.

Translating Graphics. Graphics that contain text must be translated with the rest of the site (Hopkins, 2000). This can usually be accomplished by having graphic artists retouch GIF and JPEG files by cutting the source language text, repair the background and paste in the target language text using an appropriate font. A better approach is to use pictures with layers for each component (text, shadow, borders, etc.). This approach enables the localizer to remove text, translate it, resize shadows or borders, and reinsert the translated text into the layered file. It is recommended however to avoid text-in-graphics as much as possible to avoid reprocessing images during translation.

Serving the Correct Localized Site. The process by which a specific version of a Web site is displayed could be automated through the HTTP request message that a browser makes to a server when establishing a connection, or could be left to the visitor who must select one of the languages by clicking the appropriate link on the home page. Web sites that use the later approach should make sure that the links are clearly identifiable to their multilingual visitors. One approach is to present images of flags that represent the different available versions of the site. This approach is suitable for Web sites containing country-specific content, rather than language-specific content, because flags represent countries, not languages. Using a flag from Spain to represent Spanish may alienate users in Mexico or other Spanish speaking countries. A preferred approach for language-specific sites is to let users select a language-specific version of a Web site by clicking on buttons or navigational images with the name of each available language version written in that language (Hanrahan & Kwok, 2002).

Directing Traffic to a Multilingual Site. It is important to list a multilingual site with US and international search engines. Most US search engines, such as Yahoo!, Google, MSN, and Alta Vista, maintain international directories. It is also important that a localized Web site META tags are translated so a foreign language search engines can identify them.

Content Management Workflow. Content management workflow is a potentially powerful approach for Web sites' translation and localization. Today, most companies with a need to deliver information to a global audience still rely on a manual, paper-based process for localizing their content. Human intervention is required at all stages, from the origination of source language data through translation, review, quality assurance, and final delivery of localized information. There are inherent inefficiencies to this process, which lead to longer translation turnaround times and higher localization costs (Isogen International, 2002). A content management workflow system can streamline the localization processes and reduce the cost of delivering localized information products by automating the processes of translation, review of translation for cultural context, marketing approvals, and other related steps.

Cultural Issues

While translation is an important activity of localization, it is not sufficient alone. As discussed in the Background Section, cultural differences have a profound impact on the usability, acceptability, and user performance of Web sites. We define culture as the difference among people of different countries and regions in core values, traditions, customs, norms, opinions about social and ethical issues, business and social practices, tastes, gestures, treatment of people of certain gender or age, and so on. Companies developing Web sites for different cultures must learn and understand culture differences and reflect this understanding in the design of their sites. User interfaces, visual and graphical elements, help files, documentation, and other Web site content needs to be culturally adapted for each specific country or region.

In this section we address some of the important cultural areas that need to be considered when localizing Web sites. These areas include language and cultural subtleties, local standards and conventions, graphics and design elements, color, layout, and cultural attitudes towards e-commerce.

It should be noted that the Internet is particularly viewed as a challenge to cultural identity for its potential of imposing the culture of one country upon others. In particular, many countries view the Internet as dominated by an American culture and fear the potential of "Americanization" of its younger generations. As a result some governments feel compelled to limit Internet access to its citizens in an effort to curb the undesirable cultural influence.

Language and Culture Subtleties. Lack of language and cultural subtleties can have a dramatic negative impact on a company's efforts to sell its goods and services in different countries as the following classic examples presented in Schneider (2003) illustrate. General Motors was perplexed on why its Chevrolet Nova model was not selling well in Latin America, only to discover that "no va" means "it won't go" in Spanish. Pepsi's expensive "Come Alive" advertising campaign in China was a disappointment since the message came across as, "Pepsi brings back your ancestors from the dead." Gerber baby food did not realize that in Africa many people associate the picture on a container label with its content. Therefore, when Gerber started selling their baby food using the Gerber baby pictured on the baby food jars, Africans interpreted the labels as jars of human babies. When the Totes rain gear company was about to open a German office, the company decided it would simply call the new subsidiary "Totes Deutschland," only to realize after the press releases had gone out that "Totes Deutschland" means "dead Germany" (Sprung, 1999). These examples illustrate the importance of understanding language and cultural subtleties early on to avoid costly mistakes later.

Local Standards and Conventions. Careful attention should also be given to different standards and conventions used by different countries. While the US and the United Kingdom use the English system of weights and measures, the majority of the world uses the metric system. Differences also exist in representation of dates, times, number formats, addresses, and phone numbers. The US uses the format of month, day, year, while the rest of the world uses a day, month, year format. This could lead to confusion and misinterpretation for dates such as 4/6/01. Similarly, times are represented differently in different countries. Some countries use a twelve-hour time while others use a twenty-four-hour time (also called military time). Number formats differ mainly in the use of a decimal separator. Some countries use a period to represent the separator (e.g., USA), while others (e.g., Europe) use a comma for the same purpose. Address formats also differ from one country to another. Some countries represent the address in the format of street number, street name, and city name while others use street name, street number, and city name. Phone number formats also vary in different countries, with some countries including the country and city code as part of the phone number.

Graphic and Design Elements. Web designers must take care in the selection of icons and graphic elements on Web pages. Symbols that are commonly used in some cultures may not have the same meaning or impact in other cultures. For example, the shopping cart is an understandable metaphor for US shoppers. This symbol however may cause confusion with European shoppers who shop at the market with a basket. Another classic example is the American mailbox with a little flag to indicate there is new mail. This symbol used on many sites to indicate e-mail may not be recognizable by people in other parts of the world. Some hand and finger gestures are considered obscene in certain cultures. The use of certain images is considered inappropriate in some cultures. In India, for example, it is inappropriate to use the image of a cow in a non-revered setting.

Color. Color is another aspect of Web site that designers need to consider carefully as it could be interpreted differently by different cultures. For example, while black may be considered "cool" and "edgy" in the US, it is the color of mourning and has a sinister connotation in countries of the Middle East and in Europe. On the other hand, white (and also the number four) is the symbol of death in the Japanese culture.

Layout. Beyond avoid offending visitors from different cultures, companies should thrive to do more to attract visitors by catering to their visual and shopping preferences. Different cultures are attracted by different visual elements and their arrangement on a Web page. For example in Scandinavian countries, a crisp, less cluttered, clean design feels more comfortable to the visitors of these countries than one that is busy yet more informative.

Cultural Attitude towards E-Commerce. Consumers in many countries have a completely different view on electronic commerce from that of the US For example, many shoppers prefer to pay in cash or by cash transfer instead of a credit card. Many also have reservations on using a credit card online. Companies need to find innovative and unconventional solution to attract and encourage these shoppers to conduct business with them. An example of such innovation is eS-Books venture in Japan that allows customers to order items online, and then pick them up and pay for them in cash at the local 7-Eleven store.

Legal Issues

One of the biggest obstacles to international e-commerce is the lack of a single law that governs transactions on the Internet. The Web extends a company's reach beyond a single border, making it subject to potentially numerous laws of many countries. There are however certain issues that policy makers and companies conducting international electronic commerce need to work out, any one of which can seriously threaten the success of international e-commerce. In the following we discuss five key legal issues that affect international e-commerce: jurisdiction, recognition of electronic contracts, consumer protection, conflict resolution, and privacy.

Jurisdiction. Jurisdiction refers to the question of whose rules will apply to a transaction when the buyer and seller are located in different countries (Ham & Atkinson, 2001). Traditionally the jurisdiction of courts has been defined to coincide with geographic boundaries based on the notion that consumers will shop and do business near where they live. International electronic commerce throws this notion into chaos. For example, if a person located in Country A purchases an item from a supplier in Country B by connecting to a server in Country C, it is not clear whether the laws of Country A, Country B, or Country C would apply. With so many jurisdictions involved, uncertainties arise as to the regulations and mechanisms that should apply to protect both consumers and businesses. Left unresolved, businesses and consumers are likely to shy away from international electronic transactions and stick to domestic ones.

There are two main views to the issue of jurisdiction. The first view defines jurisdiction by considering a cross-border electronic transaction as taking place in only one country. This country could be either the buyer's country (country of destination) or the seller's country (country of origin). The second view favors making the laws and regulations the same in every country through a process of harmonization.

The country of destination view is supported largely by government regulators and consumer rights advocates. This view arises from the assumption that consumers are at a disadvantage during an international transaction and therefore should be protected by the legal system of their countries. In spite of its intuitive appeal, this approach would raise the costs of international electronic commerce and therefore reduce the number of companies engaged in its conduct, since sellers need to learn and comply with myriad laws of many countries. Such an undertaking would be too expensive and never be justified by the revenue returned.

The country of origin view is favored by the seller companies. In this view, buyers assume the burden of factoring the different legal requirements of the countries they are conducting business with into their purchasing decisions. Buyers however may be intimidated by the prospect of filing a lawsuit in a foreign country to resolve disputes with sellers resulting from either innocent misunderstanding or outright fraud. For this reason, the consumer will more likely to buy an item locally at a higher price knowing that he has the assurances of his own laws and the ability to take the seller to court and win an enforceable judgment if necessary.

Proponents of harmonization seek to unify the regulatory and tax structures that govern international electronic commerce. This harmonized structure would provide a predictable environment for both buyers and sellers who engage in cross-border transactions. This approach however would require a high level of international cooperation and agreement, something that is difficult to achieve. In addition, it is generally agreed upon that flexibility, rather than the enforcement of global laws is a better model for international electronic commerce.

Given the complexity of defining jurisdiction over global e-commerce, other approaches that combine governmental and private solutions are proposed. Example of such solutions include the following:

- **Active/Passive Presence.** Under this approach, jurisdiction is determined by the sophistication of the seller's Web site to conduct electronic commerce. A static Web site is considered passive and therefore is not subject to foreign jurisdiction. A dynamic interactive Web site with information exchange is subject to foreign jurisdiction depending on the level of interactivity and commercial nature of the exchange.
- **Jurisdiction based on targeting.** Under this approach, a company would be subject to the laws of another country if it targets its marketing at the consumers in that country. The definition of targeting however can be extremely complicated, and countries must agree, therefore, on what constitutes targeting.

- **Safe Harbor Agreements.** A safe harbor agreement is an agreement negotiated between two countries to reach a middle ground when the laws and regulations of the two countries are too different. Usually this middle ground agreement is less strict than the regulations of the more stringent country but more strict that those in the less stringent country. Each country's rules and regulations still apply on domestic transactions. As a result, safe harbor agreements can achieve some of the benefits of harmonization while maintaining sovereignty.
- **Private Contractual Jurisdiction.** Under private contractual jurisdiction, the buyer and seller agree on the terms of the transaction and how to enforce the contract privately. The power of this approach is that the buyer and seller can choose any law, although it is most likely that transaction-based private contracts will favor a country of origin rule.

Through a combination of these approaches a workable system for international electronic commerce can be built. It is impossible at this point however to predict exactly how the system will look like.

Recognition of Electronic Contracts. An important step in facilitating international electronic commerce is to remove any legal obstacles to the recognition of contracts entered into by electronic means. In many countries, the law requires certain contracts to be in writing, or to be signed. Two questions arise for contracts entered into by electronic means (GIPI, 2002): (1) Is an exchange of electronic messages a "writing"? and (2) Can an electronic notation serve as a "signature?" Other related questions when entering into electronic contracts — when will an e-mail message be considered sent, and when is it received, such that a party is bound by it?

Provision to address these issues can be drawn from the Model Law on Electronic Commerce, promulgated by the United Nations' Commission on International Trade Law (UNCITRAL) in 1996. The UNCITRAL Model establishes several principles of general applicability rather than requiring amendments throughout a country's entire legal code. The following are the main provisions of the UNCITRAL model (UNICTRAL, 1996):

- **Legal recognition of data messages:** Information shall not be denied legal effect, validity, or enforceability solely on the ground that it is in electronic form. (Article 5)
- **Writing:** Anytime the law requires a writing, that requirement is met by information in electronic form if it accessible so as to be useful for subsequent reference. (Article 6)
- **Signature:** Where the law requires a signature of a person, that requirement is met in relation to a data message if a method is used to identify that person and to indicate that person's approval of the information contained in the data message, and that method is as reliable as was appropriate for the purpose for which the data message was generated or communicated, in light of the circumstances, including any relevant agreement by the parties. (Article 7)

- **Original:** An electronic data message meeting certain functional criteria can be treated as an "original." (Article 8)
- **Retention of data messages:** Where the law requires that certain documents, records or information be retained, the Model Law specifies that such requirement is met by retaining data messages, provided certain specified criteria are satisfied (Article 10)

The UNCITRAL Model law recognizes that there might be some exceptions to the use of electronic documents, such as land transactions, divorces, adoptions, wills, where the existence of a signed paper original is still desirable. The exact list of exception is left to each nation to specify, based on local considerations.

Consumer Protection. Jurisdictional issues aside, existing national legal protections available to consumers in more traditional forms of commerce should be extended to online commerce. Consumers shopping online should enjoy transparent and effective protection that is no less than the level of protection provided to them in the off-line world. Specifically online laws should be enacted for fair business, advertising, and marketing practices; clear information about an online business's identity, the goods and services that it offers, and the terms and conditions of any transaction; a transparent process for the confirmation of transactions; the right of withdrawal; liability of defective products; secure payment mechanism; protection against fraudulent charges; fair, timely, and affordable dispute resolution; and privacy protection. As an example of areas of legal protection for online commerce, see the Directive 2000/31/EC of The European Parliament and of the Council of June 8, 2000 on certain legal aspects of information society services, in particular electronic commerce, in the Internal Market (Directive on electronic commerce, 2000).

Technology could be used to support and enhance consumer protection. Electronic agents and specialized computer software can be used to identify the country, laws, regulations, protections, and remedial options that will apply to a given transaction given the particular parties involved. After such disclosure, the parties can then decide whether to continue with the transaction.

Dispute Resolution. Another important challenge to international electronic commerce is how to resolve the disputes between buyers and sellers, such as failure to deliver requested goods or services, payment disputes, etc., in an effective yet inexpensive manner. In a majority of international electronic transactions, where relatively small amounts of money are involved, recourse to the courts, even when the jurisdiction is known, is not a practical option. To help address these situations, it is better to use alternative mechanisms that allow a fast, low-cost, and easily accessible resolution to these disputes. It is important therefore for governments to encourage the use of electronic alternate dispute resolution (ADR) mechanisms, in addition to the providing recourse through the court system.

An example of a low-fee, Internet-based mechanism of alternative dispute resolution is provided by SquareTrade (Han & Atkinson, 2001). When a dispute arises between two parties, each party connects to the SquareTrade Web site to tell its side of the story and answer a multiple-choice questionnaire. Based on the input of both parties, a computer automatically generates a list of options for resolving the dispute based on historical data of similar arbitrations. If the parties can agree on one of the computer-generated resolutions, the dispute is settled. If not, a human mediator takes over for further negotiations. If a resolution still cannot be reached, the dispute goes to binding arbitration. A major appeal of this approach is it low cost and accessibility. By automating the first level of dispute resolution, the expense of having expensive arbitrators spend time on uncomplicated disputes is eliminated. In addition, it as a lot easier to access the online ADR at any time than having to take time to go to court.

On the long run, ADR may prove to be a better approach for conflict resolution than dealing with complicated jurisdictional issues of international electronic commerce. By effectively resolving a vast majority of international electronic transactions, ADR can significantly reduce the complexity of a required legal environment to resolve electronic commerce disputes.

Privacy. Privacy issues are of particularly great concern to Internet users. Internet technology allows for the easy collection and wide dissemination of personal information and online behavior. The data collected could be used in objectionable ways, from minor inconveniences such as unsolicited mail ("spam") to potentially significantly harmful uses, such as employment discrimination and blackmail.

Different countries have different views of the proper balance between consumer privacy and corporate needs and hence approach privacy in different ways. The United States, for example, favors a sector-specific approach, mandating protections for sensitive privacy areas such as medical and financial data, self regulation by Internet companies, and placing the responsibility on the Internet users to decide if the privacy policy of a company meets their comfort levels. Other countries, the European Union in particular, view privacy as a human rights issue that should be regulated by the government to protect its citizens. In 1998, the European Union adopted the Directive on Data Protection, which mandated strict privacy policies for any Web site that collects data from citizens of EU states (Directive 95/46/EC of the European Parliament, 1995). The directive provides the following principles:

1. Personal data can be collected only for specified, explicit, and legitimate purposes and not further processed in a way incompatible with those purposes
2. Personal data may be processed only upon consent
3. Individuals or companies collecting personal data must identify themselves to the subject
4. Individuals have the right to obtain confirmation that data about them are processed, to whom the data are disclosed, and the source that provided the data
5. Individuals have the right to unconditionally prohibit the use of personal data in certain circumstances (e.g., direct marketing)

These provisions of the directive are in clear conflict to the practices of US companies. Recognizing that the restrictions would negatively impact international electronic commerce of many US companies, the US government negotiated a safe harbor agreement with the European Union, which entered into effect in 2000 (International Trade Administration Electronic Task Force, 2000). The safe harbor agreement is a voluntary self-registration scheme designed to bridge the gap between the traditional US approach of sector-specific self-regulation and EU legislative, top-down approach. US companies that wish to conduct business with EU member states must certify to the US government their compliance with the safe harbor principle either through self-certification or by joining a third-party seal program that guarantees compliance with the principles.

The privacy safer harbor agreement is a good model of international cooperation to promote the growth of international electronic commerce while respecting the sovereignty of the nations involved. It provides a predictable international regulatory environment without impeding the rights of governments to legislate domestic data privacy regulation.

Technology can also play an important role in privacy protection. One of the best examples is the Platform for Privacy Preferences Project (P3P) developed by the World Wide Web Consortium (W3C) (Berkey, 2002). This technology allows companies to write privacy statements in a language readable by a Web browser. A user specifies a desired privacy level in the browser. If the privacy level of a Web site is below the predetermined level set by the user, the Web browser warns the user.

Infrastructure Issues

Limited Bandwidth. In most countries connections to the Internet have limited bandwidth not suitable to download media rich in executable content (e.g., Java applets and Flash). The design of the Web site should take this fact into consideration by, for example, offering two version of the sites, one for wide-bandwidth and another for narrow-bandwidth lines. Additionally, hosting must be regional, and mirrored to ensure acceptable download times.

Language Representation Technical Issues. Presenting information using different languages presents some technical challenges. While English characters are represented using 7 or 8 bits, some languages require 16-bit extended-character sets that require additional efforts in programming Web and database servers.

Local Connection Costs. Unlike the US, which has a flat-rate access system, local connection costs in many countries can be very high, as the cost of a local call is proportionate to the duration of the call. It is therefore important to provide Web site visitors with the information quickly, with fewer bells and whistles than, for example, a corresponding US site. Many believe that the flat-rate access system in the United States has been a key factor to the success of electronic commerce.

Digital Security Infrastructure. A fundamental requirement for the recognition of electronic documents is the ability to verify the identity of a person online and to link a document to a particular person, ensuring that a sender of a message is the person he or she claims to be. Further, a receiver must verify that a document has not been tampered with during transmission or storage. Digital signature technology using a public key infrastructure (PKI) is the most commonly used technique to ensure these requirements. PKI requires a third party to verify identities through the use of public and private "keys." A regulatory structure for licensed certificate authorities that manage the key infrastructure is therefore needed. This does not mean, however, that every nation creates their own regulatory structure for licensed certificate authorities, since creating such structures is a complex effort in an evolving industry, and carries the risk of leading to a patchwork of national systems that adversely affect international legal interoperability. A better approach is to rely on a few, larger-scale schemes that could be tailored for a region, or a particular legal system. Such an approach would lead to increased international legal interoperability.

Support Infrastructure. An often-overlooked aspect of localization is the need to provide localized e-mail and phone support. When a company commits to an international market, it has to provide a corresponding support infrastructure — staffing up with support personnel fluent in the appropriate languages to manage all aspects of customer service and marketing including e-mail, phone, and Web-based help. This may require installing localized operating systems and e-mail software for communication and support. Because of the resources required for such an undertaking, many companies do not plan for a support infrastructure and often find themselves reacting to customer service issues on a "crisis-by-crisis" basis. A better approach is to plan and budget for a support infrastructure upfront and be clear on the Web site as to what types of support are or are not provided (Yunker, 2001).

Other Issues

In this chapter we focused on language, cultural, legal, and infrastructure issues as the main barriers to international e-commerce. There are however many other barriers to global e-commerce (Turban et al., 2004). They include other legal issues, financial issues, and security issues.

Other Legal Issues. Other legal issues not addressed in this chapter include export/import regulations, intellectual property enforcement, contract laws, tax laws, and cross-border transactions laws.

Financial Issues. Financial barriers to international e-commerce include custom duties, taxes, tariffs, currency exchange, and banking and electronic payment systems.

Security Issues. Security issues include identification and authentication of buyers and sellers, digital signatures, digital certificates, trust, and encryption standards.

Conclusion

Companies seriously targeting customers outside their countries and building Web sites for that purpose should give careful consideration to language, cultural, legal, and infrastructure issues. This chapter presented a framework for identifying and proposing solutions to address these issues.

It is important to note that the issues of international electronic commerce are both numerous and complex. It is likely that many issues have not been considered because they are not foreseen. As international electronic commerce continues to grow, new challenges, particularly legal ones, will most likely emerge, and some existing ones will likely disappear. As innovations on the Internet are occurring at lightning speeds, it is important for companies and international regulatory bodies to continually address these issues and develop strategies on how best to deal with them.

References

Barber, W., & Badre, A. (1998). Culturability: The merging of culture and usability. In *Proceedings of the 4th conference on human factors and the web*. Retrieved October 10, 2004, from *http://www.research.att.com/conf/hfweb/proceedings/barber*

Berkey, J.O. (2002, January 1). *Outline of international e-commerce regulatory issues*. Retrieved March 22, 2004, from *http://www.un.int/unitar/intel_nct_campus/ecom_regulatoryissues.pdf*

Bert, E. (2000). *A practical guide to localization*. Philadelphia: John Benjamins.

Del Galdo, E. M., & Nielsen, J. (Eds.). (1996). *International user interfaces*. New York: John Wiley.

Directive 2000/31/EC of The european parliament and of the council. (2000, June 8). Retrieved March 24, 2004, from *http://europa.eu.int/ISPO/ecommerce/legal/documents/2000_31ec/2000_31ec_en.pdf*

Directive 95/46/EC of the European parliament. (1995, October 24). Retrieved January 19, 2004, from *http://europa.eu.int/smartapi/cgi/sga_doc?smartapi!celexapi!prod!CELEXnumdoc&lg=en&numdoc=31995L0046&model=guichett*

Evers, V., & Day, D. (1997). The role of culture in interface acceptance. In *Proceedings of Human Computer Interaction, Interact'97* (pp. 260-267). London: Chapman & Hall.

Ferraro, G. (1990). *The cultural dimension of international business* (4th ed.). New York: Prentice Hall.

Forrester Research. (2000, June). *The multilingual site blueprint*. Retrieved January 19, 2004, from *http://www.forrester.com*

GIPI. (2002, May 21). *The Regulatory framework for e-commerce –International legislative practice*. Retrieved March 24, 2004, from *http://www.internetpolicy.net/principles/framework.pdf*

Glasheen, C., Gantz, J., & Emberley, D. (2004). *Worldwide Internet Usage and Commerce 2004-2007 Forecast: Internet Commerce Market.* Retrieved from *http://www.idc.com/getdoc.jsp?containerId=30949*

Global Reach (2004, September 30). *Global Internet statistics.* Retrieved October 4, 2004, from *http://global-reach.biz/globstats*

Ham, S., & Atkinson, R.D. (2001, March 15). *A third way framework for global e-commerce.* Retrieved March 22, 2004, from *http://www.ndol.org/documents/global_ecommerce.pdf*

Hanrahan, M., & Kwok, W. (2002, February 20). *Globalizing the web: how to refine the international user experience to create business results.* Retrieve February 23, 2004, from *http://www.ion-global.com/insights/pdf/globalizing.web_ion.global.pdf*

Harkus, S. (2001). *Writing for translation.* Retrieved, January 14, 2004, from *http://www.multilingualwebmaster.com/library/writing-TR.html*

Heuberger, A. (2001). *Machine translation vs. translation memory.* Retrieved January 14, 2004, from *http://www.multilingualwebmaster.com/library/mt-vs-tm.html*

Hofstede, G. (1997). *Cultures and organizations: Software of the mind.* New York: McGraw-Hill.

Hopkins, Jr., R. (1996). *Website translation: A primer for webmasters, authors and owners.* Retrieved January 19, 2004, from *http://glreach.com/eng/ed/art/trans.html*

International Trade Administration Electronic Task Force. (2000, July 21). Retrieved March 24, 2004, from *http://www.ita.doc.gov/td/ecom/menu.html*

Isogen International. (2002, December 17). *Reducing localization costs with xml-based technology.* Retrieved March 1, 2004, from *http://www.isogen.com/downloads/white_papers/xml_localization_costs.pdf*

Kalakota, R., & Robinson, M. (2001). *E-business 2.0: Roadmap for success.* New York: Addison-Wesley.

Marcus, A., & West Gould, E. (2000). Crosscurrents: Cultural dimensions and global Web user-interface design. *ACM Interactions*, *2*(4), 32-46.

Morrison, T., Conaway, W. A., & Douress, J. J. (1997). *Dun & Bradstreet's guide to doing business around the world.* Prentice-Hall.

Morrison, T., Conaway, W. A., Borden, G. A., & Koehler, H. (1995). *Kiss, bow, or shake hands: How to do business in 60 countries.* Adams Media.

Napier, H. A., Judd, P. J., Rivers, O. N., & Wagner, S. W. (2001). *Creating a winning e-business.* Cambridge: Course Technology.

Nielsen, J. (2000). *Designing Web usability.* Indianapolis, IN: New Riders Publishing.

Oz, E. (2004). *Management information systems* (4th ed.). Cambridge: Course Technology.

Perinotti, T. (2001). *Internationalization and localization tasks.* Retrieved January 22, 2004, from *http://tgpconsulting.com/articles/Intl-task.htm*

Schneider, G. P. (2003). *Electronic commerce* (4th annual edition). Cambridge: Course Technology.

Schwartz, H. (2000, September). *Going global: Hungry for new markets*. Retrieved February 18, 2004, from *http://www.webtechniques.com/archives/2000/09/schwartz/*

Sheppard, C., & Scholtz, J. (1999). In *Proceedings of the 5th conference on human factors and the web*. Retrieved October 10, 2004, from *http://zing.ncsl.nist.gov/hfweb/proceedings/sheppard/index.html*

Sprung, R. (1999, September). *What's in a name*. Retrieved October 10, 2004, from *http://www.devicelink.com/pmpn/archive/99/09/002.html*

Turban, E., King, D., Lee, J., & Viehland, D. (2004). *Electronic commerce 2004: A managerial perspective*. Upper Saddle River, NJ: Prentice Hall.

UNICTRAL. (1996). *UNCITRAL Model law on electronic commerce*. Retrieved March 24, 2004, from *http://www.uncitral.org/english/texts/electcom/ml-ecomm.htm*

Worldwide ecommerce growth. (2001, November 23). Retrieved February 23, 2004, from *http://www.global-reach.biz/globstats/index.php3*

Yunker, J. (2001). *Secrets of web site globalization*. Retrieved January 14, 2004, from *http://www.multilingualwebmaster.com/library/secrets.html*

Chapter IV

E-Commerce Infrastructure and Economic Impacts in Developing Countries: Case of India

Varadharajan Sridhar, Management Development Institute, India

Kala Seetharam Sridhar, National Institute of Public Finance and Policy, India

Abstract

This chapter presents a conceptual model that explains how e-commerce adoption in developing countries is affected by various infrastructure enablers and socio-economic variables. It describes the status of infrastructure enabler variables such as computer and Internet penetration, quality and speed of Internet connectivity, security infrastructure, online payment mechanisms and dispute resolution mechanisms in India and their impact on e-commerce adoption. Furthermore the chapter highlights the relationship between e-commerce adoption and various socio-economic variables such as prices, market reach, disposable income level, and cultural orientation of consumers. The chapter discusses the taxation of e-commerce, taking into account the complexity of the tax structure in India. A couple of mini-cases exemplify the utility of e-commerce in some practical applications. With this review of e-commerce adoption, stakeholders such as the government, the policy makers and industries will be able identify ways to nurture the positive effects and mitigate the negative effects to sustain the growth of e-commerce in many developing countries such as India.

Introduction

As with income, there are huge disparities in the adoption of e-commerce across developed and developing countries. Consider this: while total online retail sales for 2002 for the United States was $43.47 billion, it was just $15 billion in Asia-Pacific region, and a mere $4 million for Africa! Business-to-business (B2B) e-commerce transactions that contributed to more than 93% of US e-commerce revenue, amounted to $995 billion. In the Asia-Pacific region, B2B accounted for $120 billion, and in Africa, a mere $0.5 billion (UNCTAD, 2003).

Five years after Amazon opened its virtual store in the US, consumer e-commerce made its debut in India in 1999. This is when Fabmart and Rediff started their e-tailing operations, selling merchandise such as music CDs and books. Business-to-Consumer (B2C) electronic commerce in India is currently estimated to be about US$ 460 million and is expected to cross US$ 1 billion soon.

The growing importance of e-business in India provides motivation for work reported in this chapter. From our review of the literature, we note that there is lack of sufficient research regarding e-commerce and related issues, especially in developing countries (Sridhar & Sridhar, 2004). Primarily because the phenomenon is new and evolving, there is little quantitative data apart from whatever is found for the US and other OECD countries. In this chapter, we propose to answer the following questions:

1. What are the infrastructure enablers of e-business and their status in a developing country like India that has been at the forefront of the global software industry since the last decade?
2. What is the relationship between e-business and socio-economic variables that affect the development and sustenance of e-business in developing countries such as India?

It is relatively well-known that infrastructure enablers that determine the extent of e-commerce adoption include computer and Internet penetration, quality and speed of Internet connectivity, existence of security infrastructure, and existence of online payment mechanisms. We develop a conceptual model in this chapter that highlights the detailed relationships between these various variables. Besides describing the status of the state-of-the-art infrastructure that enables e-commerce in the Indian context, we highlight the impact of e-business on prices, market reach, and the government's tax base, using relevant cases. Note that, given the data limitations of this rapidly expanding field, we embark upon a qualitative rather than a quantitative approach in explaining the effect of various factors on e-business adoption.

Next section summarizes the small, but growing literature in the area. Then we suggest a causal model that highlights factors affecting e-commerce adoption. Subsequently, we elaborate on the infrastructure enablers, their importance for e-commerce adoption in the context of developing countries and their status in India. We then discuss the socio-economic factors relevant in developing countries that affect e-commerce adoption. Here we discuss the impact of e-commerce on the nature of markets, and several issues that arise in the taxation of e-commerce. The last section presents concluding remarks and prescribes future research directions.

Literature Review

The body of formal literature dealing with e-commerce in the context of developing countries is relatively small, but growing. We divide the literature in this area into those that explore the infrastructure aspects and those that deal with economic and taxation issues.

Literature Dealing with Infrastructure Aspects of E-Commerce

Using a case study from Thailand, Gray and Sanzogni (2003) describe areas that are prerequisites for the introduction of electronic commerce, namely telecommunications, Internet penetration and development, and technology parks. Muthitacharoen and Palvia (2002) find that acceptance of Internet stores is lower in developing countries where consumers still heavily depend on conventional stores. They recommend that attributes such as effective laws, and availability of credit cards have to be improved for better diffusion of e-commerce. Pavlou (2003) argues that consumer trust and risk constitute a tremendous barrier to online transactions and suggests ways by which online retailers can build trust to positively influence consumers purchase intentions. Other informal studies in the context of developing countries confirm this.

Perceptions of the characteristics of Web-based e-commerce and their effect on use intentions were studied among consumers in the US and India by Belanger et al. (2002). They point out that particular characteristics such as image, trustworthiness and relative advantage vary across countries and thus affect use intentions differently.

UNCTAD (2003) examines the scope of e-commerce in the marketing of agricultural commodities in developing economies. It points out that e-commerce can reduce the number of intermediaries in the agricultural supply chain and hence improve efficiencies and prices. Further, it points out that Business Process Outsourcing (BPO) services have the potential to become a larger part of B2B e-commerce in developing countries such as India.

Review of Literature: Economic Aspects and Taxation of E-Commerce

Panagariya (2000) discusses the main economic issues raised by e-commerce for the WTO and developing countries. He suggests that e-commerce be classified as trade in services with General Agreement on Trade in Services (GATS) applied since it would be fair to developing countries.

A large number of studies explore various aspects and implications of not taxing e-commerce, domestically and internationally, mostly using data from the US (Bruce & Fox, 2000; Goolsbee, 2000; Goolsbee & Zittrain, 1999) or EU or OECD countries (McLure, 2003). We find that none of these studies highlight the potential impacts of taxation pertaining to e-commerce in the context of developing countries.

Studies do, in fact, look at the impact of electronic commerce on tariff revenue, in the context of developed and developing countries. Teltscher (2002) presents data on potential revenue losses from import duties on a number of products that have been traded physically in the past but are increasingly being imported digitally. This study finds that while developing countries account for only 18.5% of world imports of digitized products, their share in tariff revenue resulting from these imports is roughly 65%. The study reports that the absolute value of the tariff revenue loss amounts to over US $600 million for developing countries compared to US $265 million for the developed countries.

Satapathy (2001) points out that in the two categories of printed matter and software, most products such as books and newspapers are totally exempt from customs duty in India, and so the fiscal losses (Teltscher, 2002) finds could be exaggerated.

While Teltscher (2002) takes into account only *goods*, Mattoo et al. (2001) primarily focus on the electronic supply of products, that is, *electronic services.* According to their estimates, tariff revenue currently collected from printed matter, recorded tapes, CDs, and packaged software represents on average, less than 1% of total tariff revenue, and a small 0.03% of total fiscal revenue. This study is optimistic and points out that these revenue losses, while small, have to be weighed against the revenue gains from increasing productivity due to e-commerce.

While the existing studies examine various aspects of e-commerce, we note that both qualitative and quantitative literature on e-commerce in developing countries is limited.

In the light of the literature review, we view the primary contributions of this chapter as being two-fold:

1. While there is general awareness regarding the enablers of e-commerce, we develop a more detailed conceptual model that shows the interactions among the various factors. We discuss the status of various infrastructure variables and their impact on e-business adoption in India.
2. While there is a growing body of literature on taxation of e-commerce, we place this in the context of developing countries. Further, we examine the relevant tax structure in India and in this framework, explore the implications of taxing e-commerce. We summarize implications for other developing countries.

Factors Affecting E-Commerce Adoption

We provide in this section a conceptual model that enhances the understanding of various factors that affect e-commerce adoption. We broadly categorize the factors that affect e-commerce adoption as being related to: (1) infrastructure; and (2) socio-economic factors.

Since most of the e-commerce applications operate over the Internet platform, Internet penetration and quality of Internet connectivity have a direct impact on e-commerce adoption. Since, in most of the developing countries, Internet Service Provisioning (ISP) industry is still evolving, deregulation and liberalization play a major role in promoting competition in the industry which, in turn, positively affects the Internet infrastructure variables. Most of the Internet connectivity takes place through PCs, and hence PC penetration has a direct impact on Internet access.

There is a dual relationship between e-commerce adoption and online fraud. As e-commerce adoption increases, online fraud is likely to increase. At the same time, due to instances of online fraud, the e-commerce adopters are likely to shy away from it. Online security services decrease instances of fraud. Security services are promoted by deploying appropriate security infrastructure and through the implementation of appropriate legal bills/acts. Online payment mechanisms tailored to the needs of the consumers in developing countries will have positive effect on e-commerce adoption.

Further, other infrastructure such as electricity and roads also need improvement, especially in developing countries. Good roads and logistic systems (such as courier services) expedite delivery of physical goods without damage, resulting in a positive effect towards e-commerce adoption. The maturity of the IT industry can also increase the adoption of e-business, especially the BPO services. We explain this further in a later section.

We note that per capita GDP and personal disposable income directly determine accessibility to PCs and Internet infrastructure, crucial for B2C e-commerce. Further, note that the correlation between per capita GDP and education is likely to be positive and high. So any increases in per capita GDP are likely to be accompanied by increased ability to use PCs, and hence access to e-commerce services.

Further, in B2C e-commerce, the nature of products and services sold determines the success of e-commerce adoption and its continuity. For instance, travel — the sale of railway and airline tickets over the Internet — and the sale of books and CDs have been the most successful so far, for obvious reasons. They can be quickly delivered, and the final product/service is not in the least different from what is expected. Similarly, in the case of B2B e-commerce, certain industries, goods and/or services are more amenable to e-business adoption than others.

In the event of problems with the delivery/quality of the physical good delivered, the customer has to have the support of a robust dispute resolution mechanism to settle issues with the merchant. If such mechanisms are not available or possible because of the lack of laws and consumer interest groups, then the e-business adoption tends to be lower.

Governments in most of the OECD countries have given incentives to promote e-commerce activity. This also has the positive affect of increasing the tax base which encourages the government to promote this sector of the economy.

In addition to the impact various socio-demographic and technological factors have on e-commerce, e-commerce also impacts the economy in various ways. The most important way in which e-commerce can affect functioning of the economy is through its impact on prices of products and services sold.

In Figure 1, we summarize these various factors that affect e-commerce adoption through a causal model, highlighting the existence of all major cause-and-effect links, indicating the direction (cause -> effect) of each relationship. The relationship is positive (or negative) if a change in the causal factor produces a change in the same (or opposite) direction in e-commerce adoption. A closed sequence of causal links represents a causal loop. The causal loop is positive if it has all positive links or even number of negative links. Otherwise, it is a negative loop. For example, as e-commerce adaptability and adoption increase, the market size and reach increase dramatically through network externality effects, which in turn affect prices. Specifically, as the market size increases, price is lowered due to the existence of scale economies. This price advantage in turn spurs further adoption of e-commerce leading to a positive cycle of growth as indicated in the causal loop diagram.

Figure 1. Causal loop diagram of e-commerce adoption

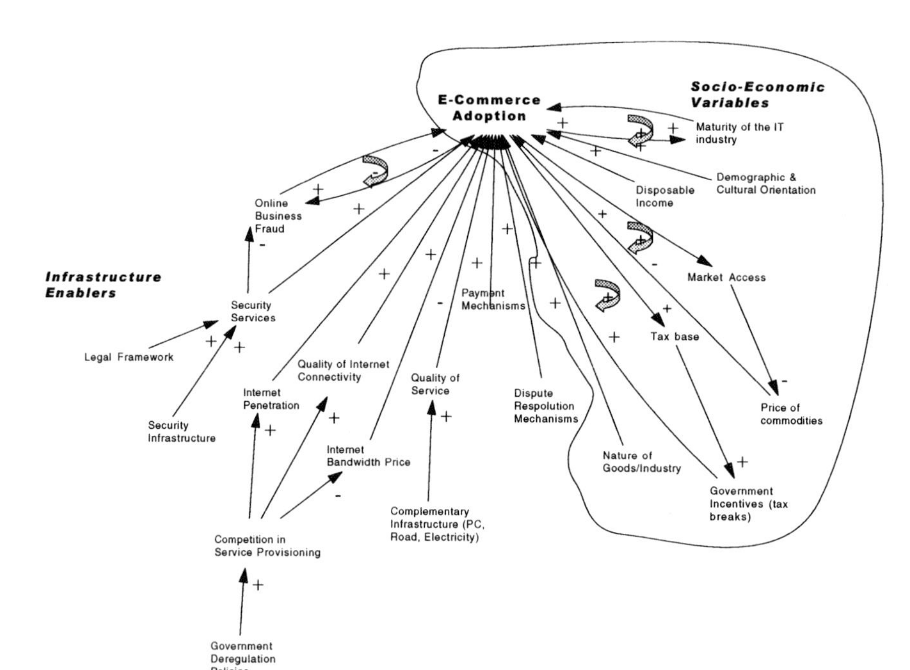

In the figure there are three positive cycles and one negative cycle. The positive cycles reinforce the extent of e-commerce adoption while negative cycles retard it. The extent of e-commerce adoption will be determined by the dominance of relevant factors and cycles.

In the following section, we illustrate the cause-effect relationships in detail.

Infrastructure Enablers and Their Impact on E-Commerce Adoption

Internet Penetration

The Internet is a key enabler and access to the Internet is crucial for adoption of e-business. Most of the developing countries lagged behind in building the Internet infrastructure. Table 1 shows the Internet and PC penetration in the world. Though Asia is catching up with Europe in number of Internet hosts, subscribers and PCs, the penetration per inhabitant is still very low.

While the Internet subscriber base is 3,262 per 10,000 inhabitants in developed countries in the year 2002, it is only 391 in developing countries (UNCTAD, 2003). The subscriber density in India stood at 159 in 2002. PC penetration in India is also very poor at 72 per 10,000 inhabitants. However, the Internet Service Provisioning (ISP) industry in India has seen a phenomenal growth of subscriber base at a Compounded Annual Growth Rate (CAGR) of 62% over the period 1998-2003, thanks to deregulation and competition introduced by the Indian government in 1998. Table 2 gives the Internet, PC and economic statistics for India for each year during 1998-2002.

Until November 1998, a monopoly government-owned operator was providing Internet services in India. The government's liberal ISP policy allowed for the market entry of an unlimited number of private ISPs with marginal license fee requirements. More than 175

Table 1. Summary of PC and Internet infrastructure in the world (2002)(Source: ITU, 2004)

Continent	Internet				Estimated PCs	
	Total Hosts	Hosts per 10,000 inhabitants	Users (in thousands)	Users per 10,000 inhabitants	Total (in thousands)	Per 100 inhabitants
Africa	243,171	3.01	9,965	123.35	9,579	1.30
Americas	122,555,360	1,450.21	217,649	2,575.99	239,717	28.95
Asia	13,390,474	37.08	211,361	584.69	157,893	4.45
Europe	18,358,407	230.38	172,918	2,169.95	167,430	21.44
Oceania	3,034,390	968.42	11,587	3,698.06	13,199	42.43

Table 2. Internet infrastructure and economic variables for India (2002) (Source: ITU, 2004)

	1998	1999	2000	2001	2002
Population (in Millions)	982.22	998.06	1012.40	1027.02	1041.85
Per Capita GDP (in 1995 US$)	430.01	453.38	464.60	482.97	474.00
Number of Telephone Mail Lines per 100 inhabitants	2.20	2.66	3.20	3.75	3.98
Number of Cellular Subscribers per 100 inhabitants	0.12	0.19	0.35	0.63	1.22
Number of PCs per 10,000 inhabitants	27.49	33.06	45.44	58.42	72.00
Number of Internet Hosts per 10,000 inhabitants	0.13	0.23	0.36	0.81	0.75
Number of Internet Users per 10,000 inhabitants	14.25	28.05	54.33	68.16	159.00

licenses were issued soon after the policy was announced and by December 2000, the Indian government had issued about 437 ISP licenses in the country. Though the government policy developed a competitive ISP market and gave a fillip to the subscriber base in the country, quality of service provisioning did not improve comparably.

Quality of Internet Services

Since the Internet is a network of networks, it is hard to define Quality of Service (QoS) for Internet access, as with other services. Typically, customer premise is connected to ISP's Point-of-Presence through a last-mile connection (i.e., leased line or dial-up) provided by the Basic Services Operator (BSO). Poor quality of the telephone lines and inadequate capacity of the switching exchanges have a negative effect on the quality of Internet connection. In India, basic telephone service provided by erstwhile monopoly government operators is notorious for poor quality, having an average of 150 faults per year per 100 main lines, one of the highest in the world (ITU, 2003). Most of the ISPs still lease transmission lines from telephone companies and hence do not have full control over the quality of service they provide. Even though the regulator, Telecom Regulatory Authority of India (TRAI), stepped in and announced guidelines for QoS of dial-up access network to the Internet, comprehensive QoS guidelines and their enforcement is still lacking in India (Sridhar & Jain, 2004).

The quality of Internet access is determined by the availability of bandwidth. While the US and other developed countries boast of a glut in Internet bandwidth, India was a bandwidth-scarce country until 2002. Internet connectivity is also expensive in developing countries. While it costs less than $4,000 per Mbps connectivity in the US, in India this cost is more than $20,000. This is mainly due to the lack of available bandwidth to meet the demand. In India, like in many developing countries, National Long Distance Service, Internet Service and International Gateway operations were provided by state-owned monopolies until 1999. The erstwhile monopolies did not have an incentive to improve the Internet infrastructure in the country. However with the recent opening up of all the above sectors for competition, bandwidth deployment is on the increase. This results in lowering Internet access prices, which will further encourage e-business adoption as depicted in the causal diagram. In fact, the bandwidth prices have reduced by more than 25% in the last year alone.

Security Infrastructure and the Legal Framework

Security remains the prime concerns of e-commerce users. Mukti (2000) specifies three types of security concerns of online shoppers: (i) payment security (ii) unauthorized access to information by hackers, and (iii) privacy of potentially confidential information. In a study (TNSI, 2002) conducted across 37 countries using over 40,000 interviews, it is reported that online payment security concerns proved to be the biggest impediment to the adoption of e-commerce.

In the transactional e-business environment, security services such as confidentiality, authentication, non-repudiation and message integrity are essential. A Public Key Infrastructure (PKI) enables users of an inherently insecure public network such as the Internet to securely and privately exchange data and money through the use of a public and private cryptographic key pair that is obtained and shared through a trusted authority. PKI automates the different processes and has been adopted by most of the countries for providing a secure Internet environment to facilitate e-business. PKI provides for a Digital Certificate that can identify an individual or organization and facilitates the conduct of secure e-business transactions (Stallings, 2004). Digital Certificates required for authentication and distribution of public keys have to be issued by Certifying Authorities (Panko, 2003). India's Information Technology Act 2000 (MLJC, 2000) specifies a broad framework for the development of Public Key Infrastructure (PKI) for conducting e-business. The Act has sections that deal with digital signatures, and the regulation of CAs.

In countries such as the US, certification services are not regulated. However, in India, the Controller of Central Certifying Authority (CCCA), set up by the Government of India, regulates certification services. The involvement of government in a PKI depends upon the extent of risk and the liability the CCCA is willing to take (Shiralkar & Vijayaraman, 2003). It is also expected that in evolving markets such as in India, a "closed loop PKI" system is adopted, in which the government reduces the risk and potential liability of a CA by acting as the highest level of CA. The Indian IT Act, among other functions, lists supervision of CAs, certifying public keys of CAs, and laying down the duties and responsibilities of CAs as the primary roles of CCA. The CCA is also entrusted with the responsibility of laying down standards to be maintained by CAs.

As e-commerce adoption increases, it is expected that there will be an increase in online fraud. As depicted in the causal model, an increase in online fraud dissuades users from engaging in e-business transactions. Indian IT Act defines various offenses including hacking, breach of confidentiality and privacy, misrepresentation, publishing false Digital Certificates and appropriate penalties for these offenses. Hacking has been adequately defined and punishment for hacking is prescribed. However, in India as in most developing countries, with the exception of a few states, government officials and police, in general, are not computer savvy. Hence cyber-crime investigations have been slow. However the Indian government created a Cyber Crime Investigation Cell within the Central Bureau of Investigation in 1999 to investigate computer crimes and frauds. The legal framework provided by IT Act and the deployment of PKI improved online security services and hence positively affected e-business adoption as shown in the causal model.

Payment Mechanisms

In Business to Consumer e-commerce, credit and debit cards are still the most accepted payment instruments. However, credit cards are much less common in India than in the developed countries. Numbers of credit and debit card holders in India are only about 12 million and 20 million respectively (out of a billion plus population). Moreover, credit/debit cards are used for less than 1% of consumer expenditure compared to 18% in case of developed economies such as the US. Hence, alternative payment mechanisms such as cash-on-delivery, pre-paid cards and debit cards have been developed. Such modes of payment require electronic transfer of funds which, in turn, require payment gateways for processing.

A payment gateway enables organizations to accept remote payments and directs them to banking systems and thus provides connectivity between Internet merchants, customers and financial institutions. The gateway (i) authenticates the parties involved in the online transaction, (ii) routes messages securely between the parties for authorization and settlement of the transaction, (iii) ensures integrity and privacy of all messages and (iv) provides administration support to the parties involved. A payment gateway is thus capable of linking the online buyer and seller, each with their respective banks/financial institutions. While B2C payment gateways deal with authorization and financial transfer of credit/debit card-based transactions, B2B gateways normally deal with inter-bank account financial transfers.

In India due to the relatively widespread use of B2C e-commerce activity, most of the foreign banks and larger Indian banks have set up their own payment gateways to link up with merchants. However, linking up inter-bank payment systems is still evolving in India. Less than 6% of banking transactions in India are electronic. The central Reserve Bank of India recently commenced live operations of the Real Time Gross Settlement (RTGS) system in March 2004, which allows settling of large value inter-bank transactions in an online, real-time mode. Electronic Clearing Systems for retail credit and debit have been deployed. These initiatives will help promote third-party payment gateways in India, and encourage e–commerce adoption.

Dispute Resolution Mechanisms

For any type of distance commerce system to succeed, there needs to be a trusted dispute resolution mechanism in place. Unfortunately, India lacks such a mechanism. Bringing suit against a merchant for a few thousand rupees may require more effort than it is worth (Van Slyke et al., 2003). Consumer rights are not as strong in India. For example, "no questions asked" return policies are much less common. In face-to-face shopping, the consumer has the ability to visually inspect merchandise. This feature is absent in distance transactions. This problem is especially acute when coupled with the weak dispute resolution system. In India, consumers cannot necessarily rely on their credit card company to nullify or suspend a transaction that is pending resolution of a dispute due to an unsatisfactory product delivery such as broken or missing parts.

Since traditional dispute resolution mechanisms cannot provide adequate redressal in e-commerce transactions where the parties do not meet face-to-face and there are differences in culture and language, Alternative Dispute Resolution (ADR) mechanisms have been devised. The main forms of ADR include arbitration, mediation, and negotiation processes that are effective in settling disputes out of court and in a less formal way than litigation in courts (UNCTAD, 2003). ADR when it takes place over computer-mediated communication is often referred to as Online Dispute Resolution (ODR). UNCTAD (2003) indicates that ODR/ADR is a process to which countries focused on expanding emerging e-commerce activities should pay particular attention. The vast majority of ODR providers are present in the U.S or Europe. UNCTAD (2003) points out that the market for ODR service in developing countries is very less. In India, there is only one listed ODR provider: cyberlaws.net (www.cybererarbitration.com). Such weak dispute resolution mechanisms have a negative effect on e-commerce adoption.

Complementary Infrastructure

For trading to take place on the Internet, other infrastructure such as electricity and roads also need improvement, especially in developing countries. It is estimated that high electricity tariffs and the need for power backups makes the cost of doing business in India about 8 times costlier.

Roads are in poor condition in India when compared to developed countries. The density of roads in 1999 was 2,561 kms when compared to 14,172 kms per million population for the United States. Only 20% of the paved roads are in good condition in India compared to the United States, United Kingdom, Italy and other developed countries (where more than 85% of paved roads are in good condition). This deters the effective delivery of goods, even if they were to be ordered through the Internet. In fact, the economic losses due to the poor condition of roads are expected to be around INR 200-300 billion a year in India.

In developing countries, where PC penetration is low, prices of PCs are high. Compounded with poor landline telecom infrastructure, there can be alternative end devices and networks that can serve as e-commerce platforms. For example in India, the cost of a PC is $800 compared to $80 for a normal GSM cell phone or $200 for a Multi Media Service capable cell phone. Developed by a team of Indian scientists and engineers, the Simputer (Simple Inexpensive Mobile People's Computer) is a hand-held device designed as a low-cost portable alternative to PCs. Equipped with a smart card reader and writer, running an open-source operating system, the Simputer can be personalized using a smart card for individual use on a changing basis (UNCTAD, 2003).

Reduced per line cost, quick deployment, better available technologies and regulatory reforms have sparked such growth of cellular services in developing countries (Sridhar

& Dutta, 2004). The total number of cellular subscribers in India is expected to increase from the existing 23 million to 40.1 million in 2006 and 72.3 million in 2011. In absolute terms, the number of mobile subscribers in India is more than the entire population of certain OECD countries such as Norway, Sweden, New Zealand, Ireland, or Switzerland! In a recent review of tariff by the International Telecommunications Union (ITU), cellular telephony tariffs are the lowest in India at $16 per month for a 300-minute basket while the same costs $96 in Mexico. All the above factors will provide the necessary impetus and critical mass for mobile commerce services (Sridhar & Dutta, 2004). For example, in India most of the mobile operators provide personal services (e-mail, jokes), financial services (banking and trading), entertainment services (news, sports, games, contests), and travel services (schedules, travel alerts). It is expected that mobile commerce services will also be extended to retail services in the near future and thus complement PC-based e-business adoption.

E-Commerce and Socio-Economic Variables

In this part, and in the various sub-sections that follow, we explore the relationship between e-commerce and socio-economic variables — primarily market access and prices, nature of good or service, purchasing power, cultural factors, and finally, taxation, all in the context of the Indian subcontinent.

Impact on Market Access and Prices

As indicated in Sridhar (2003), e-business does change a real market to be perfectly competitive, making possible the presence of large number of buyers and sellers in the market, and the availability of perfect information. E-business conducted through an Internet platform can lead to reduced transaction costs, disintermediation or the emergence of new Internet-based intermediaries, price transparency, and possible redistribution of earnings along the supply chain (UNCTAD, 2003). The process of disintermediation can save costs of distribution, allow the organization to differentiate its products and services and/or focus its attention on selected segments of the market (Whiteley, 2001). This in turn has a reinforcing positive effect on e-commerce adoption as indicated in the causal diagram.

This section will highlight the above principle through a mini-case where a large Indian agricultural conglomerate is using e-business to directly deal with suppliers in rural and remote areas of the country, eliminating middlemen and thus reducing transaction costs and improving the quality of their products along the supply chain.

Case of e-Choupal

In India, agriculture contributes to about 34% of GDP. There are more than 600,000 villages in India that have a rural population of about 70%, of which more than 65% are engaged in agriculture. Rural teledensity in India is very low. About 90,000 villages do

Figure 2. Traditional agricultural supply chain in India

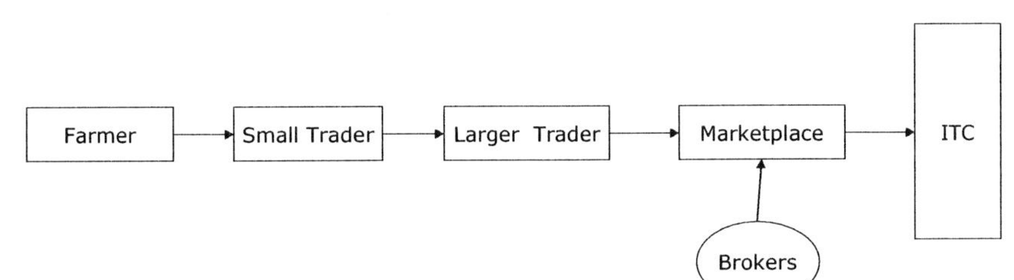

not even have a single telephone. Further, only 57% of the roads are surfaced. It is against this poor infrastructure that Indian Tobacco Company (ITC) set up e-commerce infrastructure to link up farmers in remote villages and to improve its supply chain. ITC's International Business Division exports bulk products like soybean, sugar, wheat, rice and oil worth INR (Indian rupees) 4.5 billion. Traditionally ITC has been getting the agricultural produce from village farmers using the model shown in Figure 2 (Sarvani, 2003). The farmer sold his/her agricultural produce to a small trader. The small trader then sold the products to a larger trader who then sold the produce in a "mandi" (market place). The whole procedure involved many intermediaries which resulted in procurement costs being very high. This also resulted in losses in transit and reduction in quality of the produce. There were also gaps in the information flow from the market place to the farmer. Farmers had to get weather reports and prices of commodities from the intermediaries. The intermediaries leveraged this to extract more margin and the farmers did not get their fair share of prices. ITC's main objective was to reduce the number of intermediaries in the traditional model, which in turn is expected to improve the information flow to the farmers, improve quality, and result in better prices.

In the year 2000, ITC introduced an e-business initiative called "e-Choupal" (www.echoupal.com) in rural India to link up soya and wheat growers in the villages. To emerge as a leading agri-based company in India, ITC has decided to stress Information Technology as a tool to evolve from a commodity exporter to a major player across the agricultural value chain, both as a buyer and provider of high quality goods and services. The e-Choupal model is shown in Figure 3.

ITC first opened collection centers at strategic locations in rural India to act as hubs around which the e-Choupals were built. Then ITC created the role of "sanchalak" — the convener, and assigned one of the head farmers in the village to this role. The sanchalaks were chosen carefully so that they have widespread acceptance within the farmer community in the village. A PC and VSAT (Very Small Aperture Terminal)/telephone equipment to connect to the Internet were installed in the Sanchalak's house.

Farmers grow wheat across several agro-climatic zones, producing grains of varying grades. Though these grades had the potential to meet diverse consumer preferences, all varieties were aggregated as one average quality in the market place. Now, farmers could log on through the Internet kiosks at the sanchalak's place to get prevailing market prices for their crops at home and abroad. The e-Choupal site helps the farmers to

Figure 3. The e-Choupal model (Adapted from Business World, 2003)

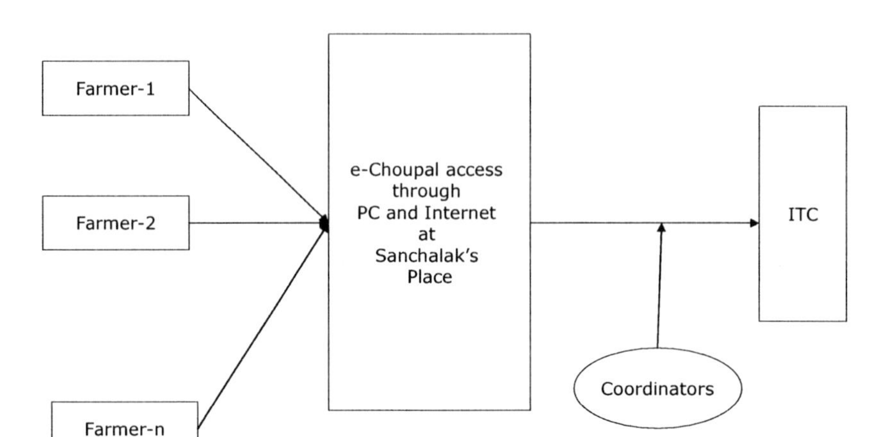

discover the best price for their quality at the village itself. The sanchalaks guide the farmers to search and select appropriate information.

After checking the prevailing prices at the e-Choupal, farmers could directly go to the collection centers to sell their produce and get their payment. The sanchalaks were paid 0.5% of the procurement price for each ton of soya procured by ITC from their choupals (Business World, 2003). ITC also devised the role of "coordinators" who act as links between farmers and the Sanchalak and also help the farmers bring their commodities to the collection centers. In this model, farmers could complete their transactions within few hours unlike many days taken previously (*Business World*, 2003). This also reduced ITC's total cost of procurement from INR 700 per ton of Soya to INR 500, a reduction of more than 25%.

Since ITC directly picks up commodities at the collection centers in each village, quality of various types of grains is maintained across the supply chain. In the very first full season of e-Choupal operations, soya farmers sold nearly 50,000 tons of their produce through the Internet platform (ITC, 2003). So far, the network of 4200 e-Choupals has empowered 2.5 million farmers in 24,000 villages across seven states of India. In the next decade, ITC plans to install 20,000 more e-Choupals, to cover over 100,000 villages in 15 states which will benefit around 25 million farmers (ITC, 2003).

ITC also started "reverse trading" to further leverage the business potential of the choupals (Business World, 2003). ITC brought in fertilizer companies to market their products and services through e-Choupals. These companies displayed their products on the site, offered special prices and also trained farmers in product usage, all in local languages. They also booked orders from farmers and delivered the products though sanchalaks. The company pay nominal fee of 10% of the value of transaction to ITC. Each sanchalak gets a commission of 5% on sales for all products sold through their choupals. Having expanded the agri commodity sourcing model through e-Choupals, ITC plans to brand the commodity exports and acts as sourcing base for global retail giants like Wal-Mart (*Economic Times*, 2004). In another innovative move, ITC began selling life insurance policies through e-Choupal network with a revenue sharing arrangement with insurance companies.

How did ITC manage the e-Choupals in the absence of complementary infrastructure facilities, such as telephone connectivity, electricity and PCs? In many villages, ITC set up on its own, Very Small Aperture Terminal (VSAT)-based Internet access kiosks at a cost of about INR 100,000 per station, thus bypassing the telecom service providers. In villages where telephone connectivity existed, the quality was very poor for Internet access. ITC upgraded the telephone line infrastructure in these places. Backup batteries are used to power the kiosks to manage electricity outage problems in villages (Business World, 2003). As per ITC's estimates, the investment could be recouped within 18 months (Sarvani, 2003). Having succeeded in e-Choupal, the positive reinforcement cycle encouraged ITC to launch other related e-business initiatives. In December 2000, the company launched "Plantersnet" (www. Plantersnet.com) to deal with coffee farmers and set up 75 kiosks to cover 6,000 coffee farmers in 125 villages (Sarvani, 2003). In November 2002, ITC created an infrastructure through 72 kiosks in remote coffee growing areas of the country and launched India's first online coffee auction system. It planned to auction 10,000 tonnes of coffee in the first year of operation. "Aquachoupal" (www.aquachoupal.com) was launched in February 2001 and set up 55 kiosks covering 10,000 shrimp farmers in more than 300 villages (Sarvani, 2003). Recently ITC announced setting up of rural development fund for INR 3.5 billion and that part of this fund will be used for improving the e-Choupal infrastructure.

The above case illustrates the penetration of ICT and associated e–business service to Indian villages made possible by the initiatives of a business entity. Since the cost of providing ICT services in rural and remote areas is high and the revenue potential is low, private companies are unwilling to roll out rural services. Hence, the responsibility for rolling out ICT in rural areas is borne by government operators. Though India witnessed an unprecedented growth in telecom penetration with teledensity shooting up from 2.86 in 2000 to 7.15 in 2004, the rural areas of the country are deprived of even basic connectivity. Nearly 85,228 Indian villages (out of 600,000) do not even have a single telephone. Rural teledensity is as low as 1.58 (per 100 inhabitants) compared to an urban teledensity of 21 (tele.net, 2004).

Following are ways by which ICT infrastructure in rural areas can be improved:

1. A regulatory driven model in which the license conditions of telecom services force the service providers to go to rural areas.

When the telecom service was privatized in 1997 in India, a rural telephony obligation was built in to the license conditions for private basic service operators. However, this model was not successful as the service providers preferred to pay the penalty instead of rolling out infrastructure in rural areas because of reasons mentioned above. Seeing this trend, the government set up a Universal Service Obligation fund in 1999, to provide subsidies to telecom operators in setting up infrastructure in rural and remote areas of the country. Despite these efforts, the private operators so far have provided only 12,581 Village Public Telephones against the license commitment of 98,000 (tele.net, 2004).

2. A socio-economic approach in which a sustainable business model is built around local communities where funding is provided either solely or jointly by a business entity.

A successful case using this approach is the Village Phone initiative in Bangladesh (Sridhar et al., 2002). Bangladesh's Grameen Bank with expertise in village-based micro-enterprise and micro-credit combined with Grameen Telecom, which provided wireless technology to provide low-cost cellular services in villages of Bangladesh. e-Choupal initiative is similar in nature where the ICT infrastructure is built by ITC (and not by the telecom service providers) to provide sustainable e-business services to the villages of India.

This case illustrates the reinforcing causal relationships between market access, price and their effect on e-business adoption.

Nature of Goods/Industry

Success of e-business depends on the nature of goods/services rendered. Most of B2C e-commerce success stories such as Amazon.com revolve around the sale of CDs and books. At Amazon, these contribute to 87% of the North American sales revenue. The rest is contributed by electronics and tools. In India, it is no different. Though some have diversified into offering groceries and jewelry, the major revenue of the popular B2C sites such as Fabmall.com, Rediff.com. and Indiatimes.com, is still derived from selling CDs, DVDs and books. For e-business to succeed, a critical mass of buyers is required. Apart from the consumer goods, goods/services that have huge demand, could be targets for porting onto e-business platforms. The following mini-case illustrates an e-commerce success story launched by the Indian Railways.

Internet Ticketing in Indian Railways

Indian Railways (IR) is the world's second-largest railway, with 6,853 stations, 63,028 kilometers of track, 37,840 passenger coaches covering over 100,000 track kilometers and 222,147 freight cars. Annually it carries some 4.83 billion passengers and 492 million tons of freight (IR, 2004). Of the 11 million passengers who climb aboard one of 8,520 trains each day, about 550,000 have reserved accommodations. Their journeys can start in any part of India and end in any other part, with travel times as long as 48 hours and distances up to several thousand kilometers. The challenge is to provide a reservation system that can support such a huge scale of operations — regardless of whether it's measured by kilometers, passenger numbers, routing complexity, or simply the sheer scale of India.

IR has been a pioneer in the use of IT in India. IT was first adopted in the 1960s when computerized passenger and freight revenue accounting, operating statistics, payroll and inventory management were introduced using computers placed in zonal data centers. The Indian Ministry of Railways established the Centre for Railway Information Systems (CRIS) in 1986 for planning and implementing all computer activities of IR. Over the years CRIS has built elaborate and well established computer systems to automate passenger reservations. The computerized Passenger Reservation System (PRS) was deployed by CRIS in 1985. Passengers could book tickets at PRS booking terminals located at station counters. PRS was implemented in five regional passenger reservation centers, each of which was a stand-alone site with its own local database. Hence reservations were only restricted to trains departing from the respective regional centers. This resulted in very long queues in station counters with average waiting time to book tickets running in to many hours.

During the mid-to-late-1990s, CRIS introduced Country Wide Enhanced Reservation and Ticketing system (CONCERT), which linked the five passenger reservation centers so that reserved tickets from any station of Indian Railways could be issued to any other station from any station counter. By 1999 all the PRS centers were networked together. The system is now handling over 800,000 reservation transactions per day. This not only offered flexibility and convenience to passengers but also resulted in drastic reduction of length of queues at booking counters. However, passengers still had to go to station counters to book their tickets.

Since trains are the most commonly used mode of transport in India, information regarding journey planning, fare enquiries, accommodation availability status, and ticket reservation status are of utmost importance to the common public. Till now, traditional methods of enquiring such as enquiry counters, announcements on TV and radio or Interactive Voice Response System on telephones were used. PRS handles 200,000 enquiries per day using the telephone-based enquiry system. However, there were shortcomings to these methods such as long waiting times in queues, and telephone line congestions.

In August 2002, IR launched the Internet Ticketing and Delivery services. The e-commerce site was planned and deployed by CRIS and Indian Railway Catering and Tourism Limited (IRCTC), a public sector company set up and fully owned by the Indian Ministry of Railways. IRCTC started providing facilities for purchasing passenger tickets through its Web site (www.irctc.co.in). This move by IRCTC has completely changed the history of train ticket booking system in India. With the implementation of Internet Ticketing and Delivery System, passengers could book their tickets from any station to any other station without even having to go to booking counters. Passengers can register and log onto the IRCTC Web site for information and booking. After finding out the availability of tickets on the travel date, passengers can fill in particulars such as name, age, preferred seat locations and payment details. Credit and debit cards are accepted for payment. The transaction is processed securely through an electronic payment gateway certified by Verisign. An e-mail conformation is sent regarding the reservation of seats along with Passenger Name Record (PNR). Tickets are delivered to the shipping address through courier for nominal charges. Passengers can also collect tickets from nominated counters. In addition to reservations, the e-commerce site handles about 700,000 passenger queries per day.

In the first 10 months of operations, the transactions grew nearly 2000%. An estimated 197,105 users are currently registered for this service. Average transaction per day is over

1,000 and is cited as the biggest e-commerce sites in Asia (outside of Japan). Most of the users (close to 63%) are youth between 18 and 25. Roughly 2% of the users are in the age group between 61 and 90 (Prasad & Sahay, 2003).

IR has already embarked on the next phase in train ticket booking. As discussed earlier, mobile services have grown tremendously in India in the last couple of years. More than 7 million text messages are being sent over Indian cellular networks per day and close to 80% of Indian mobile users use Short Message Service (SMS) in metro areas (Dutta & Sridhar, 2004). Most of the Indian mobile service providers have upgraded their networks from 2G to 2.5 G capable of providing high-speed data connectivity. General Packet Radio Service (GPRS) and Wireless Application Protocol (WAP)-enabled hand-sets are available in the Indian market. Some of the operators have already rolled out Enhanced Data Rates for Global Evolution (EDGE) services for high-speed mobile data services (Dutta & Sridhar, 2004). With these network upgrades, apart from voice and SMS, Multimedia Message Service (MMS) is also being offered by mobile service providers. Catching onto the opportunity fueled by mobile services growth, IRCTC introduced ticket booking through mobile phones in September 2004. Train tickets can be booked using either voice recognition system (http://www.irctc.co.in/ voice.html), or SMS or using enhanced data communication (http://www. irctc.co.in/reliance.html). This offers the passengers the flexibility of booking their tickets from any where from any station to any other station.

IRCTC's success in Internet Ticketing and Delivery system has prompted other service providers such as telecom companies and electric utility companies to provide payment

Table 3. List of select e-commerce companies in India (Source: http://202.41.106.14/ ~mahadev/indecom.htm)

E-Business Service	**Companies**
E-Auction	Baazee.com (B2C), auctionindia.com (B2B for industrial machinery), rosebazar.com (B2B for Flowers), teaauction.com (B2B for Tea), trade2gain.com (B2B for excess inventory)
E-Banking	Utibank.com, icici.com, pnbindia.com. statebankofindia.com
E-Education	Careerlauncher.com, egurucool.com, zeelearn.com,
E-Procurement (B2B)	Seekandsource.com, 01markets.com
E-Recruitment	Naukri.com, placementindia.com
E-Shops/E-Malls	Fabmall.com (books, music, groceries, jewelry, computers, electronics, toys, apparel), indiatimes.com (books, consumer electronics, gifts. Music, videos, magazines, apparel), floweracrossindia.com (flowers), rediff.com (books, music, gifts, jewelry, flowers, electronics, toys, computers), homeindia.com (jewelry, apparel, handicrafts, gifts), firstandsecond.com (books, music, videos)
E-Ticketing	Indiatimes.com, jetairways.com, airsahara.net, Indian-airlines.nic.in (flight tickets), irctc.co.in (train tickets),
E-Trading	ICICIDirect.com, 5paisa.com, Ittrade.com, kotakstreet.com, sharekhan.com, geogitsecurities.com
E-Travel	Etravelindia.com
Agricultural Portals	Indiaagronet.com, kisan.com, ikisan.com,
Automobile Portal	Automartindia.com, apnacar.com
B2B Exchanges	CommodityIndia.com (agricultural commodities), logisticsfocus.com (logistics), matexnet.com (machinery and materials), metaljunction.com (materials), steelexchangeindia.com (steel)

services through the Internet. This case shows that if convenience and flexibility are provided, depending on the nature of the goods or services, customers are willing to embrace e-commerce even in developing countries. Though e-business is still evolving in India, a number of companies offering a variety of innovative services have come up. Table 3 gives a representative list of such services and companies.

Purchasing Power and Its Effect on E-Commerce Adoption

Developing countries such as India have very low per capita GDP ($474 compared to the world average of $5164). The impact of increase in per capita GDP is translating to an increase in personal disposable income, and its effect on increased demand for telecom services is demonstrated by Sridhar & Sridhar (2004). The same holds true for e-commerce activity as well. With more than 70% of the population living outside metropolitan areas and close to 67% engaged in agriculture, B2C adoption limited in India. However, a recent survey in India has revealed that the percentage of middle- and high-income households who are potential adopters of e-commerce services has increased from 42% in 1992-93 to about 70% in 2001-02. It is expected that this will increase to 85% in 2006-07. With more than a billion people, this represents a potentially huge market for e-commerce services.

It is observed that national culture impacts the potential adoption of e-commerce. Tarafdar and Vaidya (2004) note that cultural preferences for physical facility-based preferences dominate over e-commerce in the Indian context. Differences in perception of e-commerce between Indian and American consumers were noted by Belanger et al. (2003). They point out that it could be due to differences in national cultural dimensions such as power distance, individualism, and long-term orientation. It is observed that majority of users of e-commerce (e.g., e-ticketing by Indian Railways) belong to the age group of 25-40. The cultural orientation of the youth is an enhancer of e-commerce. For example, in India, it is estimated that more than 22% of the population are in this age group. In 2001, an estimated 63.2 million Indian youth lived in urban areas who were potential adopters of e-commerce services (Sridhar & Dutta, 2004).

Maturity of the IT Industry and E-Business Adoption

UNCTAD (2003) shows that more investment in the IT industry in developing countries can improve their ability to advance in e-business. Dai and Kauffman (2002) cite that B2B e-markets not only support the transaction of goods, but also promote information flows along the related business processes, enabling collaboration between business partners via workflow management, project management and supply chain management. An extension of these is to provide help desk services, customer relationship management, and back office processing of business transactions. These activities, collectively referred to as IT Enabled Services (ITES), form an important part of the export-oriented, B2B e-commerce market.

Companies in developed economies such as the US and UK outsource ITES to low-wage countries such as India for cost advantages. ITES in India has registered a growth of about 71% this year, and generated a revenue of $3.4 billion. More than 170,000 are employed in this sector. It is expected that this sector will touch $57 billion by year 2008, employing more than 4 million and contributing to about 7% of India's GDP. Growth in ITES in India though is related to many factors such as availability of English speaking graduates, large technical manpower, and comparably low wages. It is also attributable to the reputation of the IT industry. Indian IT industry generates a revenue of $16 billion and is a global force in software development. It now employs more than 650,000 people and contributes to about 3% of the country's GDP. Software service companies in India have leveraged their global presence and expertise to offer superior ITES services to major clients in the US and other developed countries. Such electronic delivery of services will give a boost to India's B2B e-business activities. Relationship between maturity of the IT industry and e-commerce adoption is a two-way process. As the above e-business services grow, there is a corresponding growth in the maturity of the IT industry itself, thus resulting in a self sustainable positive loop as indicated in Figure 1.

E-Commerce and Taxation Issues

The growing volume of e-commerce transactions promises a good tax base for a developing economy such as India's that is much starved for necessary resources for spending on many useful programs. However business groups around the world argue that if electronic commerce were to be allowed to grow in an uninhibited way, there would be a general reduction in transaction costs everywhere with resulting output increases, thus widening the tax base (Bruce & Fox, 2000). For instance, states such as Virginia in the United States that host e-commerce companies such as America Online would be hesitant to tax e-commerce transactions. Similarly, strong lobbies of the IT sector in India such as the National Association of Software Services and Companies have been against taxation of e-commerce based on location of the server in India, for fear that it will nip the budding sector's growth (NASSCOM, 2004).

It is actually a myth that e-commerce is not taxed elsewhere. Bagchi (2001) points out that network service providers are required to charge a tax on consumers in several countries (e.g., Japan and Korea) though in a haphazard way. Once the issue of whether or not to tax e-commerce is sorted out, the next question is how. In fact, the question now being asked in the EU countries is precisely this. There are some problems with taxation of Internet transactions. Traditional tax models such as destination-based principles in taxation break down with e-commerce for several reasons.

First, there is the question of which jurisdiction (where the vendor is located or the one in which the customer is located) has the right to tax. Second, it is difficult for the vendor to find out where exactly the customer is located. Further, even if this is known, more frequently than not, the vendor may not have any contacts in the customer's location to collect tax (Houghton & Cornia, 2000), although it is necessary for them to track all transactions within and outside their jurisdictions. In the case of B2B transactions, it may well be the case that there could be several users of the goods and services, which they could digitally access from several locations.

Another contentious issue with international trade is that it is not clear whether e-commerce transactions are to be treated as goods or as services in the context of domestic taxation. McLure (2003) quoting from elsewhere, defines e-commerce as the use of computer networks to facilitate the production, distribution, sale and delivery of goods and services. Panagariya (2000) provides a useful rule for international trade. In international trade, there is no ambiguity if a good is ordered for over the Internet and the good were to be delivered in the conventional manner. Any such transaction would be treated as a good and GATT (WTO) trade in goods would apply. With respect to internal trade, ambiguity arises only if goods were to be *transmitted electronically* across the Internet. Satapathy (2001) gives a list of such goods, including newspapers, journals, other printed matter (which may be scanned and hence sent in digitized form, (CDs, tapes, software, etc.). Essentially then the electronic delivery of goods could be classified as services. So, based on the preceding discussion, taxation of e-commerce transactions in the Indian context is relevant as they apply to sales taxes (levied on goods) and service taxes (levied on services).

There is no ambiguity in the case of e-commerce transactions pertaining to physical goods. A problem arises if services are transmitted electronically across the Internet (for instance, downloads of music CDs and tapes from some remote server, upon payment). How can such transactions be identified and taxed, especially if the final consumers are based outside the country?

The Indian Government constituted a High Powered Committee (HPC) in December 1999 to examine the position of e-commerce transactions under existing taxation laws to determine any changes to be made and consider the possibility of taxing e-commerce transactions. The HPC submitted its report to the Central Board of Direct Taxes, Ministry of Finance, Government of India in September 2001. The Group's report highlights the importance of e-commerce in India with reduced barriers to entry, and observes that e-commerce would increase the GDP of the global economy including that of India. The Group advocates a cautious approach pertaining to taxation of e-commerce, as inappropriate taxation has the potential of harming developing countries. For instance, some existing businesses could prosper, and others could fail as a result of e-commerce, and uniform taxation could have unintended consequences.

Yet another approach to taxation of e-commerce is to tax net income derived from e-commerce. The HPC advocates this, but discourages the "base erosion approach." This is an approach where tax at low rates on all cross-border payments would be imposed for goods and services on a gross basis in lieu of an income tax. We should note, however, that taxation of income from e-commerce would be subject to similar ambiguities as those that apply to consumption — how to identify the sources of such income, if not declared?

In general, the HPC makes the assumption in its report that e-commerce will cause a significant tax base erosion for India. As the report by Nishith Desai Associates (2002) points out, however, e-commerce business methods are available to all business enterprises, and all states can potentially develop globally competitive export sectors. This means that e-commerce will lead to an increase in the market size and bring about scale economies and productivity increases, although due to data limitations, there are no studies to date that measure these impacts.

A milestone in the path of tax reform in India that is expected to be accomplished soon is introduction of the Value Added Tax (VAT). The rationale for VAT is that the current tax structures in Indian states are cascading, currently being levied at multiple levels that represent stages in the manufacture of a good (inputs, intermediate, and final goods). Further, this would be an integrated tax on goods and services. However, the introduction of VAT requires substantial work in terms of standardization of procedures, transactions, and computerization of records at various levels of sale and purchase in the various Indian states, for which some are ready, and some of which are not. The standardization is necessary to enable the granting of tax credits that have been paid for at the various intermediate levels of the manufacture of a good. Given the substantially different tax systems Indian states have, the best we can hope for is a sub-national VAT, which means each state's VAT rate and structure could be different. Further, as Bagchi (2001) points out, with the bulk of e-commerce in India being likely B2B, any tax on e-commerce will add to business costs. These would be over and above the costs businesses in India already face with poor infrastructure.

So it is neither a certain nor a clear proposition that in the Indian context, given the volume of transactions (even B2B), whether e-commerce transactions are taxable, and if so, how. We believe that the taxation of e-commerce in the context of India or other developing countries at this point, is premature, and should await the better development of supporting infrastructure. The taxation of e-commerce, in the developed Americas, OECD, and EU countries, could, however, be reality soon.

Summary and Future Directions for Research

In this chapter, we have presented a conceptual model for analyzing e-commerce adoption in the context of developing countries. We have analyzed the effect of infrastructure variables such as Internet penetration, security infrastructure, online payment systems, dispute resolution mechanisms, cyber laws, and complementary infrastructure such as roads and electricity on e-commerce adoption. We have also investigated the relationship between e-business and many economic characteristics such as market access, prices, nature of goods, and social characteristics such as demography, purchasing power and cultural orientation. We have also investigated issues pertaining to taxation of e-commerce. We provide two mini-cases to demonstrate the potential benefits and limitations of e-commerce in India. We note the effects of a matured IT industry in India on e-commerce adoption and describe how they continue to mutually enhance each other.

We have provided detailed causal relationships in our model that are empirically estimable, once data become available. Currently, the lack of reliable data does not permit empirical investigations. For example, if data on the value of B2C e-commerce transactions were to be available, we could estimate the revenue losses that are foregone by not taxing e-commerce. In the absence of reliable data on various model variables, we could use "system dynamics" methodology advocated by Forrester (1985) to simulate the cause-effect relationships and estimate the extent of e-commerce adoption, given the values of the exogenous variables. Using the simulation model, one can also perform sensitivity analysis on the extent of e-business adoption due to changes in both infrastructure and socio-economic model variables. This analysis will enhance our understanding of the interaction of the variables and help us recommend strategies for businesses, and policy directives for the governments to actively promote e-commerce adoption in the country.

Based on our exploratory work, and the model we develop, there is considerable scope for substantial increase in e-commerce activity depending on the techno-economic characteristics we highlight. These are likely to have substantial positive effects on rural communities as well, in many developing countries. Given this comprehensive view of e-commerce adoption, the major stakeholders such as the government, the policy makers and industries should strive to nurture the positive effects and mitigate the negative effects to sustain the growth of e-commerce. Substantive quantitative research in this area is possible when such data become available.

References

Bagchi, A. (2001, February 26). Taxing e-commerce: Some points to ponder. *The Economic Times*.

Belanger, F., Van Slyke, C., & Sridhar, V. (2002). Comparing the influence of perceived innovation characteristics across countries. *Proceedings of the Fifth International Conference on Electronic Commerce Research*, Montreal, Canada, October 23-27.

Bruce, D., & Fox, W. (2000). E-commerce in the context of declining state sales tax bases. *National Tax Journal, 53*(4), 1373-1388.

Business World. (2003, January 20). ITC's rural symphony. *Business World*, 30-37.

Dai, Q., & Kauffman, R. (2002). Business models for Internet based B2B electronic markets. *International Journal of Electronic Commerce, 6*(4), 41-72.

Economic Times. (2004). ITC to focus on commodities, go slow on tobacco. Retrieved July 28, 2004, from *http://www.economictimes.com*

Forrester, J. W. (1985). *Industrial dynamic.* Cambridge, MA: MIT Press.

Goolsbee, A. (2000). In a world without borders: The impact of taxes on Internet commerce. *Quarterly Journal of Economics, 115*(2), 561-576.

Goolsbee, A., & Zittrain. (1999). Evaluating the costs and benefits of taxing Internet commerce. *National Tax Journal, 52*(3), 413-428.

Gray, H., & Sanzogni, L. (2003). Technology leapfrogging in Thailand: Issues for the support of e-commerce infrastructure. *The Electronic Journal on Information Systems in Developing Countries, 16*(3), 1-26.

Houghton, K., & Cornia, G. (2000). The National Tax Association's project on electronic commerce and telecommunication taxes. *National Tax Journal, 53*(4), 1351-71.

Indian Railways (IR). (2004). Passenger reservation system. Retrieved October 1, 2004, from *http://www.indianrail.gov.in/abcrisprs.html*

Indian Tobacco Company (ITC). (2003, May). *Transforming lives and landscapes: ITC's rural development philosophy at work.* Kolkata, India: Corporate Communications Department, ITC Limited.

International Telecommunications Union (ITU). (2003). *Yearbook of statistics.* Geneva, Switzerland: ITU.

International Telecommunications Union (ITU). (2004). *Internet indicators: Hosts, users and number of PCs.* Retrieved April 1, 2004, from *http://www.itu.int*

Mattoo, A., Perez-Esteve, R., & Schuknecht, L. (2001). Electronic commerce, trade and tariff revenue: A quantitative assessment. *World Economy, 24*(7), 955-970.

McLure, C. (2003). The value added tax on electronic commerce in the European Union. *International Tax and Public Finance, 10,* 753-762.

Ministry of Law, Justice and Company Affairs (MLJC), Government of India. (2000). *The Information Technology Act 2000.*

Mukti, A. (2000). Barriers to putting business on the Internet in Malaysia. *The Electronic Journal on Information Systems in Developing Countries, 2*(6), 1-6.

Muthitacharoen, A., & Palvia, P. (2003). B2C Internet commerce: A tale of two nations. *Journal of Electronic Commerce Research, 3*(4), 201-212.

National Association of Software and Services Companies (NASSCOM). (2004). E-commerce in India. Retrieved April 15, 2004, from *http://www.nasscom.org*

Nishith Desai Associates. (2002). *Taxation of electronic ccommerce in India.* New Delhi: Taxmann Allied Services.

Panagariya, A. (2000). *E-commerce, WTO and developing countries, policy issues in international trade and commodities study series.* Geneva: United Nations Conference on Trade and Development.

Panko, R. (2003). *Corporate computer and network security.* New Delhi: Pearson Education.

Pavlou, P. (2003). Consumer acceptance of electronic commerce: Integrating trust and risk with the technology acceptance model. *International Journal of Electronic Commerce, 7*(3), 101-134.

Prasad, R., & Sahay, S. (2003). E-ticketing! Milestone in Indian Railway. *E-Commerce,* 12-18.

Sarvani, V. (2003, July). ITC's e-Choupal: Taking e-Business to farmers. *Case Folio,* 7-19.

Satapathy, C. (2001, September 29). WTO work program on e-commerce: Strategy for further negotiations. *Economic and Political Weekly,* 3665-3668.

Shiralkar, P., & Vijayaraman, B. (2003). Digital signature: Application development trends in e-Business. *Journal of Electronic Commerce Research, 4*(3), 94-101.

Sridhar, K. (2003, February 3). Why new economy lags the old. *The Business Line.*

Sridhar, K., & Sridhar, V. (2004). *Telecommunications infrastructure and economic growth: Evidence from developing countries (National Institute of Public Finance and Policy Working Paper No.14/2004)*. New Delhi: National Institute of Public Finance and Policy.

Sridhar, V., Arun, J., Chetan, R., Jayesh, E., Vipul, D., & Vishwadeep, S. (2002, January). Rural telephony: Community way. *Voice & Data*, 84-86.

Sridhar, V., & Dutta, A. (2004). Prospects of m-Commerce in developing countries. *Annual Review of Communications*, *57*, 337-341.

Sridhar, V., & Jain, P. (2004). The elusive last mile to the Internet. *Annals of Cases on Information Technology*, *6*, 540-560.

Stallings, W. (2004). *Cryptography and network security: Principles and practices*. New Delhi: Pearson Education.

Tarafdar, M., & Vaidya, S. (2004). Adoption of electronic commerce by organizations in India: Strategic and environmental imperatives. *The Electronic Journal on Information Systems in Developing Countries*, *17*(2), 1-25.

Taylor Nelson Sofres Interactive (TNSI). (2002). Global e-Commerce Report. Retrieved September 20, 2004, from *www.tns-global.com*

Tele.net. (2004). Rural rollout. *Tele.net*, *5*(8), 10-12.

Teltscher, S. (2002). Electronic commerce and development: Fiscal implications of digitized goods trading. *World Development*, *30*(7), 137-1158.

United Nations Conference on Trade and Development (UNCTAD). (2003). *E-Commerce and Development Report 2003*.

Van Slyke, C., Belanger, F., & Sridhar, V. (2003). Comparing the influence of perceived innovation characteristics across countries. *Proceedings of the Fifth International Conference on Electronic Commerce Research*, Montreal, Canada, October 23-27, 2002.

Whiteley, D. (2001). *E-commerce: Strategy, technologies and applications*. New Delhi: Tata McGraw-Hill.

Chapter V

E-Marketplace Adoption Success Factors: Challenges and Opportunities for a Small Developing Country

Andreja Pucihar, University of Maribor, Slovenia

Mateja Podlogar, University of Maribor, Slovenia

Abstract

This chapter introduces e-marketplace adoption success factors as a challenge and opportunity for a small developing country. The chapter offers insights into e-marketplace definitions with e-marketplaces' business models and business processes. Further it describes opportunities for, and threats to e-marketplaces' use. Success factors of e-marketplace adoption are divided into three groups: organizational factors, e-marketplace factors and environmental factors. The authors argue that each of these group influences significantly an organization's e-readiness for e-marketplace adoption. The importance of each of these factors is described on the basis of the results of research, conducted in 119 large organizations in Slovenia. Furthermore, the authors believe that by understanding these factors, organizations will be able to prepare better for successful e-marketplace adoption and successfully exploit important competitive advantages offered by new e-commerce business models.

Introduction

The use of the Internet and Electronic Commerce (EC) in the late 90s caused the disappearance of intermediaries between buyer and seller. The seller could sell products and services directly to the buyer without an intermediary in between (Shaffer & Zettlemeyer, 1999; Wigand & Benjamin, 1995; Choi et al., 1997). With advances in the development of electronic commerce technologies, new types of intermediaries were established. These new intermediaries were offering new value-adding services. New intermediaries invariably attract additional buyers and sellers with new services that support trading processes (Clarke, 2001; Chircu & Kauffman, 2000; DeSisto, 2000; Chircu & Kauffman, 1999; *The Economist*, 1999; Barling & Stark, 1998; Zwas, 1998). This new way of doing business describes the e-marketplace that represents one of the new business models developed in the late 90s. The e-marketplace is the result of using innovative technology in business processes.

E-marketplace might be viewed as a virtual marketplace where buyers and suppliers meet to exchange information regarding product and service offerings, and to negotiate and carry out business transactions (Archer & Gebauer, 2000). Furthermore, in the age of the Internet and with the emergence of new information and communication technologies, the e-marketplace is a Web-based information system; where multiple suppliers and multiple buyers can undertake business transactions via the Internet (Russ, 2001). The e-marketplace uses Internet technologies and standards to distribute product data and to facilitate online transactions (Segev et al., 1999).

Doing business in the e-marketplace enables sellers to enter new markets, to find new buyers, and to increase sales. Conversely, an e-marketplace gives a buyer access to a broader range of products and services offered by sellers. A buyer has the option to quickly compare various offerings by price and performance. E-marketplace services support the exchange of large amounts of data regarding supply and demand between buyers and sellers, and the implementation of the business transactions (DeSisto, 2000; Lenz, 2000).

There are different business models of e-marketplaces. In practice, we can find extensive evidence that the e-marketplace is supporting many different processes between a buyer and a seller. Some e-marketplaces support only the aggregation of supply and demand, and searching and matching of buyers or sellers. In addition, many e-marketplaces support different types of auctions and negotiations. On the other hand, few e-marketplaces support the entire trading process where business services, such as contracting, finances, logistics, insurance, legal, payments and other services are needed. Many e-marketplaces are oriented to supporting supply chains (Pucihar & Podlogar, 2003).

The appeal of doing business on the Web is clear. By bringing together large numbers of buyers and sellers and by automating transactions, e-marketplaces expand the choices available to buyers, give sellers access to new customers (buyers), and reduce transaction costs for all participants. By extracting fees for the transactions occurring within the business-to-business (B2B) marketplaces, market makers (intermediaries) can earn high revenues. Moreover, because the marketplaces are software-based — not bricks and mortar — they can gain scale with minimal additional investment, promising even more attractive margins as the markets grow (Kaplan & Sawhney, 2000).

Problem Definition

Review of the literature identifies many advantages and opportunities from the field of electronic commerce and e-marketplaces (Podlogar & Pucihar, 2004; Pucihar & Podlogar, 2004). In the commercial arena, we can find many cases of successful and unsuccessful use of the e-marketplace, for both buyers and sellers. Unsuccessful use of the e-marketplace might be caused by improper services of e-marketplace providers. If services of e-marketplace providers do not add value to buyer or seller, then in the long term they will not choose the e-marketplace as a way of doing business; and the e-marketplace will be forced to close its business since it will not attract a sufficient critical mass of companies.

Without a sufficient number of sellers, buyers cannot choose among different sellers. The same rule also applies to buyers - if there are not enough buyers on the e-marketplace, then sellers do not have sufficient interest to join the e-marketplace, since they will not have parties to whom they could sell their products or services.

The direction in which the e-marketplace will evolve is still uncertain, because of the quickly changing environment and fast development of e-commerce technologies. In the long term, only the e-marketplace that will offer the most value-adding services will survive in the turbulent global economic environment. As electronic commerce connects markets to one global market, organizations will have to be able to compete on the global market. They will try to gain competitive advantage in different ways, and one of the ways will be by doing business in the e-marketplace.

An organization will adopt an e-marketplace only if sees sufficient benefits that might be gained from the value-added services. E-marketplace adoption is related to business process re-engineering, connecting of information systems, gaining new knowledge with training and learning, and making investments in new technologies and software (Bakos, 1991; Archer & Gebauer, 2000). Organizations may enter the e-marketplace as a buyer or a seller (supplier).

An organization must take very careful steps to prepare itself for a successful entrance in to an e-marketplace. Only this kind of approach will enable the organization to gain full use of the available opportunities, and thereby bring the anticipated business results (Archer and Gebauer, 2000). For that reason it is important to research the factors that organizations need to consider to successfully adopt the e-marketplace. By knowing what these factors are, organizations will be able to better prepare for successful e-marketplace adoption and therefore be able to compete in the global marketplace (Pucihar, 2002).

For that reason we investigated the importance of organizational factors, e-marketplace requirements and business environmental factors in the research that was completed with 119 large organizations in Slovenia. The importance of these factors is presented further in the chapter.

Doing business in an e-marketplace in Slovenia is not, as yet, very widespread. However, there are some indices from various organizations showing their interest in adoption and developing a use of the e-marketplace. Since Slovenia has become a member of the European Union on May 1, 2004, this research is especially important for organizations in Slovenia, who will soon have the opportunity to operate and compete on the common European and global market.

Background: Prior Research

On the basis of prior published research results and interviews with industry experts, we identified three groups of factors that organizations need to consider when making decisions regarding e-marketplace adoption (Pucihar, 2002):

- Organizational factors,
- E-marketplace factors,
- Environmental factors.

Organizational Factors

Organization of the IS Departments and the Level of Use of Information Technology

The use of information technology and correct organization of the information systems department in an organization are important for the achievement of gaining competitive advantage in the e-marketplace. Moreover, they enhance the possibility for more successful business in the e-marketplace (Auer & Reponen, 1997; Bharadwaj, 2000; Grewal et al., 2001).

Awareness of Benefits of E-Marketplace Use

The principal benefit of adopting the e-marketplace is lower transaction costs for all participants. This leads to a more effective business. Doing business in the e-marketplace enables a company to simplify business processes, and to facilitate a faster search of new buyers and suppliers, and make a faster entrance into new markets (Porter & Millar, 1985; Malone et al., 1987; Bakos, 1991; Steinfield & Caby, 1993; Streeter et al., 1996).

The Level of Top Management Support in the Process of Organization's E-Marketplace Adoption

Top management support is most important for successful implementation of electronic commerce projects. It is important that top management underpins the initiative for the e-marketplace adoption and assigns a responsible person for work on this project. It is also important that the top management tracks and encourages the work on the project (Turban et al., 2002; Ramsdell, 2000).

Formulated Electronic Commerce Strategy in Organization

A formulated electronic commerce strategy in the organization is very important as electronic commerce is a strategic issue for today's businesses operating in competitive markets. Electronic commerce is of critical importance for most organizations. It enables them to compete in global competitive markets and furthermore enables their survival (Turban et al., 2002).

Training for the Use of E-Marketplace Interface and Programs

Whilst e-marketplace adoption is pursued, organizations are also facing new programs and interfaces that enable doing business in the e-marketplace. For successful use of programs and interfaces of the e-marketplace it is important to provide training and education for users (Grewal et al., 2001).

Awareness of and Possibility to Make Extra Investments, Needed for E-Marketplace Adoption

Whilst adopting the e-marketplace, organizations may face extra costs, such as new technology and programs needed to connect to the e-marketplace, necessary training and education providing for users, and necessary business process re-engineering. Organizations should be aware of possible extra costs and must have the possibility to make the necessary extra investments (Bakos, 1991; Archer & Gebauer, 2000).

Ability to Setup and Maintain the Electronic Catalogue of Products and Services

If organizations do not have an electronic catalogue of its products and services, it may be a limitation for e-marketplace adoption. Design and maintenance of an electronic catalogue of products and services require investments, time and knowledge. The issue is much more complex, if the seller has a wide range of products or services with different characteristics, which need to be described in a catalogue (Choudhury et al., 1998; Enterworks, 2000; Mello, 2001).

Ability to Ensure and Provide the Standardized Form of Data

Organizations need to assure the relevance of the data, and provide it in a timely manner, and in a standardized form. Only this will enable them to do business with other organizations electronically. Use of standards for electronic commerce is very important because it enables organizations to exchange business documents. Formulation of standards and use of standards are also found important by the European Commission (Dai & Kauffman, 2001; Esichaikul & Chavananon, 2001; Buescher & Vittet-Philippe, 2000.)

Readiness of Business Processes for Inter-Organizational Relationship

Organizations need to organize business processes in a way that will enable them to connect with other organizations. It is important that organizations are flexible and know how to take advantage of new business models that are the results of electronic commerce (Esichaikul & Chavananon, 2001).

E-Marketplace Factors

E-Marketplace Services

The e-marketplace provides different services to buyers and suppliers in the e-marketplace. The most commonly used services in today's marketplaces are related to maintaining a product electronic catalogue, negotiating support and performing online auctions (Lefebvre et al., 2001). The aim of intermediaries, i.e., e-marketplace service providers, is to provide a wide range of services to all the participants in the e-marketplace. Among others, these services include product development, logistics and insurance services, payments and similar. Also very important are services that ensure a higher level of trust between buyers and sellers in the e-marketplace (Bailey & Bakos, 1997; Bakos, 1998; Archer & Gebauer, 2000; Durlacher, 2000; Mello, 2001). There are few marketplaces operating today that offer such a wide range of services. Most are oriented in providing services for matching buyers and sellers, negotiation and auction services (Dai & Kauffman, 2001).

Way of E-Marketplace Adoption

While considering e-marketplace adoption, an organization has two options. One is to create its own e-marketplace or, second, to adopt existing e-marketplaces available in the market. We differentiate several e-marketplace business models, such as e-marketplaces controlled by buyers or sellers, and independent or neutral e-marketplaces (Berryman et al., 1998; Kambil & van Heck, 2002).

Lack of trust may be one of major factors inhibiting organizations to adopt the e-marketplace. High levels of trust between buyers and sellers are especially important in Internet commerce and e-marketplaces, where buyers and suppliers might meet for the first time (Lee & Clark, 1996a, 1996b; Ba et al., 1998, 1999a, 1999b).

Trading Mechanisms

Buying and selling in e-marketplaces may be conducted through electronic catalogues, where products and services are described in detail. Prices are generally defined and vary according to ordered quantity. Another way of trading is by various types of auctions. Auction may be run by seller or buyer (reverse auction). Buyer's auctions are most frequently used on e-marketplaces. Using reverse auctions, buyers may achieve price decreases of up to 30%. Important factors that influence successfully conducted auction are trust and quality of products and services (Turban et al., 2002; Klein & O'Keefe, 1999).

Environmental Factors

Existence of Trust Between Business Partners

Business environments, where organizations are doing business, influence an organization's intention of e-marketplaces adoption. If buyers have strong relationships with existing suppliers, and between them also exists a strong level of trust, then they may not want to exploit possible benefits of new e-marketplaces. In some cases buyers would rather keep strong relationships with an existing supplier, with which they have good business experiences (Bakos, 1997; Dai & Kauffman, 2001).

Various characteristics of business, such as innovations, use of new information technologies, information exchange, trust, flexibility and others, lead buyers to cooperate with fewer suppliers. Such organizations usually do not decide to adopt e-marketplaces, where they have the possibilities to get new, cheaper and more efficient suppliers (Bakos, 1997).

Use of Electronic Commerce Between Business Partners

Electronic commerce enables stronger relationships between buyers and suppliers and new ways of buying from existing suppliers (Bakos & Brynjollfson, 1997; Kraut et al., 1998; Pucihar, 1999). Older technologies and information systems were harder to connect. In many cases, the solution that worked for connecting with one organization did not work with another organization. These solutions were very expensive. Today, the problem with connectivity of different technologies has disappeared. The Internet became the integrator of different technologies. E-marketplaces enable buyers and sellers to enter new global markets and explore new opportunities.

Encouragements from Business Environment

Organizations are influenced by the dynamic environment (Grewal et al., 2001). Important large buyers have a greater influence on business relations. We may find many cases, even in Slovenia, where the buyer invites suppliers to join the e-marketplace and negotiates for business with other potential suppliers.

The government has a strong role in promoting encouragement and spreading the benefits of electronic commerce (Gricar, 2001; European Commission, 2000). The result of the research that was carried out among organizations in Slovenia has shown that government's activities have been an important role in accelerating electronic commerce (Pucihar, 1999).

Many important e-commerce solutions have been implemented in Slovenia's organizations and government institutions. Since the year 2000, Slovenia has had the law of e-commerce and e-signature that regulates e-commerce. Its aim is the spreading of e-commerce (Gricar, 2001). The European Commission recommends that government should enable self-regulation in the field of e-commerce (European Commission, 2000).

E-Marketplace Definitions

E-Marketplace Business Models

The e-marketplace is a virtual marketplace where buyers and suppliers meet to exchange information about product and service offerings, and to negotiate and carry out business transactions (Archer & Gebauer, 2000). Furthermore, in the age of the Internet and with the emergence of new information and communication technologies, the e-marketplace is a Web-based information system, where multiple suppliers and multiple buyers can undertake business transactions via the Internet (Russ, 2001). The e-marketplace uses Internet technologies and standards to distribute product data and to facilitate online transactions (Segev et al., 1999).

An intermediary can provide four important mechanisms that cause marketplaces to add value (Christiaanse et al., 2001; Bailey & Bakos, 1997):

- Aggregating together a large number of buyers and sellers,
- Matching buyers and sellers to negotiate prices on a dynamic and real-time basis,
- Ensuring trust among participants by maintaining a neutral position, and
- Facilitating market operations by supporting certain transaction phases.

Intermediary functions may support a multiplicity of activities, including brokerage — auction payments, logistics, legal, consulting, or may support inter-company communications through third-party, inter-organizational systems and related systems (Bakos, 1991; Choudhury et al., 1998; Segev et al., 1999; Archer & Gebauer, 2000; Kaplan & Sawhney, 2000; Grewal et al., 2001; Russ, 2001).

The success of the e-marketplace depends on the perceived net benefit of buyers and suppliers. From the viewpoint of transaction cost economics, information technology helps to reduce transaction costs, risks, and coordinating costs of e-marketplaces (Clemons et al., 1993).

There are many different types of e-marketplaces operating today. We may divide e-marketplaces into those controlled by sellers, those controlled by buyers and those controlled by neutral third parties (Berryman et al., 1998):

- E-marketplaces controlled by sellers are usually set up by a single vendor seeking many buyers. Its aim is to create or retain value and market power in any transaction.
- E-marketplaces controlled by buyers are set up by or for one or more buyers with the aim of shifting power and value on the marketplace to the buyer's side. Many involve an intermediary, but some particularly strong buyers have developed e-marketplaces for themselves. Such e-marketplaces are also called consortium e-marketplaces.
- Neutral or independent e-marketplaces are set up by third-party intermediaries to match many buyers to many sellers.

E-Marketplaces might be also divided into horizontal and vertical marketplaces. A horizontal marketplace addresses a specific function (e.g., human resources, office supplies) and serves a wide range of industries, while a vertical marketplace focuses on a wide range of functionalities in a specific industry such as chemicals, steel or automotive (Baldi & Borgman, 2001; Ramsdell, 2000).

We may also classify marketplaces into four categories (Kaplan & Sawhney, 2000):

- MRO (Maintenance Raw and Operations) hubs are horizontal e-markets that enable systematic sourcing of operating inputs,
- Yield managers are horizontal e-markets that enable spot sourcing of operating inputs,
- Exchanges are vertical e-markets that enable spot sourcing of manufacturing inputs, and
- Catalogue hubs are vertical e-markets that enable systematic sourcing of manufacturing inputs.

The appeal of doing business on the Web is clear. By bringing together large numbers of buyers and sellers and by automating transactions, e-marketplaces expand the choices available to buyers, give sellers access to new customers (buyers), and reduce transaction costs for all participants. By extracting fees for the transactions occurring within the business-to-business (B2B) marketplaces, market makers (intermediaries) can earn vast revenues (Kaplan & Sawhney, 2000).

E-Marketplace Business Processes

Traditional markets and e-marketplaces consist of many trading processes that create value for both buyers and sellers. For successful e-marketplace operation, trading processes must be transformed from traditional business exchanges to e-marketplace exchanges in an efficient manner. The biggest challenge is how to successfully translate current business processes from a traditional to a virtual environment (Kambil & van Heck, 2002). This is possible by information technology that enables business process integration in the e-marketplace.

These processes that must be partly or fully integrated to e-marketplaces in order to achieve their successful operation may be divided into processes related directly to trade executing and processes that enhance trust among trading partners (Kambil & van Heck, 2000).

Processes related directly to trade executing are as follows (Kambil & van Heck, 2000):

- Search processes for discovering and comparing trading opportunities,
- Pricing processes for discovering prices,
- Logistics processes to coordinate transfer of goods and services,
- Payment and settlement processes to transfer funds from buyer to seller, and
- Authentication processes to verify the quality of goods or services sold and credibility of trading partners.

Processes for enhancing trust among trading partners and legitimizing the trade are as follows (Kambil & van Heck, 2000):

- Product representation processes for presenting products and services in the e-marketplace,
- Regulation processes to set market rules and social principles of e-marketplace trading,
- Risk-management processes to reduce risks in transaction processes between trading partners,
- Influence processes to ensure that commitments among trading partners are met, amd
- Dispute-resolution processes to resolve conflicts among trading partners and market makers.

E-marketplace processes are shown in Figure 1.

Simplified trading processes between buyers and suppliers are shown in Figure 2. In this trading case, the supplier creates a catalogue where buyers search for products and

Figure 1. E-marketplace processes

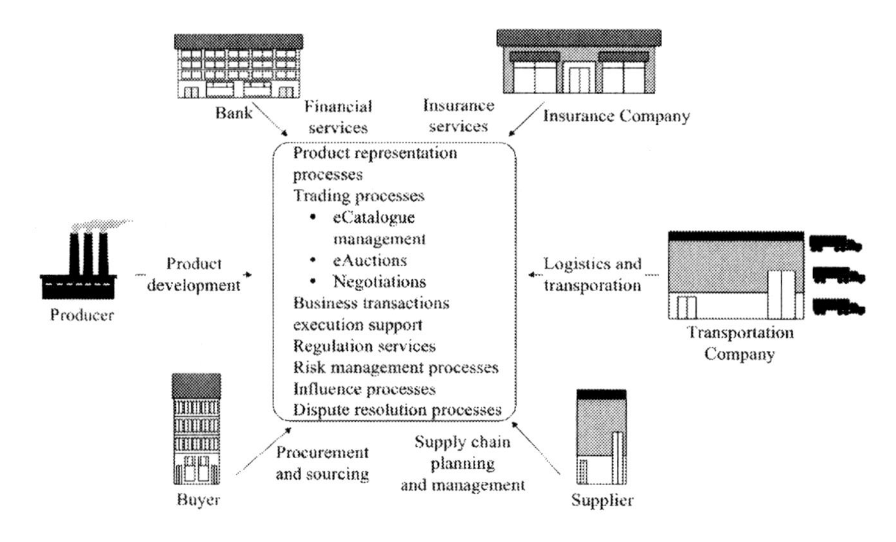

Figure 2. Trading processes

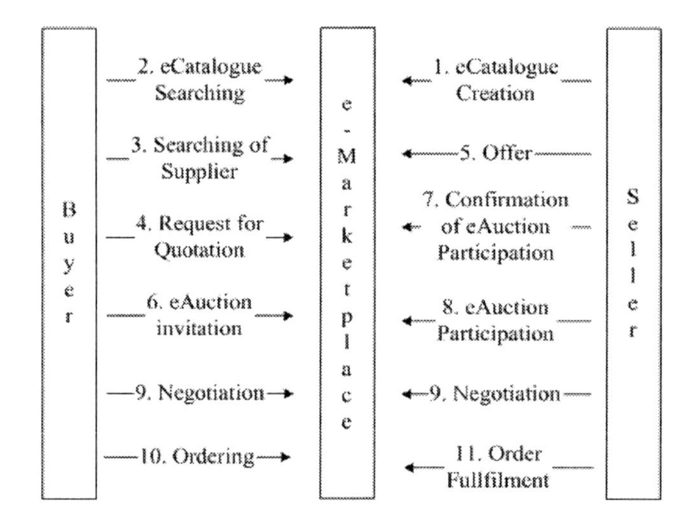

services. The buyer creates requests for quotation for several possible suppliers. Then suppliers send their offer. Selected suppliers are than invited to participate at auction with their offers. After the auction is closed, the buyer awards the supplier and the bidding process is started. This is followed by ordering and the supplier takes care of order fulfillment.

Of course this is not the only possible trading process. There are many options available to buyers and suppliers. It is also possible that the supplier starts an electronic auction in order to discard unnecessary inventories or to increase sales.

Opportunities and Threats of E-Marketplace Use

Opportunities of E-Marketplace Use

Use of the e-marketplace enables many benefits to be gained by organizations. Some benefits for buyers and sellers are described in Table 1 (Bakos, 1998; Berryman et al., 1998; Archer & Gebauer, 1999; Dai and Kauffman 2001; Turban et al., 2002).

Threats of E-Marketplace Use

As well as benefits that organizations may acquire, they may also be faced threats while using the e-marketplace. Some threats for the buyers and sellers are described in Table 2 (Lee & Clark, 1996a, 1996b; Bakos, 1997; Archer & Gebauer 2000; Dai & Kauffman, 2001; Lefebvre et al., 2001; Mello, 2001; Turban et al., 2002).

Table 1. Possible benefits of e-marketplace use

Possible benefits of e-marketplace use for sellers:	**Possible benefits of e-marketplace use for buyers:**	**Possible benefits of e-marketplace use for sellers and buyers:**
Easy and fast entrance to new markets	Easy and fast entrance to new markets	Lower transaction costs
Selling at every time	Easy and fast comparison between offers of many sellers	Better co-operation between buyers and sellers in product development processes
Physical store not needed anymore	Fast selecting of best offer	Better image of organization
Less mistakes in order management and fulfillment	Fast purchasing without long term relationships contracts	Lower marketing costs
Faster detection of new buyers for lower costs	Lower negotiation cycles	Faster time of new product to the market
Possibilities for sales increasing	Lower procurement costs	Lower production time
Lowering of superfluous inventory	Ordering at any time	More efficient production
More efficient production planning	Order status tracking	Lower inventory
More efficient product development	Easy changing of orders	Simplifying of business processes
	Consolidation of minor value orders	Better customer relationship management
		Less mistakes

Table 2. Possible threats of e-marketplace use

Possible threats of e-marketplace use for sellers:	Possible threats of e-marketplace use for buyers:	Possible threats of e-marketplace use for sellers and buyers
Loss of direct relationship with buyers and possibility of loss of existing buyers	Loss of direct relationship with suppliers	Lack of trust between buyers and suppliers, while they first meet in e-marketplace
Stronger negotiation power of buyers	Possible unreliability of unknown sellers	Lack of trust to e-marketplace intermediary
Reverse auctions (sellers are competing for business with price decreasing)	Lack of trust in products and services quality	Lack of services that would support whole business processes
Payment of commission to the intermediary (also in trading with existing buyers)	Uncoordinated different ways of supplying	
Buyers have strong relationship with existing suppliers (are not ready to get new ones)		
A lot of time, knowledge and funds necessary for e-catalogue creation and management		

Use of E-Marketplace in a Developing Country: Situation in Slovenia

Currently there are no business-to-business e-marketplaces operating in Slovenia. However, we can observe some interest expressed by organizations, such as Telekom Slovenije, various information systems providers and some government institutions (Government Center of Informatics), to establish the e-marketplace. In addition, some companies are looking for opportunities to establish private e-marketplaces to involve all their partners along the supply chain. In Slovenia there are several information systems providers that would like to sell their software, designed to operate in e-marketplaces. Conversely, we have organizations that have already joined e-marketplaces outside Slovenia. Some of them have already awarded and won some business contracts (Pucihar, 2002).

In the Slovenian market we may find e-marketplaces at an early stage of development. These e-marketplaces are mostly supporting business-to-consumer or consumer-to-consumer e-commerce processes. In most cases these are Web stores or Web malls providing some services that are typical for e-marketplaces: such as multi-vendor e-catalogues that enable comparison between products and services of several sellers, secure payments over the Internet, Web-based bulletin boards for supply and demand announcements, and electronic auctions support.

Since the year 2001 there are several human resource sites — recruitment agencies — present on the Web. They operate as intermediaries between organizations and people searching for a job. The same situation is also in the car industry and real estates services. In each case, intermediaries are also integrated in specialized business portals for various industries. These agencies intermediaries represent a basic horizontal e-marketplace. With their services they could serve various vertical e-marketplaces.

One of the major factors for slow development of e-marketplaces in Slovenia is the lack of awareness of possible benefits that might be gained by use of e-marketplaces. While considering e-marketplace adoption, organizations are also concerned by:

- Short-term benefits of e-marketplace use,
- Extra costs and extra work,
- Lack of trust and security of doing business on e-marketplaces,
- Business partners are not yet ready for doing business of e-marketplace or they currently do not use them,
- Lack of knowledge about the most suitable e-marketplace, and
- Unavailability of standardized data of products and services for e-catalogues.

We have previously stated that some organizations already have experienced doing business on e-marketplaces. Most of these organizations are production exporting organizations. The usual case in Slovenia is that most organizations enter the e-marketplace by request of an important large buyer that wanted to achieve the best offer between several suppliers. The first winning case of the e-marketplace use in organizations in Slovenia was achieved in 2002 by the organization Iskraemeco. This case is described in the next section.

There are few cases in organizations in Slovenia of entering e-marketplaces for e-procurement or e-sourcing purposes. This way of entering the e-marketplace is the most frequent in other countries. Sourcing and procurement in e-marketplaces enable buyers to gain extra savings. We may find the reasons different in Slovenia due to the lack of awareness of e-marketplaces.

E-marketplaces are becoming more and more important for organizations in Slovenia. On May 1, 2004 Slovenia became a member of European Union. From that moment, our organizations have to be able to compete in the global market. Organizations will try to make savings in every step of their business. One of the possible ways is also doing business in e-marketplaces. Successful cases of e-marketplace adoption might encourage other organizations in Slovenia.

E-Readiness Factors for Successful E-Marketplace Adoption

Research Methodology

The research was carried out in 119 large organizations that had the highest revenues and also a Web page in Slovenia during May 2002. To be qualified as a large organization, it has to fulfil at least two of the following conditions based on the 51st article of Organizations Act:

- More than 250 employees,
- Annual revenue more that 16.8 million EUR, and
- Average asset value of more than 8.4 million EUR.

Another requirement was that the organization had its own Web site.

The main reason for making such choices lies in the assumption that e-marketplace adoption is associated with extra expenses and extra investments in new knowledge, business processes re-engineering and new technologies, and it is expected that large organizations have more resources available for such purpose. The research was based on a questionnaire.

The questionnaire was designed on the basis of a research model that was derived from theoretical knowledge derived prior to the research, and the use of case studies. Opinions of experienced experts and information technology and Internet service providers in this field were also considered. The research model consists of organizational factors, e-marketplace factors and environmental factors that an organization needs to consider while adopting e-marketplace. The research model is shown in Figure 3.

The questionnaire was divided into the following sections:

- Respondents and organizations' data,
- Experiences with e-commerce and e-marketplaces use,
- Factors for successful e-marketplace adoption:
    - Organizational factors
    - E-marketplace factors
    - Environmental factors.

Figure 3. Research model

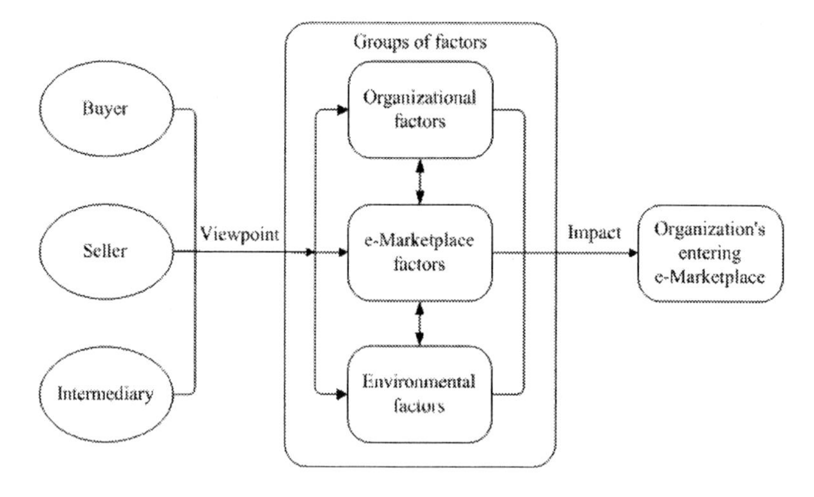

In order to ensure that the responses reflect the situation in each organization, letters with the questionnaires were addressed to the chief executive officers of 250 organizations. They were asked to distribute the questionnaires to the chief procurement officer, chief sales officer and chief information officer. A total of 119 useful responses were obtained and analyzed.

E-Commerce and E-Marketplace Use Experiences

Most of the respondents are IT personnel (42%), 24.4% of respondents work in the field of procurement and 17.7% of them work in sales. Most of respondents are very experienced. Almost a half of them have more than 10 years of experience in their field of work (46.2%). Almost a third of respondents have 5 to 10 years of experience in their field of work (31.9%). Most of the organizations sell their products and services in Slovenia (45.4%) and European countries (33.6%). The same situation is for procurement: most organizations buy products and services in Slovenia (45.6%) and European countries (37.4%).

The research results show that few organizations have any knowledge and experiences of e-marketplaces. Only 23.5% of companies cited at least one business-to-business e-marketplace that is of interest to their organization. Only 11% of the organizations already have the experiences of doing business in the e-marketplace. The research results show that 2.5% of the organizations already have experiences of buying, 3.4% of selling, 3.4% of service providing and 1.7% of buying and selling in the e-marketplace. Three quarters (75.6%) of the organizations have no experiences, and 13.4% of the organizations did not answer the question (Pucihar, 2002).

The research results are in line with our expectations. E-marketplaces are new ways of conducting electronic commerce and it has been developing very fast over the last few years. For this reason it is very satisfying to identify that some organizations in Slovenia have already been doing business in the e-marketplace.

We may find even more positive results when we examine organizations' intention to adopt e-marketplaces in the future. More than half of the organizations (56.3%) are thinking about adopting e-marketplaces in the near future. Over one tenth (11.8%) of the organizations intends to enter an e-marketplace by 2003 and more than a quarter (26.9%) of them in the future. Some organizations already know how they will adopt the e-marketplace: 6.7% of them will enter as sellers, 2.5% as buyers and 7.6% as a buyer and a seller. Less than 1% of the organizations (0.8%) intend to enter the e-marketplace as service providers. Only 11.8% of organizations don't think about adoption of e-marketplace in the future. Almost a quarter (21.8%) of the organizations has no opinion about adopting the e-marketplace in the future, while one tenth (10.1%) of the organizations did not answer the question (Pucihar, 2002).

Organizational Factors for Successful E-Marketplace Adoption

"Relevant organization of the IS department" and a "high level of IT use" are both important factors for successful e-marketplaces adoption. Only 32.7% of respondents confirmed that the information technology support group in their organization is large enough to offer effective support to the users. Almost half of the respondents (45.4%) mean that in organization they have enough knowledge about information technology. Many more respondents (69.8%) confirmed that in their organization they follow the development of and possibilities for information technology use (Pucihar, 2002).

Few respondents have the awareness of the benefits of e-marketplace use. It may be one of the reasons for the low level of e-marketplace use in large organizations in Slovenia. Less than half of the respondents (44.6%) disagree that in their organization they are awarded for successfully doing business on e-marketplaces. In addition, many respondents (35.8%) disagree that they are aware of possible troubles when doing business on e-marketplace. A good third of the respondents (37.8%) disagree that employees in their organization know the benefits of doing business on the e-marketplace. Almost half of the respondents (49.6%) also disagree that the organization's employees know what e-marketplaces are the most suitable for their organization to adopt (Pucihar, 2002).

Top management must support the use of new business models enabled by modern IT. Nowadays it may be the only way for their organization to gain important competitive advantage in the global marketplace. More than half of the respondents agree that the top management in their organization supports new ways of business (67%) and investments in electronic business (54.6%) (Pucihar, 2002).

Many of the respondents could not define or answer if costs of preparing for, and adoption of, the e-marketplace are low or high. This may explain the stated costs being neither high nor low for e-marketplace services (34.5%), cost of e-marketplace fees (32.8%), costs of setting up and maintaining an electronic catalogue of products and/or services (40.3%) and IT costs (31.9%). The highest number of the respondents thinks that training and educational costs are high (44.5%). Their answers could be attributed to the lack of experiences of using e-marketplaces (Pucihar, 2002).

The majority of respondents agree that they are able to provide the standardized form of data. Almost half of respondents (46.2%) agree that characteristics of their products and services are easy to describe. More than half of respondents (63%) agree that in their organization they are able to offer standardized data about products and services (Pucihar, 2002).

A majority of the respondents marked each of the following organizational factors as important for successful adoption of the e-marketplace (Pucihar, 2002):

- Top management support of new information technology and ways of business (87.3%)
- Formulated e-commerce strategy (82.3%)
- Trained employees with knowledge (84.4%)
- Modern information technology infrastructure (81.2%)
- Readiness of business processes to connect with business partners (80.5%)
- Information technology department support in the organization (77.9%)
- Use of enterprise resource planning system that enables them to connect with business partners (76.3%)
- Defined position of electronic commerce executive in the organization (77.1%)
- Availability of an electronic catalogue of products and services (67%)
- Experiences with electronic commerce (67%)

The importance of organizational factors for a successful entry in the e-marketplace is shown in the Table 3.

Table 3. Organizational factors

Organizational factors	The level of importance of factors in %						
	Not important at all				Very important	No answer	Average value
	1	2	3	4	5	0	
Top management support to new information technology and ways of business	0.1	0.1	8.5	30.5	56.8	4.2	4.5
Formulated e-commerce strategy	0.1	0.8	13.6	33.1	49.2	3.4	4.4
Trained employees with knowledge	0.1	1.7	9.3	40.7	44.1	4.2	4.3
Modern information technology infrastructure	0.1	1.7	12.0	41.0	40.2	5.1	4.3
Readiness of business processes to connect with business partners	0.8	0.8	11.9	35.6	44.9	5.9	4.3
Information technology department support in the organization	0.8	2.5	14.4	38.1	39.8	4.2	4.2
Use of enterprise resource planning system that enables connecting with business partners	1.7	2.5	14.4	33.1	43.2	5.1	4.2
Defined position of electronic commerce executive in organization	0.1	7.6	11.0	39.8	37.3	4.2	4.1
Availability of electronic catalogue of products and services	0.1	5.1	16.9	33.9	33.1	11.0	4.1
Experiences with electronic commerce	0.1	3.4	26.3	40.7	26.3	3.4	3.9

E-Marketplace Factors for Successful E-Marketplace Adoption

Trading Mechanisms

As respondents do not have many experiences with doing business on e-marketplaces, they have low estimations about the importance of e-marketplace trading mechanisms. Over one third of respondents marked that they would preferably to use an electronic catalogue as a trading mechanism (33.9 %), followed by buyer's auction (22.1 %) and seller's auction (14.4 %).

Way of E-Marketplace Adoption

The majority of respondents would enter e-marketplace with supply chain partners (53.7%). The second choice is to enter independent e-marketplace (30.5%), followed by entering e-marketplace with key industry players (competitors) (28.5%). Less interest is shown in the creating of the organization's own e-marketplace (16.5%).

E-Marketplace Services

A majority of respondents marked that matching buyers and suppliers (79.5%), electronic catalogue management (72.6%), payments (67%), shipment insurance (53.4%) and trans-

Table 4. Importance of e-marketplace factors

e-Marketplace factors	The level of importance of factors in %						
	Not important at all				Very important	No answer	Average value
	1	2	3	4	5	0	
Trading mechanisms							
Electronic catalogue	14.4	16.9	12.7	25.4	8.5	22.0	3.0
Buyer's auctions	16.9	19.5	16.1	15.3	6.8	25.4	2.7
Seller's auctions	18.6	21.2	22.0	7.6	6.8	23.7	2.5
Way of e-marketplace adoption							
Enter e-marketplace with supply chain partners	2.5	5.9	13.4	31.9	21.8	24.4	3.9
Enter e-marketplace with key industry players (competitors)	3.4	13.4	27.7	21.8	6.7	26.9	3.2
Enter independent e-marketplace	8.5	14.4	20.3	20.3	10.2	26.35.8	3.1
Create own e-marketplace	25.2	13.6	13.6	10.7	5.8	31.1	2.4
e-Marketplace Services							
Matching buyers and suppliers	0.1	0.1	9.4	42.7	36.8	11.1	4.3
Electronic catalogue management	0.1	0.9	14.5	31.6	41.0	12.0	4.3
Payments	1.7	5.9	9.3	26.3	40.7	16.1	4.2
Shipment insurance	2.5	8.5	20.3	32.2	21.2	15.5	3.7
Transportation	3.4	10.2	20.3	28.0	22.0	16.1	3.7
Negotiations	4.2	14.4	33.9	18.6	14.4	14.4	3.3
Auctions	4.2	21.2	28.0	22.9	8.5	15.3	3.1

portation (50%) are the most valuable services provided at e-marketplaces. Respondents shown less interest in negotiation support (33%) and different types of auctions (31.4%).

Detailed opinions of respondents about e-marketplace factors are described in Table 4.

Environmental Factors for Successful E-Marketplace Adoption

Existence of Trust Between Business Partners

A majority of respondents agree that between them and their suppliers exists a high level of trust (72%). Also majority of respondents agree that between them and their buyers exists a high level of trust (76.1%). Almost 80% of respondents confirmed that their organization has strong relationship with existing suppliers (79.7%) and existing buyers (81.2%).

Encouragements from Business Environment

A majority of respondents disagree that their important buyers (44%) and suppliers (53.4%) already do their business on e-marketplaces. Also, a majority of respondents disagree that Slovenian government encourages business-to-business e-commerce (42.4%).

Use of Electronic Commerce Between Business Partners

A majority of respondents disagree that they use e-commerce with their suppliers (42.4%) and buyers (40.2%).

Organization's Relationship to Business Environment

A majority of respondents are interested in spreading their business to new markets (94.9%). More than one third of respondents agree that their buyers enforce a way of business (36.7%). The majority of respondents agree that they usually get new business with personal contacts (73.7%).

Detailed opinions of respondents regarding business environmental factors are described in Table 5.

Table 5. Environmental factors

Environmental factors	The level of agreeing in %						
	Strongly disagree				Strongly agree	No answer	Average value
	1	2	3	4	5	0	
Existence of trust between business partners							
Our organization has strong relationship with existing buyers	0.1	2.6	12.8	37.6	43.6	3.4	4.3
Our organization has strong relationship with existing suppliers	0.1	1.7	15.3	44.1	35.6	3.4	4.2
Between us and buyers exists high level of trust	0.9	0.1	17.9	53.0	23.1	5.1	4.0
Between us and suppliers exists high level of trust	0.1	1.7	22.0	51.7	20.3	4.2	3.9
Encouragements from business environment							
Slovenian government encourages business-to-business e-commerce	12.7	29.7	29.7	19.5	5.9	2.5	2.9
Our important buyers already do their business on e-marketplaces	25.4	18.6	16.1	7.6	2.5	29.7	2.2
Our important suppliers already do their business on e-marketplaces	22.9	30.5	13.6	5.1	1.7	26.3	2.1
Use of e-commerce between business partners							
We use e-commerce with our suppliers	12.7	29.7	29.7	19.5	5.9	2.5	2.8
W use e-commerce with our buyers	12.8	27.4	35.9	15.4	4.3	4.3	2.7
Organization's relationship to business environment							
We are interested in spreading our business to new markets	0.1	0.9	0.9	27.4	67.5	3.4	4.7
We usually get new business with personal contacts	1.7	5.1	15.3	44.9	28.8	4.2	4.0
Our buyers enforce a way of business with our organization	8.5	19.7	27.4	23.9	12.8	7.7	3.1

Model Discussion

Preconditions for the successful adoption of e-marketplace are awareness of the opportunities and threats of e-marketplace adoption. By knowing these factors, organizations are able to design further necessary steps for efficient use of an e-marketplace in order to improve competitiveness and achieve a competitive advantage in the global market.

While considering e-marketplace adoption, organizations need to be adequately organized. Research results have shown that nearly all investigated organizational factors are important for efficient adoption of e-marketplaces. These factors are: top management support of new information technology and new ways of business, information technology department support in the organization, formulated e-commerce strategy, defined position of electronic commerce executive in organization, trained employees with knowledge, modern information technology infrastructure, readiness of business processes to connect with business partners, use of enterprise resource planning system that enables connecting with business partners, availability of products and services, and experiences with electronic commerce. By considering all of these factors, organizations can achieve internal readiness for efficient e-marketplace adoption.

Another step is to define goals for e-marketplace adoption. It has to be defined as to why the e-marketplace will be used — either for procurement or for sales processes, for searching for new buyers or suppliers or for joining existing supply-chain partners in order to improve business processes and achieve lower transaction costs. By knowing these factors, an organization is able to define the way of entering the e-marketplace and choose an optimum e-marketplace to achieve their goals. While considering e-marketplace adoption, the organization has to be informed by trading mechanisms and services available at selected e-marketplaces.

The environment where an organization operates may influence the organization to adopt or not adopt an e-marketplace. If the organization has a strong relationship and good trust with existing business partners, then there less possibility that they will consider e-marketplace adoption. An organization needs to evaluate its relationship with current business partners and estimate where additional costs savings might be achieved. Especially important are encouragements from the business operating environment: such as government activities giving awareness and thus creating and using e-marketplace for its business.

Model of factors that need to be considered for successful e-marketplace adoption are shown in Table 6.

Table 6. Model of factors for efficient e-marketplace adoption

Awareness of Possible Opportunities and Threats of e-Marketplace Adoption		
Organizational factors	**e-Marketplace Factors**	**Environmental Factors**
Top management support to new information technology and ways of business	**Way of e-marketplace adoption**	**Existence of trust between business partners**
Formulated e-commerce strategy	Enter e-marketplace with supply chain partners	Strong relationship with existing buyers
Trained employees with knowledge	Enter e-marketplace with key industry players (competitors)	Strong relationship with existing suppliers
Modern information technology infrastructure	Enter independent e-marketplace	Existing of high level of trust to buyers
Readiness of business processes to connect with business partners	Create own e-marketplace	Existing of high level of trust to sellers
Information technology department support in the organization	**Trading mechanisms**	**Encouragements from business environment**
Use of enterprise resource planning system that enables connecting with business partners	e-Catalogue	Government encourages business-to-business e-commerce
Defined position of e-commerce executive in organization	Buyer's auctions	Buyers already do their business on e-marketplaces
Availability of e-catalogue of products and services	Seller's auctions	Suppliers already do their business on e-marketplaces
Experiences with e-commerce	**e-Marketplace services**	**Use of e-commerce between business programs**
	Matching buyers and suppliers	E-commerce use with suppliers
	e-Catalogue management	E-commerce use with buyers
	Payments	**Organization's relationship to business environment**
	Shipment insurance	Interest in spreading our business to new markets
	Transportation	Getting new business with personal contacts
	Negotiations	Buyers enforce a way of business with our organization
	Auctions	

Success Story of the Slovene Supplier: An Invitation Which Led to an Unexpected Win

For the encouragement of other organizations to start using e-marketplaces for international business, the first success story of Slovene supplier is presented in this chapter (Grièar et al., 2004).

The company Iskraemeco from Kranj in Slovenia is one of the world's leading producers in the field of measuring and managing electrical energy. The company produces equipment and systems for energy measurement and management segmentation. At the end of 2001, the company was invited by the company CLP Power Hong Kong Limited (CLP Power) to join the marketplace Freemarkets and participate at the reverse auction with other invited suppliers. CLP Power provides a safe and reliable electricity supply to two million domestic and commercial customers, serving approximately 80% of Hong Kong's population. The company decided to supply measuring equipment for the next three years from the most suitable supplier.

As Iskraemeco previously had a business relationship with CLP Power Company, they decided to participate at the auction. First, they had to prepare everything necessary for the registration to the marketplace Free-markets. After the registration, they were trained for using the marketplace bidding software BidWare to participate at the auction. The auction started on January 22, 2002 at 7 a.m. and finished at 10:56 a.m. The auction took place in real time and participating suppliers were able to compare their offers with others. Each supplier had to be very well prepared for the auction and needed to have the appropriate data available in real time for decision making about possible price decreasing. Two weeks after the auction's closing time, the suppliers were selected and contracts signed. Iskraemeco won two- thirds of the business.

Iskraemeco Public relations department reported that the company will use e-marketplaces for creating global relationships. This brings benefits especially for companies that are currently not leading and well known in their industry. These companies may compete on an equal footing at reasonable costs for the business with well-known companies all over the world. In Iskraemeco they are aware of that e-marketplaces allows benefits to buyers. Buyers get the most advantages with reverse auctions, where possible suppliers under huge pressure compete for the business with price decreasing. But on the other hand, the lowest price is not always the key element for awarding the suppliers. Buyers usually require other important elements to be considered in the first offer of each supplier. On the basis of first offers, the buyer estimates supplier's qualifications for the auction.

In the case outlined above, price was not the only key element for winning the business. There were also other important elements needed, such as quality of products and services, timing for order fulfillment and after-sales services. Compared to other competing suppliers in this auction, Iskraemeco did not offer the lowest price but they were close to the middle among other offers. But their offer had better fulfilled other required elements when compared with others and ultimately that was the reason why they were awarded the business.

Iskraemeco Company keeps its eyes open and looks for opportunities by regularly participating in other auctions when they are invited.

Recommendations

If an organization wants to adopt an e-marketplace successfully, it is important that it has a clearly defined strategy of e-marketplace adoption. It needs to know what the goals of doing business in the e-marketplace are. It is important that the organization has a clear vision, whether they will enter the e-marketplace as a buyer or a seller. Because there are many e-marketplaces present in the global market, it is important that the organization knows e-marketplaces that are relevant to its business mission and its future goals. The organization needs to investigate what is the process of entering the e-marketplace, what types of services are available, what is the price to subscribe and use a service, etc.

It is also important that the top management supports the project of e-marketplace adoption and that they have a clear vision and strategy how they will exploit the benefits of this new way of doing business. A defined position of electronic commerce executive in an organization might elevate the importance of electronic commerce and electronic business for the organization. It is also important that employees are educated and trained for such a way of doing business. Usually the e-marketplace providers offer all the necessary training on how to use different e-marketplace software.

It is important that an organization is aware of costs that will incurred by e-marketplace adoption. In addition to the direct cost such us e-marketplace annual fees and service fees, organizations will also be faced with various indirect costs, such us training costs, investments in new technology or possibly investments in setting up and maintaining e-catalogue of products and services. Only if the organization has a strong vision for adopting the e-marketplace and if it will be aware of additional costs that may appear, will it be able to assure the necessary resources. Every organization has to calculate expected benefits and costs, because in any case, expected benefits have to be higher than expected costs to make a choice to adopt the e-marketplace.

The availability of an electronic commerce catalogue of product and services is important. Organizations that intend to adopt the e-marketplace must be able to offer standardized electronic data of their products and services. The use of electronic commerce standards enables the connection of different information systems. This is especially important in the case of e-marketplace adoption, because many organizations that are using different information systems need to communicate with each other.

Almost all e-marketplaces operate on Web-based information systems, thus it is important that modern information technology infrastructure, which enables Internet commerce, is in place in the organization. Many e-marketplaces operate on the basis of information systems integration, this being especially valid for consortia-based e-marketplaces. As many organizations have implemented enterprise resource planning systems, it is important that they are able to connect their systems with other information systems of business partners. The use of electronic commerce standards and different interfaces for business-to-business systems integration make integration processes much easier.

It is important that organizations have their business processes ready to be connected with business partners. They must change and modify business processes to the new way of doing business. The necessary changes of business processes are dependent of the type of e-marketplace that the organization is willing to adopt. For example, an organization will have to make more changes if it decides to enter a private or consortia-based e-marketplace than third-party e-marketplaces. To enter a private or consortia-based e-marketplace, an organization has to consider changing their business processes, because the next step will usually be the integration of information systems. To enter an independent, third-party e-marketplace, it is usually enough to use the Web browser to make an order or to participate in an online auction.

E-marketplaces are new business models that are developing and changing very rapidly. Therefore, it is important that the business environment encourages organizations to use such models to gain competitive advantage in the global market. Governments need to play a key role in such activities. In some cases governments establish e-marketplaces for simplifying the trading processes with national companies, especially in the segment of public procurement.

Current Status and Future Developments

In June 2003 there were approximately 1,000 public e-marketplaces worldwide; with 500 of these active in Europe. E-marketplaces are not equally distributed over all industries. The most important groups for vertical marketplaces are building and construction, logistics and agriculture (e-Business W@tch, 2003).

According to e-Business W@atch research, large organizations use e-marketplaces in different ways than smaller ones. Larger organizations are more active and more often take the initiative. They are more frequently using online catalogues for selling and buying and are more active in taking the initiative of offering or initiating requests for quotation. Large organizations are also above-average users of consortia marketplaces, which are operated by group of companies — buyers or sellers. Larger organizations are still stronger users of e-marketplaces than SMEs.

The use of e-marketplaces as a new way of conducting business is not without problems. Many open issues remain in the use of e-auctions and related online negotiations. Many organizations, especially smaller ones, fear unfair competition by the initiators of auctions — often the buyers. One often-stated issue is so-called fake or proxy bidding, where the initiator of a procurement auction either provides bids himself or via proxy participants to drive the price down. A second way can be the setting of unrealistically low target prices. But the question remains how often such situations really happen in e-marketplaces. Such kind of opinion might be the result of lack of trust and awareness of e-marketplace use (e-Business W@tch, 2003). Invariably, however, e-marketplace providers try to ensure fair trading with different mechanisms and regulations.

It is expected that through globalization and market consolidation, organizations will find themselves within a strong competitive environment. Costs savings will become one of their important targets. In order to achieve and remain competitive, organizations will try to use various internet business models. One of them is e-marketplaces.

The situation in Slovenia shows that interest from organizations in e-marketplaces has increased in the last three years. Some organizations that are highly export-oriented were already encouraged by buyers to participate at reverse auctions. In the last year, growth of interest has been shown in e-marketplace use for procurement processes. Organizations, especially those with foreign investors or ownership, show interest in searching for new, cheaper, more reactive and flexible suppliers.

Although in many developing countries current use of e-marketplaces is not widely spread as yet, it is expected that e-marketplace use will increase in the coming years. One of the main reasons for such a situation is often lack of knowledge and awareness about possible benefits of e-marketplace use. The higher level of awareness and knowledge could be achieved by many activities — projects from government and other public and private institutions. In this process universities also could play a significant role, especially with establishing of laboratories for collaboration between organizations, students, information technology and e-marketplace providers. With such a kind of collaboration the knowledge is transferred between collaborating parties.

E-marketplaces offer many advantages: especially for organizations from developing countries. In many cases such organizations are more flexible than others in order to survive in the market. By suitable use of e-marketplaces, they can compete in a global market and achieve important competitive advantage from their offers and cost savings.

References

Archer, N., & Gebauer, J. (2000). Managing in the context of the new electronic marketplace. *Proceedings 1st World Congress on the Management of Electronic Commerce,* Hamilton, Ontario, Canada, January 19-21.

Auer, T., & Reponen, T. (1997). Information system strategy formation embedded into a continuous organizational learning process. *Information Resource Management Journal, 10*(Spring), 32-43.

Ba, S., Stinchcombe, M., Whinston, A. B., & Zhang, H. (1999a). Trusted third parties in the electronic marketplace. W. D. Haseman & D. Nazareth (Eds.), *Proceedings of the Fifth Americas Conference on Information Systems (AIS-99),* Milwaukee, WI (pp. 238-240).

Ba, S., Whinston, A. B., & Zhang, H. (1998). The design of a trusted third party for electronic commerce transactions. E. D. Hoadley & I. Benbasat (Eds.), *Proceedings of the 4th Americas Conference on Information Systems (AIS-98),* Baltimore (pp. 269-271).

Ba, S., Whinston, A. B., & Zhang, H. (1999b). *The impact of authentication and reputation on competition on the electronic markets: Electronic commerce in Europe* (ed. M. Fehimovic). London: World Markets Research Center.

Bailey, J., & Bakos, J. Y. (1997, Spring). An exploratory study of the emerging role of electronic intermediaries. *International Journal of Electronic Commerce,* 1(3), 7-20.

Bakos, J. Y. (1991). A strategic analysis of electronic marketplaces. *MIS Quarterly,* September, 295-310.

Bakos, J. Y. (1997). Reducing buyer search costs: Implications for electronic marketplaces. *Management Science, 43*(12), 1676-1692.

Bakos, J. Y. (1998, August). The emerging role of electronic marketplaces on the Internet. Association for Computing Machinery, *Communications of the ACM,* New York.

Bakos, J. Y., & Brynjolfsson, E. (1993). Information technology, incentives, and the optimal number of suppliers. *Journal of Management Information Systems, 10*(2), 37-53.

Baldi, S., & Borgman, H. P. (2001). Consortium-based B2B e-Marketplaces – A case study in the automotive industry. In B. O'Keefe, C. Loebbecke, J. Gricar, A. Pucihar, & G. Lenart (Ed.), *14th Bled Electronic Commerce Conference: e-Everything: e-Commerce, e-Government, e-Household, e-Democracy,* Bled, Slovenia, June 25-26 (pp. 629-645).

Barling, B., & Stark, H. (1998). Business-to-business electronic commerce: Opening the market. *Ovum,* June/August/October. Retrieved from *http://www.ovum.com/*

Berryman, K., Harrington, L. F., Layton-Rodin, D., & Rerolle, V. (1998). Electronic commerce: Three emerging strategies. *The McKinsey Quarterly, 1.* Retrieved from *http://www.mckinseyquarterly.com/electron/elco98.asp*

Bharadwaj, A.S. (2000). A resource-based perspective on information technology capability and firm performance: An empirical investigation. *MIS Quarterly, 24*(1), 169-196.

Buescher, R., & Vittet-Philippe, P. (2000). *B2B e-Commerce: Impact on enterprise policy – A first assessment.* DG Enterprise e-Business Report No. 1, European Commission, Enterprise Directorate-General, July.

Chircu, A.M., & Kauffman, R.J. (1999, May). Strategies for Internet middlemen in the intermediation/disintermediation/reintermediation cycle. *Electronic Markets, 8*(2), 109-117.

Choi, S.Y., Stahl, D., & Whinston, A. (1997). *The economics of electronic commerce.* Indianapolis, IN: Macmillan Technical Publishing.

Choudhury, V., Hartzel, K. S., & Konsynski, B. R. (1998, December). Uses and consequences of electronic markets: An empirical investigation in the aircraft parts industry. *MIS Quarterly, 22*(4), 471-507.

Christiaanse, E., Sinnecker, R., & Mossinkoff, M. (2001). The impact of B2B exchanges on brick and mortar intermediaries: The Elemica Case. In S. Smithson, J. Gricar, M. Podlogar, & S. Avegerinou (Eds.), *Proceedings of the 9th ECIS Conference: Global Co-operation in the New Millennium,* Bled, Slovenia, June 27-29 (pp. 422-432).

Clarke, R. (2001). Towards a taxonomy of B2B e-Commerce schemes. *14th Bled Electronic Commerce Conference: e-Everything: e-Commerce, e-Government, e-Household, e-Democracy,* Bled, Slovenia, June 25-26.

Clemons, E.K., Reddi, S.P., & Row, M.C. (1993). The impact of information technology on the organization of economic activity: The "move to the middle" hypothesis. *Journal of Management Information Systems, 10*(2), 9-35.

Dai, Q., & Kauffman, R. J. (2001). Business models for Internet-based e-Procurement systems and B2B electronic market: An exploratory assessment. The *Thirty-Fourth Annual Hawai'i International Conference on System Sciences (HICSS-34),* Maui, HI, January 3-6.

DeSisto, R. (2000). e-Marketplaces: Place your bets of fold your cards. *Gartner Symposium ITXPO 2000,* Insight for the Connected World. Walt Disney World, Orlando, Florida, October 16-20.

Durlacher. (2000). Business to business e-commerce. Investment perspective. Durlacher Research Ltd. Retrieved December 7, 2001, from *http://www.durlacher.com/fr-research.htm*

e-Business W@tch. (2003). *The European eBusiness Report 2002/2003. A portrait of e-business in 15 sectors of the EU economy.* European Commission, Enterprise Directorate General.

The Economist. (1999). The rise of the Infomediary. Retrieved June 26 from *http://www.economist.com/editorial/freeforall/19990626/su3300.html*

Enterworks. (2000). Challenges and solutions in supplier enablement: Opportunities for leveraging e-marketplaces to your advantage. Executive Overview, Enterworks Inc. Retrieved November 28, 2001, from *www.enterworks.com*

Enterworks. (2001). Nine things you need to know as an electronic exchange supplier. Enterworks Inc. Retrieved November 28, 2001, from *www.enterworks.com*

Esichaikul, V., & Chavananon, S. (2001). Electronic commerce and electronic business implementation success factors. *14th Bled Electronic Commerce Conference: e-Everything: e-Commerce, e-Government, e-Household, e-Democracy,* Bled, Slovenia, June 25-26.

European Commission. (2000). EcaTT Final Report, *Electronic Commerce and Telework Trends: Benchmarking Progress on New Ways of Working and New Forms of Business across Europa.* EMPIRICA, Project EcaTT98, EP29299, July.

Grewal, R., Comer, J. M., & Mehta, R. (2001). An investigation into the antecedents of organizational participation in business-to-business electronic markets. *Journal of Marketing, 65*(July), 17-33.

Gricar, J. (2001). Uvodnik. Organizacija. Revija za management, informatiko in kadre. Tematska stevilka: Izboljsanje konkurencnosti z elektronskim poslovanjem: eEvorpa – eSlovenija. *Letnik 34,* stevilka 3, 118-119.

Gricar, J., Lenart, G., & Pucihar, A. (2004). E-business W@tch reports as a resource in student projects - A Slovenian case study. The European e-business report: A portrait of e-business in 10 sectors of the EU economy, (Enterprise Publications). 2004 edition. Luxembourg: Office for Official Publications of the European Communities, 223-225.

Kambil, A., & van Heck, E. (2002). *Making markets. How firms can design and profit from online auctions and exchanges.* Boston: Harvard Business School.

Kaplan, S., & Sawhney, M. (2000, May-June). e-Hubs: The new B2B marketplaces. *Harvard Business Review*, 97-103.

Klein, S., & O'Keefe, R. M. (1999, Spring). The impact of the Web on auctions: Some empirical evidence and theoretical considerations. *International Journal of electronic Commerce*, 3(3), 7-20.

Kraut, R., Steinfield, C., Chan, A., Butler, B., & Hoag, A. (1998). Coordination and virtualization: The role of electronic networks and personal relationships. *Journal of Computer-Mediated Communication, 3*(4). Retrieved from *http://www.ascusc.org/jcmc/vol3/issue4/kraut.htm*

Lee, H. G., & Clark, T. H. (1996a, Fall). Impacts of electronic marketplace on transaction cost and market structure. *International Journal of Electronic Commerce, 1*(1), 127-149.

Lee, H. G ,& Clark, T. H. (1996b, Winter). Market process reengineering through electronic market systems: Opportunities and challenges. *Journal of Management Information Systems, 13*(3), 113-136.

Lefebvre, L. A., Cassivi, L., & Lefebvre, E. (2001). Business-to-business e-commerce: A transition model. The *34th Annual Hawai'i International Conference on System Sciences (HICSS-34)*, Maui, HI, January 3-6.

Lenz, C. (2000). E-marketplaces: How are they changing your business environment? *Gartner Symposium ITXPO 2000*, Insight for the Connected World. Walt Disney World, Orlando, FL.

Malone, T. W., Yates, J., & Benjamin, R. I. (1987, June). Electronic markets and electronic hierarchies. *Communications of the ACM, 30*(6).

Mello, A. (2000). How e-Markets Fail Suppliers. *ZDNet,* September 26. Retrieved October 4, 2001, from *http://techupdate.zdnet.com/techupdate/stories/main/0,14179,2814612,00.html?chkpt=zdnnecrttu*

Podlogar, M., & Pucihar, A. (2004). Electronic commerce in procurement process: Experiences and the frequency of use. V. Florjancic, Joze (ur.), Pütz, Karl (ur.). Informatics and management: Selected topics. Frankfurt am Main [etc.]: P. Lang, 279-295.

Porter, M. E., & Millar, V. E. (1985, July-August). How information gives you competitive advantage. *Harvard Business Review, 63*(4), 149-160.

Pucihar, A. (2002). *Entering e-markets success factors.* Doctoral dissertation. Kranj: University of Maribor.

Pucihar, A., & Podlogar, M. (2003). Influence of business environment on successful entering e-marketplace: Case in large organizations in Slovenia. V: Jasková, Mária (ur.). ECON '03 : [selected research papers], (Research works proceedings, Vol. 10, 2003, Economics set). Ostrava: Technical University of Ostrava, Faculty of Economics, 371-379.

Pucihar, A., & Podlogar, M. (2004). Organizational perspective on successful entering electronic marketplaces. V: Florjancic, Joze (ed.), Pütz, Karl (ur.). Informatics and management: selected topics. Frankfurt am Main [etc.]: P. Lang, pg. 311-327.

Ramsdell, G. (2000). The real business of B2B. Retrieved from *http://mckinseyquarterly.com/electron/rebu00.asp*

Russ, N. (2001). *E-marketplaces: New challenges for enterprise policy, competition and standardisation.* eEurope go Digital. Workshop Report, Brussels, April 23-24.

Segev, A., Gebauer, J., & Färber (1999). Internet based electronic markets. *Electronic Markets, 9*(3), 138-146.

Shaffer, G., & Zettlemeyer, F. (1999). *The Internet as a meduim for marketing communications: Channel conflict over the provision of information.* Working Paper, June.

Steinfield, C., & Caby, L. (1993). Strategic organizational applications of videotex mong varying network configurations. *Telematics and Informatics, 10*(2), 119-129.

Streeter, L. A., Kraut, R. E., Lucas, H. C., & Caby, L. (1996). How open data networks influence business performance and market structure. *Communications of ACM, 39*(7), 62-73.

Turban, E., King, D., Lee, J., Warkentin, M., & Chung, M. H. (2002). *Electronic commerce 2002. A managerial perspective.* Upper Saddle River, NJ: Prentice Hall, Pearson Education.

Wigand, R. T., & Benjamin, R. I. (1995). Electronic commerce: Effects on electronic markets. *Journal of Computer-Mediated Communication, 1*(3). Retrieved from *http://www.ascusc.org/jcmc/vol1/issue3/wigand.html*

Zwass, V. (1998). Structure and macro-level impacts of electronic commerce: From technological infrastructure to electronic marketplaces. In K. E. Kendall (Ed.), *Emerging information technologies.* Thousand Oaks, CA: Sage Publications. Retrieved from *http://www.mhhe.com/business/mis/zwass/ecpaper.html*

Chapter VI

E-Commerce Activity, Opportunities, and Strategies in Latin America

Raul Gouvea, University of New Mexico, USA

Dante Di Gregorio, University of New Mexico, USA

Suleiman K. Kassicieh, University of New Mexico, USA

Abstract

In this chapter, we analyze international differences in e-commerce activity and strategies within Latin America and between Latin and non-Latin markets. Starting at a macro-level, we compare e-commerce activity and Internet-related developments in Latin America with developments in other regions. Moving to a closer focus on Latin America, we then discuss the evolution of e-commerce within Latin American countries, with an emphasis on Brazil, Chile, and Mexico. Finally, we identify prevalent business models and analyze the strategies of key players in this sector, and discuss the outlook for the future.

Introduction: Global E-Commerce

E-commerce has been described by different authors from different perspectives. These definitions center on many issues such as:

1. Distribution channels and the ability to reach a much wider audience than one is able to reach with brick and mortar or even phone order business models. This model is used to explain many business-to-consumer (B2C) models.
2. Supply chain coupling between suppliers and manufacturers giving new importance to the idea of just-in-time supply models and zero inventory mechanisms. This has been widely used to explain business-to-business (B2B) models.
3. Consolidation of customer purchasing power models whereby customers (both individuals and companies) are helped by visiting one portal where they can link to many different Web sites. This has given the power of negotiating better prices to small groups who now can consolidate their purchases, thereby reducing transaction cost and making profit possible to the consolidator. This explains many intermediary Web sites from travel to books to auction sites.
4. Ability to use the Internet to push information that can be helpful to customers who in turn get faster information and have less paperwork to go through to purchase new products. This is used mostly in technology areas such as software and hardware where changes are fast and customers seek information on new products.
5. Government processes from tax filing to permit requests are being fast tracked through Internet procedures increasing efficiency of business and individuals through their dealings with the government. Government purchasing is being also streamlined through e-commerce dealings with suppliers.
6. Services are becoming more dependent on the Internet to supply information, reducing the need to deal with customers via customer support. As security of the Internet grows, more financial transactions are done electronically, thereby reducing the cost of service and subsequently reducing the cost to customers of service provisions.

In this chapter, we analyze international differences in e-commerce activity and strategies within Latin America and between Latin and non-Latin markets. The chapter proceeds in four sections, as follows. First, we present an overview of the evolution of global e-commerce, with a brief profile of all major regions. Second, we adopt a macro-level perspective to compare major Latin American countries to other regions of the world in terms of e-commerce activity and the factors that determine e-readiness. In the third section, we focus more specifically on opportunities and firm strategies in the Latin American e-business environment. The breadth of our coverage of e-commerce in the region is proportional to the market size in each country (i.e., since Brazil, Chile, and Mexico jointly account for approximately 75 to 90% of the regional market in this sector, these countries account for a proportional amount of coverage in this chapter). Finally, we draw conclusions regarding the evolutionary path of e-business in Latin America, and identify challenges at the national level as well as at the firm level.

World E-Commerce Activities

The pace of change in Latin American e-commerce has been difficult to fully understand in the current competitive climate. Mirroring industry events elsewhere in the world, there has been substantial volatility among the leading e-commerce ventures in Latin America, with a high rate of organizational mortality, frequent changes in market structure, and seemingly limitless possibilities for future growth, albeit with ambiguous profit potential. Existing firms and dot-com start-ups have aggressively entered the arena, and many firms from both camps have already been forced to exit or to undertake radical strategic changes.

The use of e-commerce in different parts of the world has shown staggering numbers. Global e-commerce estimates in 2000 expected transactions to reach US$6.7829 trillion by 2004, from US$657 million in 2000 (Forrester Research, 2001). In most countries and regions, e-commerce markets are expanding rapidly. The data in Table 1 indicate that North America, Asia, and Western Europe are presently the dominant e-commerce markets, and are likely to continue to dominate.

The growth of e-commerce is substantial in Latin America and other emerging markets but it is still merely noise compared to the developed countries. Latin American e-commerce is expected to grow from $3 billion in 2000 to $81 billion (a 2700% increase) in 2004 but that is still just above 2% of the e-commerce transactions in the world. This raises several questions regarding the actions that policy makers in these countries should take to expand the e-commerce capabilities of the countries involved and what role entrepreneurs may play in this area so that they can complement the policy makers' activities by creating new companies that can enhance their countries' capabilities in this area. The Internet is drastically transforming the nature, geographic boundaries, timing, and scope of global competition, thus fundamentally changing the "mental geography" of international business (Ingari, 1999; Hof, 1999). The globalization process and the Internet are feeding on one another, expanding the frontiers of globalization, and rapidly changing

Table 1. Worldwide e-commerce growth (US$ billions)

Year	2000	2001	2002	2003	2004	% of total, 2004
North America	509	908	1,495	2,339	3,456	12.8
Asia Pacific	53	11	286	724	1,649	8.0
Western Europe	87	194	422	853	1,533	6.0
Latin America	3	7	14	31	81	2.4
Other countries	3	6	13	31	68	2.4
Total	**657**	**1,233**	**2,231**	**3,979**	**6,789**	**8.6%**

Source: Forrester Research, Inc. (2001)

the demographics of the Web (Daily, 1999; Eltoweissy et al., 1999). English is still the lingua franca of the Internet, where 96% of all e-commerce transactions are conducted in English (Dogar & Power, 2000; Pickering, 1999). However, in a recent study by Global Reach the English-speaking population on the internet reached 288 million people out of a population of 508 million, whereas non-English speakers with Internet access reached 516 million from the population of 5.8 billion people (Global Reach, 2004). Most Internet users now have a first language other than English. Nielsen Net Ratings (2003) reported that the US has 29% of the Internet access, with Europe, Asia-Pacific and Latin America at 23%, 13% and 2% respectively with the rest of the world at the remaining 33%.

In order to globalize, e-companies are finding it increasingly important to localize the content of their offerings (Schibsted, 1999; Zeff, 1999). More than any other industry, the development of strategic alliances between e-companies from developed nations with e-companies from developing nations is necessary to foster localization of content. As a result, most e-MNCs have bought regional companies or engaged in partnering approaches, and are establishing multilingual sites (Kotha, 1998).

One of the problem areas that affect Internet use is the digital divide between developed and less-developed countries. Figure 1 shows the difference in users among the world regions. It highlights the imbalance in Internet use. There is cause for concern that many parts of the world will be left out of the e-commerce phenomenon. It also shows the projected number of online users by the year 2003. North America, Western Europe and Asia are poised to expand their positions as the largest players on the Web. Emerging Web frontiers such as Latin America and Eastern Europe are also showing promising expansion potential.

It should be noted here that the actual numbers as reported by Global Reach (2004) for users in the different world regions are shown in Table 2.

At the onset of this new millennium, the globalization process is permeating all segments of business transactions (Kurzweil, 1999). One of the key features of the globalization process has been the advent of the Internet, which is launching a global remapping

Figure 1. Projected number of online users by region (millions of users)

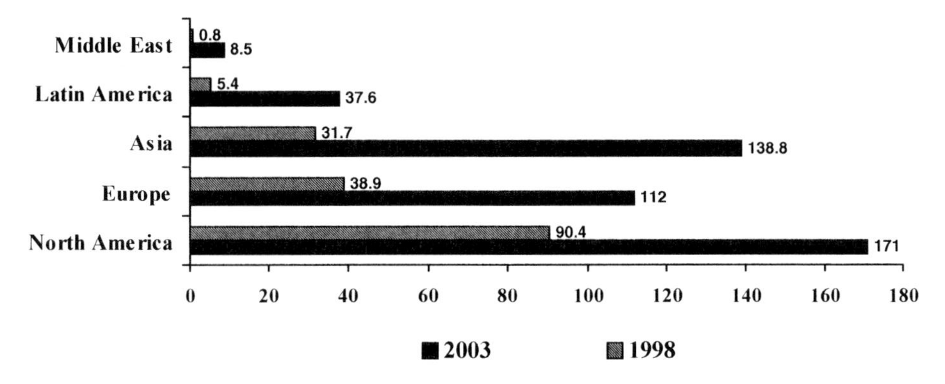

Source: Weyer (2000, p. 69)

Table 2. Internet users in different world regions in 2004

Region	Number of Users (2004) in millions
Middle East	10.5
Latin America	53
Asia	281.5
Europe	202.4
North America	205

Source: ClickZ Stats (2004)

process. This process is redefining space, mass, and time barriers. It is modifying the fabric of global business, boosting competition at the global level, and forcing companies to redesign their domestic and global strategies, business models, and corporate structures (Kotler, 1999; Westland, 1999). The Internet is also creating new industries and modifying existing ones. These transformations are making the global economy look like a cluster of digital metropolises that are erasing national boundaries (Gross, 1999; Rayport, 1999).

The New Economy or digital economy has created an explosion of economic growth and productivity, albeit accompanied by an equally impressive explosion in volatility. In this digital economy, the ability and capacity to use and manipulate information now measures the economic might of an economy or business cluster (Mandel, 1999). This represents a drastic shift from physical capital investment to intellectual capital (Byrne, 1999).

The Internet has had a substantial impact on the international business environment. Companies have been forced to rewrite business strategies in many industries across the globe, by shrinking business response times, and shortening service and product life cycles. The Internet is also fostering a diffusion of technology that is taking place at Internet time, leading to acceleration in the rate of innovation (Lucier & Torsilieri, 2000). As has been the case with other disruptive technologies such as railroads, telephone, electric power, radio & television, jet travel and microelectronics, the Internet is boosting the value of innovation (Mandel, 1999).

This new e-business environment places emphasis on delivery. In the Internet age, core competencies like speed, product quality, and quality service are key to a company's success in the global e-marketplace (Drucker, 1999). The Internet is also raising expectations since it lowers costs in supply and demand chains, opening markets previously forbidden to the average customer (Kotler, 1999).

The Internet has also boosted trade across boundaries by making more and better information available. This has leveled the playing field for companies of different sizes. Small and medium-sized companies don't face the same capital constraints to reaching a global market as they did prior to the Internet (Yoffie & Cusumano, 1999). Moreover, niche markets can now enjoy economies-of-scale at the global level by tapping into global market segmentation. The Internet has made identification of these market segments much easier. The creation of infomediaries is fostering development of new industries that can bring the global Web-marketplace together.

The nature of the Internet also makes it easier to expand the share of services such as banking, gambling, consulting, retailing, and education in global trade (Mandel, 1999). On the investment side, the Internet facilitates the integration of companies' activities at the global level, fostering innovations and integration of the supply chain, and therefore cutting costs.

This global electronic shopping center is open 24 hours, 7 days a week, and the e-consumer is an interactive consumer. The added use of e-mail creates tremendous interactivity. The Internet offers a unique opportunity to close the Awareness-Interest-Desire-Action cycle (AIDA). Companies are gearing themselves up to identify, communicate and deliver customer value (Ingari, 1999). These e-consumers are clustered in digital communities around the world, bringing new meaning to global segmentation of markets. These digital communities or clusters also largely impact development of e-commerce trends and bring together like-minded consumers from across the globe as never before (Gross, 1999). Thus, the Internet intensifies and changes the nature of communication between companies and customers. The new e-business paradigm changed to "we sell what we can deliver" (Drucker, 1999).

The Internet holds the promise of narrowing the gap between the "haves" and "have-nots" or between developed and less-developed countries. However, the global digital divide is already imposing a new form of colonization on emerging business frontiers not yet connected to the Web. Despite all of the promising talks of globalization, there are 2 billion people throughout the world that have never even made a phone call.

E-Commerce in Regions Other Than Latin America

Prior to exploring the case of Latin America in depth, it may be useful to provide greater context via an overview of e-commerce in two other regions: Europe and Asia.

The European economy entered the new millennium with renewed economic confidence. The "euro-sclerosis" days are apparently over. Inflexible labor markets, lackluster economic growth, and anti-business government policies are no longer permeating the European economy. The launch of a common currency is unifying fragmented markets, and facilitating the emergence of venture capitalists, an essential feature of the New Economy (Halper, 2000). However, Europe still lags behind the US when it comes to the Internet age. Only 10% of Europeans are connected to the World Wide Web, and Europe grossed US$19 billion in e-commerce for 1999, only 20% of the US volume (Fairlamb & Edmondson, 2000; Weier, 2000).

Table 3. New old world - 1999

Percent of Population with:			Online Sales
Country	Cell Phones	Internet Access	(US$ billions)
France	28%	10%	9.2
Germany	21%	15%	16.3
Sweden	55%	48%	86.0
Britain	32%	23%	26.0
USA	**25%**	**43%**	**112.0**

Source: Baker and Echikson (2000, p. eb44)

Across Europe, however, one finds different levels of Internet penetration. Scandinavians now rival the US in per capita usage of the Web. Germans are Europe's e-commerce leaders, accounting for 15.5% of the global share, compared to England's 6.6% and France's 3.8% of global share. England, meanwhile, is investing in skill advancement and upgrading the quality of its infrastructure, as creation of a computer literate population is a main concern of the British government (Weier, 2000).

Despite their relative lag, European companies are quickly developing new technologies and new product technologies. For instance, the development of mobile mini-portals, or phones that carry Internet services, is the new challenge (Mandel, 1999). These smart phones are linking customers to the Web 24 hours a day. Europe's common standard for cellular communication (GSM) is now in use from China to South Africa. By spreading their cellular standard, US cellular companies are edged out by European companies like Nokia and Vodafone, which are thriving in the wireless Net environment. By 2003 it is expected that 248 million Europeans will carry cell phones, up from 141 million in 1999 (Baker & Echikson, 2000).

Growth of the Web in Europe still faces a number of challenges. First, the increasing unification of European markets enhances European ethnocentrism, which may have a negative impact on the goal of building a global village. Second, in many countries, phone calls are charged by the minute, discouraging use of the Internet. Third, despite creation of the Euro, companies are still dealing with fragmented markets, several languages, and a different set of laws. Fourth, the extensive use of English has previously discouraged Europeans from using the Web. Fifth, in order for Europe to act as an electronic shopping center, import regulations have to be standardized. Currently, as a result of cumbersome customs procedures, most European Web sites are focusing on domestic shoppers. Sixth, Europeans are more willing to impose governmental control over the Web than Americans are. Issues such as consumer protection and privacy are becoming the focus of many lawmakers in Europe.

Asian countries are also rapidly narrowing the Internet gap. The convergence of PCs, cellular phones, and e-business is sending waves throughout Asia. The Asians have woken up to the potential benefits of information technology on productivity and consumer spending. Asia's immense untapped markets could turn into an "Eldorado" for e-commerce companies (Weier, 2000). With 2.7 billion people, close to half the world's

Table 4. Technology tigers

Country	USA	Singapore	Hong Kong	Taiwan
GDP per capita	$30,500	$28,400	$23,200	$14,400
Internet Technology Spending per capita	$ 1,200	$ 600	$ 300	$ 100
Internet Technology Spending (% of GDP)	4.2%	2.2%	1.3%	1.0%
Internet users (% of PCs)	55.0%	55.0%	42.0%	42.0%
Internet Users (% of population)	23.0%	17.0%	11.0%	5.0%

Source: International Data Corporation in Arnold (1999)

population, and a young population with a rising disposable income, Asia has all the ingredients for booming e-commerce (Arnold, 1999). According to some estimates, Asian e-commerce could amount to US$32 billion by 2003. In China alone, e-commerce is expected to increase from US$8 million in 1999 to about US$3.8 billion by 2003.

The Asian e-race is leading a number of countries to build technological parks, such as the Malaysia Multimedia Super Corridor. Singapore is planning to build a Science Hub and intends to build a fully wired society (Bremmer & Ihlman, 2000). These efforts could help narrow the substantial gap that currently exists between the US and some of the leading Asian Internet players, as shown in Table 4.

Like the Europeans, Asians are looking for alternative Web designs. In Japan the advent of smart-phones, or Internet-ready cellular phones, is dramatically changing the status of the Internet in Japan. Japan is emerging as the technological leader in wireless Net communications. In a country with 17 million computers, and 50 million cellular phones, wireless technology looks like the most efficient way to access the Web. By 2001, it is expected that 20 million smart phones will be in operation in Japan. Japan is also leading the global wireless industry by being the first country to jump into the third generation of mobile telephony or 3G (Kunii, 1999). This new technology will allow for video-conferencing, Web surfing, and a number of other applications. However, some argue that the "speed game" imposed by the Internet culture may not go along with Japanese corporate culture. Could we see the emergence of e-Keiretsus or e-Chaebols down the road?

E-Asia faces numerous challenges (Weier, 2000). First, credit cards are not widely used in Asia. Second, the vastness of Asia and the heterogeneous nature of its infrastructure imposes limitations on distribution. Third, narrow and illiquid markets hamper the emergence of venture capitalists. Fourth, Asia is not a homogeneous business environment and regulation intensity varies from country to country. Fifth, foreign direct investment restrictions in some countries, like China, hinder the Web's progress in Asia. Sixth, a lack of research-leader universities limits internal development of applicable technologies. Seventh, the absence of a risk-loving corporate culture makes the Internet less attractive to traditional Asian industries. Lastly, the anarchic nature of the Internet negatively influences the foundations of many authoritarian regimes in Asia.

Table 5. Breaking down the barriers to the new economy (1999)

Country	Net Population (million)	PC Penetration (% of households)
Japan	20.0	30.0%
South Korea	7.8	23.0%
Taiwan	4.2	35.0%
India	2.1	2.5%
China	10.0	1.7%

Source: Bremmer and Ihlwan (2000, p. 91)

Comparing Latin American E-Readiness to Other Regions

Some of the issues that require examination in the context of e-commerce capabilities center around the economic power gained through connectivity so that countries without the infrastructure investments in communications will continually fall behind the other countries, increasing thereby the wealth gap between nations. As e-commerce develops, the ability of any consumer in any country can theoretically become the same as any other person irrespective of where they are. This, however, assumes that goods can move freely and more importantly that consumers have equal access. Equal access, however, depends on the availability of many of the complimentary support services such as package delivery services (such as DHL, Fedex, etc.), quick clearance through customs, freedom from excessive tariffs or taxation and secure, verifiable payment systems.

Another important set of considerations surround the policies of developing countries in building communications infrastructure and allowing consumers to have access to the Internet at a reasonable cost. This will probably be a more pressing issue for countries if they have the companies that can benefit from the Internet (net gain in exports) in trade rather than a net loss (more imports). As exports grow, there are incentives for countries to support the communications infrastructure because that supports economic development and a higher tax base. There is a "chicken-and-egg" problem because companies will not have e-commerce sites if they cannot provide high-speed communications to customers and governments will not build infrastructure if it is not needed. The recommendation for countries, though, is to build the infrastructure because that serves companies with potential Internet presence as well as builds new markets for consumers.

The above-mentioned factors, including information and communication technology infrastructure, logistics infrastructure, institutional environment, and human capital jointly determine a country's "e-readiness," or the extent to which conditions in a given country are ripe for developing a healthy e-business sector. In the following tables, we present comparative data of e-business activity and e-readiness indicators for major Latin American markets, as compared with other emerging markets as well as developed markets.

Table 6. Comparative data: E-commerce and Internet usage

	E-Commerce	B2B sites	B2C sites	Internet	Internet	Internet	Users, as	Hosts per	E-Commerce
Country	Market 2001 $mil	Yahoo 11/2001	Yahoo 11/2001	Hosts 1/2001	Hosts 1/95	Users 1999	% of Pop. 1999	1000 Pop. 2001	Sites per Million Pop
Brazil	1,758	406	315	876,596	800	3,500,000	2.08%	5.22	4.29
Mexico	1,357	354	463	559,165	6,656	1,822,000	1.89%	5.79	8.46
Argentina	371	161	209	270,275	1,262	900,000	2.46%	7.39	10.11
Chile		92	117	74,708	0	700,000	4.66%	4.97	13.92
Colombia		42	64	46,819	1,127	664,000	1.60%	1.13	2.55
Venezuela		62	86	16,154	529	525,000	2.21%	0.68	6.24
Peru		50	101	10,705	171	400,000	1.59%	0.42	5.98
Uruguay		23	10	54,065	172	300,000	9.06%	16.32	9.96
Costa Rica		58	326	7,357	798	150,000	4.18%	2.05	106.99
Bolivia		19	23	1,324	0	78,000	0.96%	0.16	5.16
Guatemala		27	26	5,603	0	65,000	0.59%	0.51	4.78
Panama		31	56	15,084	17	45,000	1.60%	5.37	30.95
El Salvador		10	18	577	0	40,000	0.65%	0.09	4.55

The data in these tables comes from sources including the World Bank, the Heritage Foundation, the Yahoo internet directory (i.e., for the number of B2B and B2C sites listed in Yahoo for each country, and a per capita measure of such sites) and eMarketer, with the estimates of national market size being derived from the authors' computed averages

Table 6. Comparative data: E-commerce and Internet usage (cont.)

	E-Commerce	B2B sites	B2C sites	Internet	Internet	Internet	Users, as	Hosts per	E-Commerce
Country	**Market 2001 $mil**	**Yahoo 11/2001**	**Yahoo 11/2001**	**Hosts 1/2001**	**Hosts 1/95**	**Users 1999**	**% of Pop. 1999**	**1000 Pop. 2001**	**Sites per Million Pop**
Ecuador		50	99	2,636	325	35,000	0.28%	0.21	12.00
Dominican Republic		58	56	7,907	0	25,000	0.30%	0.94	13.56
Nicaragua		6	12	1,400	49	20,000	0.41%	0.28	3.66
Honduras		13	24	128	0	20,000	0.32%	0.02	5.86
Belize		31	112	293	0	12,000	4.86%	1.19	579.30
Haiti		6	4	1	0	6,000	0.08%	0.00	1.28
Latin America		**1,499**	**2,121**	**1,950,797**	**11,906**	**9,307,000**	**1.93%**	**4.05**	**7.51**
European Union (15)		**26,000**	**24,922**	**12,607,544**	**930,456**	**59,675,000**	**15.89%**	**33.58**	**135.63**
G7 Countries		**386,570**	**226,972**	**16,586,950**	**893,615**	**151,430,000**	**21.85%**	**23.93**	**885.14**
United States	147,000	355,246	197,767	2,267,089	37,615	74,100,000	26.63%	8.15	1987.61
Japan	29,650	763	533	4,640,863	96,632	27,060,000	21.38%	36.67	10.24
Germany	15,005	1,228	681	2,163,326	207,717	14,400,000	17.54%	26.35	23.25

of estimates made by consulting firms such as IDC, Forrester Research, Morgan Stanley Dean Witter, Boston Consulting Group, and eMarketer (to improve accuracy and reliability, only countries for which multiple estimates were available are included). Although estimates of market size in each country are to be viewed as educated guesses at best, these estimates match up quite well with other indicators of e-commerce and

Table 6. Comparative data: E-commerce and Internet usage (cont.)

	E-Commerce	B2B sites	B2C sites	Internet	Internet	Internet	Users, as	Hosts per	E-Commerce
Country	**Market 2001 $mil**	**Yahoo 11/2001**	**Yahoo 11/2001**	**Hosts 1/2001**	**Hosts 1/95**	**Users 1999**	**% of Pop. 1999**	**1000 Pop. 2001**	**Sites per Million Pop**
United Kingdom	12,560	16,643	16,634	2,291,369	241,191	12,500,000	21.01%	38.51	559.27
Canada		10,419	9,124	2,364,014	186,722	11,000,000	36.08%	77.53	640.94
Italy	3,200	1,001	942	1,630,526	30,697	7,000,000	12.14%	28.29	33.71
France	4,190	1,270	1,291	1,229,763	93,041	5,370,000	9.16%	20.98	43.69
Eastern Europe		**920**	**843**	**904,630**	**38,710**	**5,693,000**	**5.22%**	**8.29**	**16.16**
Poland		134	134	371,943	11,477	2,100,000	5.43%	9.62	6.93
Czech Republic		206	210	153,902	11,580	700,000	6.81%	14.97	40.47
Hungary	30	65	58	158,732	8,506	600,000	5.96%	15.77	12.22
Romania		114	80	41,326	597	600,000	2.67%	1.84	8.64
Slovakia		29	41	36,680	1,414	600,000	11.12%	6.80	12.97
Slovenia		83	63	23,594	1,773	250,000	12.59%	11.88	73.53
Bulgaria		104	64	17,166	144	235,000	2.86%	2.09	20.47
Croatia		56	62	23,814	1,090	200,000	4.48%	5.33	26.43

Internet activity, such as the number of e-commerce sites per 1,000 inhabitants in each country, as derived from the count of sites on the Yahoo directory.

As shown in the tables, Latin American countries fall behind developed countries in all indicators of e-commerce activity and e-readiness. A comparison between Latin America and other emerging markets, however, produces more mixed results. In general, Latin

Table 6. Comparative data: E-commerce and Internet usage (cont.)

Country	E-Commerce Market 2001 $mil	B2B sites Yahoo 11/2001	B2C sites Yahoo 11/2001	Internet Hosts 1/2001	Internet Hosts 1/95	Internet Users 1999	Users, as % of Pop. 1999	Hosts per 1000 Pop. 2001	E-Commerce Sites per Million Pop
Estonia		40	37	40,094	1,396	200,000	13.87%	27.80	53.38
Latvia		49	54	19,059	612	105,000	4.32%	7.84	42.37
Lithuania		40	40	18,320	121	103,000	2.78%	4.95	21.63
Select Asian Emerging Mkts		**11,194**	**5,650**	**682,059**	**22,822**	**26,360,000**	**0.96%**	**0.25**	**6.14**
South Korea	3,750	1,149	413	397,809	18,049	10,860,000	23.18%	8.49	33.33
China	9,337	2,017	616	70,391	569	8,900,000	0.71%	0.06	2.10
India	221	4,618	2,258	35,810	359	2,800,000	0.28%	0.04	6.89
Malaysia		1,438	669	68,248	1,606	1,500,000	6.61%	3.01	92.78
Indonesia		523	445	26,727	177	900,000	0.43%	0.13	4.68
Thailand		847	811	63,447	1,728	800,000	1.33%	1.05	27.52
Philippines		490	356	19,448	334	500,000	0.67%	0.26	11.39
Vietnam		112	82	179	0	100,000	0.13%	0.00	2.50

countries lag most Eastern European countries (e.g., Czech Republic, Hungary, Slovakia, Slovenia, and Estonia) and some Asian countries (e.g., South Korea and Malaysia) in terms of actual usage of the Internet for e-commerce and other purposes. The second table provides clues as to why Latin countries lag these other emerging markets. Most importantly, Latin countries generally lag the other emerging markets listed in economic

Table 7. Comparative data: E-readiness indicators

	GDP per Cap	Illiteracy	Life Exp	Pop. 02	Rural Pop	Econ Freedom	Tel /1000	2002 Cost of	2002 TV	High Tech	Trade, as %
Country	$ in 2002	% of youth	2002	mil	% of tot	Heritage	2002	Local Call $	Sets per 1000	Exports, %	of GDP 2002
Brazil	7,450	5.5	69	174.5	18	3.10	223	0.03	349	19.0	24.3
Mexico	8,800	3.5	74	100.8	25	2.90	147	0.16	282	21.0	52.4
Argentina	10,190	1.5	74	36.5	12	3.48	219	0.03	326	7.0	33.7
Chile	9,420	1.0	76	15.6	14	1.91	230	0.10	523	3.0	55.2
Colombia	6,150	2.5	72	43.7	24	3.13	179	0.03	303	7.0	30.6
Venezuela	5,220	1.5	74	25.1	13	4.18	113	0.04	186	3.0	41.0
Peru	4,880	3.0	70	26.7	27	2.83	66	0.08	172	2.0	26.9
Uruguay	7,710	1.0	75	3.4	8	2.55	280	0.17	530	3.0	31.5
Costa Rica	8,560	1.5	78	3.9	40	2.71	251	0.03	231	37.0	73.8
Bolivia	2,390	2.5	64	8.8	37	2.59	68	0.09	121	7.0	39.5
Guatemala	4,030	20.0	65	12.0	60	3.16	71	0.07	145	7.0	35.7
Panama	6,060	3.0	75	2.9	43	2.83	122	0.12	191	1.0	31.1
El Salvador	4,790	11.0	70	6.4	38	2.24	103	0.07	233	6.0	57.3
Ecuador	3,340	4.0	70	12.8	36	3.60	110	0.03	237	7.0	47.1
Dom. Republic	6,270	8.5	67	8.6	33	3.51	110	0.06	..	1.0	65.0
Nicaragua	2,350	13.5	69	5.3	43	2.94	32	0.08	123	5.0	59.7
Honduras	2,540	11.0	66	6.8	45	3.53	48	0.06	119	2.0	64.1
Belize	..	..	..	..	..	2.69	..	..	..	..	..
Haiti	1,610	33.5	52	8.3	63	3.78	16	..	6	..	41.0
Latin America	**6,950**	**4.5**	**71**	**524.9**	**24**	**2.97**	**168**	**0.06**	**289**	**16.0**	**41.2**
European Union	**25,700**	**0.5**	**78**	**305.5**	**22**	**2.10**	**555**	**0.13**	**597**	**17.0**	**56.3**
G7 Countries	**28,635**	**0.0**	**79**	**705.9**	**20**	**2.15**	**590**	**0.10**	**736**	**21.0**	**41.1**
United States	36,110	0.0	77	288.4	22	1.85	646	0.00	938	32.0	18.3
Japan	27,380	0.0	82	127.2	21	2.53	558	0.07	785	24.0	18.9
Germany	26,980	0.0	78	82.5	12	2.03	651	0.09	661	17.0	55.8
United Kingdom	26,580	0.0	77	59.2	10	1.79	591	0.18	950	31.0	39.9
Canada	28,930	0.0	79	31.4	21	1.98	635	..	691	14.0	67.1
Italy	26,170	0.0	78	57.7	33	2.26	481	0.11	494	9.0	41.7
France	27,040	0.0	79	59.5	24	2.63	569	0.12	632	21.0	46.2

development, as measured by GDP per capita on a purchasing power parity basis, as well as in literacy rates and in the importance of trade for the national economy. The gap between Latin countries and other emerging markets is smaller or non-existent in other indicators of e-readiness, such as economic freedom (as measured by the Heritage Foundation's annual index of economic freedom), urbanization, life expectancy, and technology infrastructure.

Table 7. Comparative data: E-readiness indicators (cont.)

	GDP per Cap	Illiteracy	Life Exp	Pop. 02	Rural Pop	Econ Freedom	Tel /1000	2002 Cost of	2002 TV	High Tech	Trade, as %
Country	$ in 2002	% of youth	2002	mil	% of tot	Heritage	2002	Local Call $	Sets per 1000	Exports, %	of GDP 2002
Eastern Europe	**10,716**	**0.0**	**72**	**262.1**	**33**	**2.64**	**296**	**0.12**	**478**	**10.0**	**98.8**
Poland	10,450	0.0	74	38.6	37	2.81	295	0.08	422	3.0	50.9
Czech Republic	14,920	0.0	75	10.2	25	2.39	362	0.13	538	14.0	113.9
Hungary	13,070	0.0	72	10.2	35	2.60	361	0.13	475	25.0	109.3
Romania	6,490	0.0	70	22.3	45	3.66	194	0.11	697	3.0	69.3
Slovak Republic	12,590	0.0	73	5.4	42	2.44	268	0.12	409	3.0	130.2
Slovenia	18,480	0.0	76	2.0	51	2.75	506	0.07	366	3.0	92.9
Bulgaria	7,030	0.0	72	8.0	32	3.08	368	0.02	453	12.0	88.1
Croatia	10,000	0.0	74	4.5	41	3.11	417	0.09	293	12.0	69.6
Estonia	11,630	0.0	71	1.4	31	1.76	351	0.09	502	4.0	156.7
Latvia	9,190	0.0	70	2.3	40	2.36	301	0.11	850	4.0	75.4
Lithuania	10,190	0.0	73	3.5	31	2.19	270	0.14	487	5.0	96.4
Asia	**6,167**	**2.0**	**70**	**2,834.5**	**56**	**3.33**	**140**	**0.03**	**230**	**29.0**	**92.4**
South Korea	16,960	0.0	74	47.6	17	2.69	489	0.03	363	32.0	66.0
China	4,520	1.0	71	1,280.4	62	3.64	167	0.03	350	23.0	49.0
India	2,650	..	63	1,048.6	72	3.53	40	0.02	83	5.0	..
Malaysia	8,500	3.0	73	24.3	41	3.16	190	0.03	210	58.0	182.4
Indonesia	3,070	1.5	67	211.7	57	3.76	37	0.03	153	16.0	51.1
Thailand	6,890	2.0	69	61.6	80	2.86	105	0.07	300	31.0	105.6
Philippines	4,450	5.0	70	79.9	40	3.05	42	0.00	182	65.0	91.7
Vietnam	2,300	..	70	80.4	75	3.93	48	0.02	197	..	101.3

In addition to differences between regions in e-commerce and e-readiness, there is substantial within-region variation for both developed and emerging market regions. Italy, for instance, stands out as having a similar level of economic development and infrastructure as other developed countries, yet exhibits substantially lower rates of Internet usage and market potential. Turning to Latin America, in absolute terms, Brazil and Mexico stand out as accounting for the majority of Internet usage and e-commerce in the region. Although Mexico was an earlier mover in the adoption of Internet technologies, Brazil has since surpassed Mexico. However, there is some indication that the gap is narrowing once again. In terms of the intensity of Internet usage and e-commerce, Brazil and Mexico remain prominent players in the region, joined by Argentina, Uruguay, Chile, Venezuela, Costa Rica, Panama, Belize, Peru, and Colombia. Beyond Panama and Costa Rica, Central American nations have been among the slowest adopters of e-commerce and the Internet. These intra-regional differences are largely the result of the same socio-economic and infrastructure-related issues addressed above, such as income levels and literacy rates.

A Closer Look at E-Commerce in Latin American Markets

We now turn to focus on e-commerce and Internet activity in specific Latin American countries, as well as the competitive actions of firms operating in this sector.

Latin American Internet users form about 2% of the world Internet users (Nielsen Net Ratings, 2003). Latin American countries are joining the Web at different speeds and intensities with the number of Internet users in the region growing from 2 million in 1997 to 60 million in 2004 (World Bank, 2001; www.eMarketer.com). Furthermore, the regional online population is expected to reach 7% of the Latin American population. Another more forward-looking estimate forecasts 77 million users by 2005 (www.nua.com; Jupiter Media Metrix). It is important to emphasize that Latin American countries are facing diverse levels of Internet penetration. For instance, in 2003, Chile had the highest Internet penetration rate, close to 22%, compared to Brazil's 6%. E-commerce should increase from US$36.2 million in 1997 to an expected US$8.0 billion by 2003 (www.nua.com).

Emarketer (2002) estimates that Argentina, Brazil and Mexico account for 65% of Latin American trade. Chile is home to a smaller population, leading to a quicker saturation than Argentina, Brazil, and Mexico. This has been a change from 2000 when Brazil had two-thirds of the e-commerce activities, followed by Mexico with 10% (Almeida, 2000). Argentina, Brazil, Chile, Colombia, Mexico, and Venezuela constitute the core of Latin America's e-commerce (Zelner, 2000).

However, in coming years, the relative importance of Brazil should diminish. The region's business-to-business (B2B) and business-to-consumer (B2C) e-commerce numbers are expected to reach US$8 billion combined by 2003 (Latin Trade, 1999). Brazilian e-commerce is expected to increase from US$121 million in 2000 to US$4.3 billion by 2003, and Mexican e-commerce from US$25 million to US$1.5 billion. Chile should expand B2B to US$9 billion in 2004, and B2C to US$400 million, and Argentina's e-commerce should expand from US$15 million to US$1.1 billion in the same period. It is important to highlight though, that the majority of this e-commerce is still heavily directed at US Web sites, which makes development of Latin American e-commerce even more appealing to US investors and companies (Latin Trade, 1999; Somagi, 1999; www.eBusinessforum.com). Table 8 indicates the percent of online spending directed at domestic rather than foreign sites.

Table 8. Domestic online spending in Latin American countries in 2000

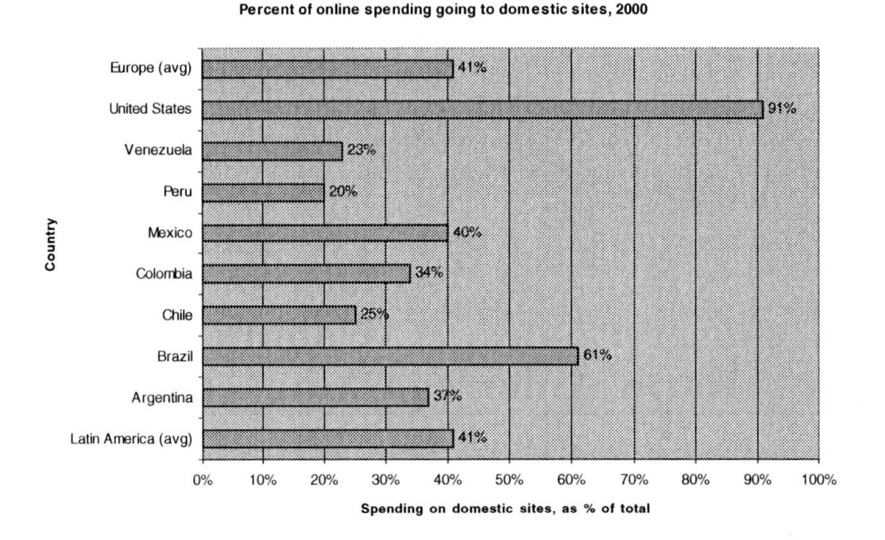

Source: eMarketer, 2001

One interesting characteristic of Latin America e-commerce is its high level of concentration. For instance, the top 20 Latin American Web sites account for 73% of e-commerce sales (Cavalcanti, 2000). In addition, brick & mortar companies reportedly control 98% of e-commerce revenues in Latin America (Negocios Exame, 2000).

Business-to-business e-commerce is poised to grow faster in the region than business-to-consumer e-commerce. A number of companies operating in Latin America are now realizing the potential for cost reductions through use of the Web. Multi-nationals like Volkswagen are driving these changes in Mexico and in Brazil. For instance, Volkswagen built a Dealer Communication System extranet in Mexico, linking its dealerships to Volkswagen's factory in Puebla. In Brazil, Volkswagen made use of its extranet system mandatory to all of its 738 Brazilian and Argentine suppliers as early as 1999 (Case, 1999).

The Latin American Web sector is going through changes at e-speed. Foreign investment is pouring in and changing the profile of the industry while opening new markets in a short period of time. Multi-nationals like the Spanish Telefónica and Microsoft are racing to connect Latin American customers (Dorschner, 1999). The convergence of Web-related technologies at the global level is reflected in the region, where phone lines, TV cable, and Internet access could be provided by a handful of companies in the near future. BellSouth, for instance, is expanding its involvement in Latin America, going beyond wireless services by planning to offer Internet services in ten Latin American countries.

Latin American and multi-national companies are also rushing to consolidate their position in the industry. Grupo Carso, the holding company for Teléfonos de México, acquired Prodigy and has developed a joint venture with Microsoft. The Spanish company Telefónica's Terra spin-out bought several international Service Providers (ISPs) and portals, such as the Brazilian company ZAZ and the Mexican company, Infosel, throughout Latin America. Terra Networks is investing US$600 million to develop its network in Brazil, Mexico, Peru, Chile and Guatemala, while Microsoft is developing a partnership with GloboCabo in Brazil to launch a local ISP (Dorschner, 1999; Vargas, 2000).

The lack of state-of-the-art telecommunications infrastructure is leading to another race parallel to the evolution of the Internet. Establishing more capable, more reliable and less costly phone connections would eliminate one of the region's main bottlenecks for growth of the Internet in Latin America. Chile is the only exception in the region. The country shows the most modern telecommunication infrastructure in the region. Several multi-nationals are rushing to wire the region with high-speed broadband cable (Dorschner, 1999). Telefónica is spending US$1.7 billion to lay undersea fiber-optic cable connecting Brazil and Argentina to Miami. AT&T is developing a joint venture with FirstCom, creating AT&T Latin America, to wire a number of Latin American markets.

Cyber-pioneers are appearing all over Latin America. Market-oriented reforms such as trade liberalization and privatization are having a substantial impact on Internet expansion in Latin America. Privatization of the telecommunications industry in several Latin American countries has had substantial impacts on the allocation of phone lines per capita, quality of service, and reliability. For instance, in Brazil the number of fixed lines per 100 inhabitants doubled in the second half of the 1990s, from 8.4 fixed lines to 16 fixed lines per 100 inhabitants, or a total of 25.8 million fixed phone lines for 1999. Still, Chile boasts the highest number of telephone main lines per capita in the region, 23.6 per 100 people. In the cellular industry the changes have been more substantial. From 1994 to 1999, cellular phones increased their reach from 0.5% of the Brazilian population to 7.2% of the Brazilian population. Fiber optics usage has also increased in the same period from less than 4,000 kilometers in 1995 to about 40,000 kilometers by 1999 (Gurovitz, 1999).

Despite such recent efforts, Latin American countries lag well behind the US and Europe in some vital aspects of the Internet environment. In terms of number of PCs, the US leads with 145 million units, followed by Japan with 30 million units. In Latin America, Brazil has 6.8 million units, followed by Mexico with 3.6 million, Argentina with 1.6 million, and Chile with 0.9 million units. In terms of PCs per capita, the US has 41 PCs per 100 inhabitants compared to Brazil's 4 PC's per 100 inhabitants and Argentina's 5 PCs per 100 inhabitants (Lopes, 1999).

The private sector is not the only sector in Latin America affected by the impact of the Internet. For instance, low levels of efficiency and transparency have always characterized business transactions between the government and private sector in the region. The Mexican private sector is utilizing e-commerce to combat red tape and lack of efficiency in government procurement. "Compranet," a B2G (business-to-government) Web site, is being used to source US$20 billion worth of products and services for the Mexican government (Conger, 2001). In Chile "ComprasChile" is reaching more than 120 public entities, making transactions between the Chilean private sector and state agencies more transparent and efficient. In another example, Brazilians are filing their income taxes online, increasing efficiency and reducing red tape. These are just some examples on how the Internet is substantially changing the Latin American business environment.

The Latin American e-commerce industry still faces a number of challenges, many of which are similar to those faced by Asian countries. First, low credit card ownership hampers e-commerce. In addition, Latin Americans' fear of credit card fraud keeps many e-customers from shopping on the Web. Only international credit cards have been accepted, keeping a number of potential buyers outside the e-commerce loop. Second, shipping costs are high and customs crossing can be cumbersome and lengthy. Third, Latin Americans like to have a personal touch when shopping, which e-commerce lacks. Fourth, phone calls are charged by the minute, making access and use less attractive. Fifth, the integration of shipping services, inventory management and customer service is still in its infancy. Sixth, low computer penetration limits the impact of the Internet. These barriers are creating a number of opportunities for companies that can expedite and facilitate e-commerce in Latin America (Caturla, 2000a, b). Companies like shippers can facilitate and expedite the e-commerce process by brokering payments or integrating the whole process.

Table 9 provides an overview of many of the firms active in the Latin American e-commerce sector. Firms are listed according to their market scope and region of origin. This table is not intended to be exhaustive, but rather to provide an overview of many of the major players in this sector and their strategic orientation. Also, there is substantial flux not only in the market participants, but also in each firm's strategic orientation, as firms attempt to re-invent themselves (e.g., from a portal to a marketplace to an e-service provider) in response to competitive dynamics and trends in business models. Furthermore, consolidation has transformed local players into regional or global players, as occurred through Terra's acquisition of Infosel, a leading ISP in Mexico. This means that the barriers between the various types of firms present in the market are in constant flux and are transcended by multiple firms.

Table 9. Partial list of leading e-commerce firms in Latin America

Predominantly National B2Cs, Portals & ISPs	Regional B2Cs, Portals, & ISPs	National B2Bs, Marketplaces, e-services	Non-Latin firms with substantial investments in the region	Non-Latin firms serving the market w/little customization
dot-coms: Saraiva, Globo, decompras, Yupi, alo.com, todito.com, lineinvest.com, UOL, BOL, iG.com, Cade, Shoptime.com, Celebrando.com **Bricks/Mortar:** Pao de Acucar, Sanborns, Elektra, Liverpool, Palacio de Hierro, Gandhi, Martí	Starmedia, Submarino, Patagon, deremate, Telmex/T1msn, El sitio, Qbueno, viajo.com, Latin Stocks.com, sportsya.com	Latinexus/ Cemex, Artikos, Eficentrum, Banco do Brasil, Itau, Unibanco, Bradesco, Asista.com, Mercantil.com, Estrutura.net, Exiros	AOL Latin America, Terra Lycos, Microsoft, Volkswagen, General Motors, Citibank, Santander, Banco Bilbao Vizcaya, Telefonica, Ebay/ mercadolibre, Yahoo!, Foodtrader	Virtually endless number of firms (Amazon.com, Travelocity, Priceline, etc.)

In Table 9, a few trends can be identified through a closer look at the firms included in the table. First, although the B2B sector may at first appear more amenable to a pan-regional approach than B2C, there are very few B2B players that are truly regional in scope (two exceptions are Mercantil.com and Acquira.com). Instead, major B2B players tend to have either a national focus (e.g., the many Brazilian banks that are pioneers in online banking and e-services and the Chilean Cienpunto) or have been consolidated into global networks (e.g., eBay's majority acquisition of a regional player, mercadolibre.com). Among the few highly visible regional marketplaces that have been created, such as Latinexus, the market scope has tended to revert to focus predominantly on a single country or has broadened to a global perspective. Latinexus, for instance, was a prominent start-up created by leading Mexican and Brazilian industrial and financial giants, including Cemex and Alfa from Mexico and Bradespar and Votorantim from Brazil. The original stated strategy of Latinexus was to leverage the tremendous procurement needs of these giant firms to provide similar procurement services for smaller firms throughout the region. More recently, Latinexus has been taken over by one of its founders, Cemex, in order to provide supply chain management for Cemex as well as to service external customers.

Second, while there are a large number of strictly national firms in the sector, there are commonalities among these firms. Many of the large, general B2C and portal firms that have a national focus are located in Brazil (e.g., Saraiva.com, Shoptime), whereas firms in this sector originating from other countries have tended to broaden their scope to the regional level (e.g., deremate/arremate, Patagon). Another important segment among national B2C companies consists of established brick-and-mortar firms that utilize online operations to complement their core business, such as Falabella, Almacenes Paris and Ripley, Sanborn's, Palacio de Hierro, Liverpool, and Martí, all of which are located in Mexico. These firms are much more likely to remain national in scope, mirroring their physical scope.

Third, although firms in the above table are distinguished by originating from within Latin America or from outside the region, this distinction is not clear in many cases. On the one hand, many firms from within Latin America (e.g., Infosel, mercadolibre, Cade) have been acquired by global players. In rarer instances, the opposite has occurred, such as in the case of Grupo Carso's acquisition of the Prodigy ISP and CompUSA. Moreover, multiple dot-coms founded specifically to serve the Latin American market were initially located in Miami or other locations in the US (e.g., Starmedia, Qbueno), and the vast majority of dot-coms located within Latin America have been dependent on external providers of financing and other resources. These firms have generally turned to US-based venture capitalists and equity markets, including the NASDAQ, rather than or in addition to seeking local sources of funding.

Conclusion

Continued volatility is perhaps the only identifiable certainty regarding the future of e-commerce in Latin America. Nonetheless, our analysis allows us to draw the following conclusions regarding the regional e-commerce environment at present, as well as more tentative projections regarding future developments.

Latin American E-Markets are not Homogenous

Latin America countries show dramatic differences in their business and economic environments. The region is divided not only by two languages, Portuguese and Spanish, but also by different levels of economic development and social environments (Robles, Simon & Haar, 2003). In the case of e-commerce, different levels of economic development have affected the speed and expansion of e-commerce in the region. For instance, Chile shows the most developed infrastructure in the region and the most dynamic economy, explaining the higher rates of connectivity statistics, such as telephone main lines per capita, number of personal computers per capita, and Internet users as a percentage of the population. In addition, Chile's earlier adoption of market-oriented reforms has resulted in a more efficient and more stable business environment. Countries like Argentina, Brazil and Mexico face much different economic and social conditions. In these countries the digital divide is much more pronounced than in Chile. These countries also show a much worse income distribution than Chile's, resulting in lower levels of connectivity, such as telephone lines per capita, number of personal computers per capita, and Internet users as a percentage of the population. These are issues that need to be considered when approaching Latin American markets.

Argentina, Brazil, Chile, Colombia, Mexico, and Venezuela form the core of Latin America's e-commerce. By 2004, the region should have close to 60 million Internet users, with Argentina, Brazil, and Mexico, representing close to 65% of the Latin America's total users. These countries show the largest potential for e-commerce in the region. Chile's small population will lead the industry to face saturation sooner than in these other countries.

Privacy laws are not the same throughout the Latin American markets, either. Several countries in the region did not have legislation regulating the collection or even application of personally identifiable data on the Internet. Recently, however, several countries are implementing specific data legislation. In this regard, in 1999 Chile was the first country in the region to implement a data protection law. In 2000, Mexico implemented the Mexican E-Commerce Act. Brazil, on the other hand, has had a data privacy bill waiting for approval by its Congress since 1996. Latin American privacy laws are being drafted after the European Privacy Directive. This in the future will facilitate e-commerce transactions in the region and between Latin America and the European Union.

The digital divide has held the region back so far, but may disappear. Academics and policy makers have been concerned that a digital divide will not only prevent developing countries from obtaining social welfare gains associated with technological advancement, but rather that differential access to technology will actually exacerbate the gap in standards of living between developed and developing nations.

The digital divide presents two challenges to Latin American economies and businesses. First, although e-commerce has grown in the region, it accounts for a much smaller share of the regional economy in Latin America than in the United States or

European Union. If the assumptions that e-commerce adoption leads to enhanced allocative efficiency and greater entrepreneurial opportunities for value creation are valid (Amit & Zott, 2001), then the existence of a digital divide would appear to hinder the region's economic development and provide regional firms a competitive disadvantage in global markets. Second, even if new technologies were to eventually diffuse throughout the region, early-mover advantages would accrue to firms from outside the region, and Latin American firms would then suffer from a competitive disadvantage in local markets as well.

To date, large Latin American firms have been better positioned than smaller rivals to profit from e-business. Although a digital divide separates Latin America from other parts of the world, the largest Latin American firms have been able to overcome and arguably profit from the digital divide. Whereas the region as a whole lags in Internet access, infrastructure development, and online transactions, the largest Latin American firms do not appear to lag behind their global counterparts in technology adoption and e-commerce activity. As shown in a preceding section, a greater share of large Brazilian, Chilean and Mexican firms have e-commerce enabled Web sites than do large Italian, Canadian, Japanese, and French firms. Many of these firms compete in global markets and obtain capital as well as knowledge-based resources from global sources. Also, they often control key distribution channels in local markets, which gives them an advantage over both small, local rivals, as well as foreign market entrants. Even if they only reach parity with global competitors in terms of technology adoption, they will be accorded a technology-based competitive advantage in their home markets.

The future expansion of Latin American e-commerce faces a number of challenges. Among these challenges, several stem from features of the industry structure (Manaut, 2000). First, low penetration and the small number of Internet users in Latin American countries is not allowing e-commerce companies to generate sufficient economies of scale, with the possible exception of Brazilian firms. This has forced many companies to broaden their scope to the regional level, only to find that they lack the resources and market knowledge to succeed elsewhere in the region. Logistics in Latin American e-commerce is still a major bottleneck for building an efficient supply chain. Integration of shipping services, inventory management and customer service is still in its infancy. Lack of state-of-the-art, locally based logistics is attracting foreign companies like FedEx and leading to creation of private carriers like the Mexican Estafeta and the Argentine Macri. Customs crossings, however, are cumbersome and lengthy, and shipping rates are still very high. E-commerce safety is still of paramount importance to companies and consumers in Latin America. The absence of a consumer protection code is a barrier to further expansion of Latin American e-commerce. However, some changes are already taking place. In the fall of 2000, Mexican Web e-commerce contracts were given the same validity as regular non-virtual commerce contracts. Other aspects of physical and information infrastructure are inadequate and/or costly. For instance, phone calls are charged by the minute in most Latin countries, which particularly inhibits browsing and B2C transactions. Finally, the limited local content of Latin American e-commerce sites entices Latin e-customers to shop in better-developed, non-Latin American sites. Overall, regional Latin American e-commerce firms face competing pressures to standardize in order to achieve the economies of scale necessary to compete against global rivals, while at the same time needing to localize in multiple national institutional environments in order to provide the responsiveness of a local firm.

References

Amit R., & Zott, C. (2001). Value creation in e-business. *Strategic Management Journal, 22*, 493-520.

Arnold, W. (1999, September 19). Where the start-up dance is still hard to do. *The New York Times*, Section 3, BU 1(13).

Baker, S., & Echikson, W. (2000, February 7). Europe's Internet bash. *Business Week – Ebiz*, eb40-eb44.

Bremmer, B., & Ihlwan, M. (2000, January 31). Edging toward the information age. *Business Week*, 90-91.

Byrne, J. (1999, October 4). The search for the young and gifted. *Business Week*, 108-116.

Case, B. (1999, June). Extranet exertions. *Latin Trade*, 42.

Caturla, F. (2000a). Acesso gratis a Web chega a Argentina. *Gazeta Mercantil Latino-Americana, 4*(192), 13.

Caturla, F. (2000b). Dobra o comercio virtual na Argentina. *Gazeta Latino-Americana, 4*(194), 9.

ClickZ Stats. (2004). Global online populations. Retrieved from *www.clickz.com/stats*

Conger, L. (2001, July 30). Compra virtual reduz farude no Mexico. *Gazeta Mercantil, 10.*

Daily, J. (1999, November). A world of opportunity. *Business 2.0*, 1.

Dogar, R., & Power, C. (2000, December-February). Mapping a virtual planet. *Newsweek*, Special Edition, 79-81.

Dorschner, J. (1999, November 21). The new Internet frontier. *The Herald*, E1(2).

Drucker, P. (1999). Can e-commerce deliver? *The Economist – The World in 2000*, 95.

Eltoweissy et al. (1999). Enabling global e-commerce: A technology perspective. *Proceedings of the Association for Global Business,* Las Vegas, November 18-21 (pp. 124-134).

Emarketer. (2002). Latin American Net population to soar. Retrieved from *www.emarketer.com*

Fairlamb, D., & Edmondson, G. (2000, January 31). Work in progress. *Business Week*, 80-87.

Global Reach. (2004). Global Internet statistics (by Language). Retrieved from *www.global-reach.biz/globstats*

Gurovitz, H. (1999a). Brasil digital. Brasil em Exame 1999, Edicao 700, 10-12.

Gurovitz, H. (1999b). Planeta E. *Exame, 32*(12), 149-159.

Halper, M. (2000, February). Europe's new capital. *Business 2.0*, 172-182.

Hof, R. (1999, October 4). A new era of bright hopes and terrible fears. *Business Week*, 84-98.

Ingari, F. (1999, Fourth Quarter). The Internet as a marketing medium. *Business & Strategy*, 6-8.

Kotha, S. (1998). Competing on the Internet: The case of Amazon.com. *European Management Journal*, *16*(2), 212-222.

Kotler, P. (1999). *Kotler on marketing.* New York: The Free Press.

Kunii, I. (1999, October 18). I'm online all the time and it's free. *Business Week*, 30.

Kurzweil, R. (1999, December). The Web within us. *Business 2.0,* 173-174.

Latin Trade. (1999, June). The outer limits of e-commerce. *Latin Trade*, 36-54.

Lopes, M. (1999). "Por Que o Vale do Silicio nao e Aqui." *Brazil em Exame*, 40-42.

Lucier, C., & Torsilieri, J. (2000, First Quarter). The trillion-dollar rate to "E". *Business & Strategy*, *18*, 6-14.

Manaut, S. (2000, May 22-28). Obstaculos travam e-commerce na Regiao. *Gazeta Mercantil Latino-Americana, 7.*

Mandel, M. (1999, October 4). The Internet economy, the world's next growth engine. *Business Week*, 72-77.

Nielsen Net Ratings. (2003). Global Net population increases. Retrieved from *www.nua.com/surveys*

Pickering, C. (1999, December). Brain Waves. *Business 2.0*, 175.

Rayport, J. (1999, Third Quarter). The truth about Internet business nodels. *Business & Strategy*, 16, 5-7.

Robles, F., Simon, F., & Haar, J. (2003). *Winning strategies for the new Latin markets.* New York: Prentice Hall.

Schibsted, E. (1999, November). All the world in stages. *Business 2.0*, 45-49.

Somagi, L. (1999). Quem Arrisca Petisca. *Brazil em Exame*, 32-38.

Vargas et al. (2000, January). Internet: Vamos Ficar Para Tras? *Amanha*, *150*, 19.

Westland, J., & Clark, T. (1999). *Global electronic commerce.* Cambridge, MA: The MIT Press.

Weyer, M. (2000, First Quarter). Globalism vs nationalism vs e-business. *Business & Strategy*, *18*, 63-80.

Yoffie, D., & Cusumano, M. (1999, January-February). Judo strategy: The competitive dynamics of Internet time. *Harvard Business Review*, 71-81.

Zeff, R. (1999). *Advertising on the Internet.* New York: John Wiley.

Chapter VII

Assessing Job Seekers' Acceptance of Online Recruitment in Egypt

Nahed Azab, Middlesex University, UK

Abstract

The development in information and communication technology had an implication on Internet adoption growth rate and especially on Web technology use. The evolution of the online recruitment industry closely parallels such development worldwide. Egypt, is no exception. This trend started in 1998 and proved to be an efficient way to provide better interaction and information exchange between job seekers or "candidates" and employers. This case investigates online recruitment in Egypt, and reveals the factors contributing to the success of such business. A model is presented using the Technology Acceptance Model (TAM), discussing the main constructs affecting candidates in embracing this relatively new technology. By identifying the aspects that shape candidates' preference to use Web-based recruitment over traditional off-line recruitment, agencies could gain an insight of the candidates' behavior in a developing country like Egypt. This, in turn, could be compared with the corresponding behavior in developed countries. Such comparison will help in identifying similarities and differences in candidates' perception towards online recruitment in both cases. As a result, appropriate business strategies stemming from the culture and social and economic variables of any country could be set in order to boost the diffusion and efficiency of online recruitment business.

Background

The Internet and the Web have revolutionized the way people communicate. This revolution has not only had a radical change on ways of doing business, but also created new business models that did not exist in the past. The range of Internet applications grew fast due to emerging technologies and competitive business pressures. Since 1995, the Internet has witnessed a rapid development of innovative applications. Among these applications are online recruitment, which was adopted by many users worldwide due to the high potentials it brings to both job seekers and employers.

As for job seekers, online recruitment sites allow them to post their CVs and avail them for companies looking for employees. They also let them check vacant jobs and apply for what they consider to be appropriate. Most of the online recruitment agencies offer other services for job seekers like tips to better write a CV, ways of enhancing their interview's performance, information regarding the job market, etc.

Employers have also realized the benefits of online recruitment agencies as they allow them to advertise for vacancies in a media much cheaper than other traditional media, facilitate focusing on the right candidates without the hassle of going through an enormous number of CVs. Online recruitment sites allow them to filter the applicants according to any criteria they choose. This feature minimizes the search process, and provides them with the information they need instantly and dynamically. Employers can also have access to more detailed information about each candidate because many online recruitment agencies validate the CVs sent by job seekers and ensure their consistency and completeness before accepting them and making them available for employers.

Online recruitment allows both job seekers and employers to look for their needs at a time convenient to them, provides them with a broader exposure of the job market, resulting in a much higher probability for a better match.

Developments in online recruitment features are always progressing as online recruitment agencies are always competing to offer better and different services to attract more users.

Online recruitment business started in Egypt when three graduate students realized the absence of this service in Egypt and the potential it could bring when provided. They designed the first online recruitment site in Egypt, "*CareerEgypt*" (which has been changed now to *CareerMideast*) and launched it in 1998. The success realized by *CareerEgypt* encouraged other players to get into the business and to make continuous innovations and developments in this direction.

This case discusses online recruitment business in Egypt. It first gives a quick summary of main information about Egypt, basic social and economic indicators, the history of the Internet in Egypt and its status in the present, and the situation of e-business in it. The country overview will provide readers with a general idea to put the research topic in context. The study aims to investigate the reasons that helped the success of online recruitment in Egypt and states the most famous sites in this domain. The case will also test the extent to which job seekers accept this technology by applying an extended version of the Technology Acceptance Model (TAM) created by F. Davis and associates in 1989. TAM proved to be very effective in measuring users acceptance of emerging technologies.

Egypt's Overview

Egypt is at the heart of the Middle East North Africa (MENA) region. It occupies the Northeast corner of Africa constituting the continent's gateway into Asia (Figure 1).

Most of the population (95%) lives in the narrow banks of the Nile River. Egypt has the highest population in the Arab world (70.5 million—*Central Authority for Public Mobilization and Statistics, CAPMAS, January 2004*), and among African nations, is second only to Nigeria. Cairo, the capital, is the continent's most populous city with 9.83 million capita (*Wholesome Words, 2003*). The total area of Egypt is approximately 1 million km^2. Arabic is the official language, but English and French are widely spoken.

A recent statistic performed by the Central Authority for Public Mobilization and Statistics (CAPMAS) in January 2004 revealed that unemployment percentage in Egypt reached 10.7% (2,300,000), which is considered to be on the high end of the acceptable average compared to other nations worldwide. Table 1 shows the main estimates for this study.

According to the latest estimates provided by the World Bank in 2003, the GDP (Gross Domestic Product) in Egypt per capita is US$1,219 per year. This value is considered very low when compared to its correspondent in developed countries (GDP per capita in United Kingdom is 30,354, and in United States is 37,457).

The Ministry of Foreign Trade evaluated the inflation rate at 3.2% for fiscal year 2002/2003, total exports at US$8,205 million, and total imports at US$14,779 million. This results in a negative trade balance of US$6,574 million.

Figure 1. Map of Egypt

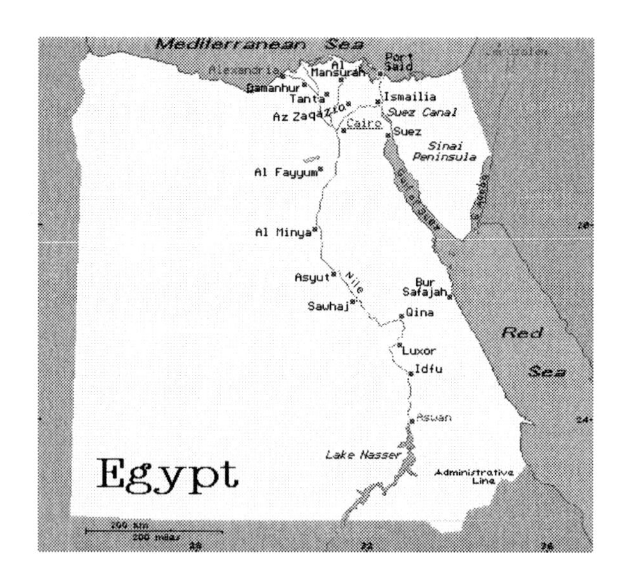

Table 1. Egypt statistic results (Central Authority for Public Mobilization and Statistics, CAPMAS, January 2004)

Population	70,547,718
Population growth (per annum)	1.96% (1.99% in 2002)
Population between age 15 to 65	40,588,695 (59.1%)
Working population	20,703,000 (20,176,000 in 2002)
Adult illiteracy (above age 10)	12,818,607 (28.59%) Males: 17% Females: 40.65%
Unemployment	2,300,000 (10.7%)

Internet in Egypt

Internet services started in Egypt in October 1993. A 9.6 K link carrying Internet traffic was established between the Egyptian Universities network (EUN) and the European Academic and Research Network (EARN) in France. At that time, the estimated number of users was 2,000.

The creation of public-private partnerships in 1996 resulted in the establishment of 12 privately-owned Internet Service Providers (ISPs) providing wide-scale Internet services using the infrastructure delivered by the state-owned telecom operator, Arab Republic of Egypt National Telecommunications Organizations (ARENTO), and through the international gateway of the Information and Decision Support Cabinet (IDSC). The competition between the ISPs lead not only to successive reductions in the average cost of Internet access, but also to a wide variety of Internet access options to address different segments in the society.

The Internet Society of Egypt (ISE) was formed in 1997 as an initiative to act as a regulatory body for several issues concerning the Internet in Egypt. ISE helped in building awareness of the Internet in Egypt. It was also responsible of establishing a code of ethics and policies for Internet use and in regulating ISPs.

In 1998, the Internet market attracted 120,000 users, and the number of ISPs reached 20. The Ministry of Communications and Information Technology (MCIT) was formed in 1999. Its main objective was to boost the telecommunication sector and to develop Egypt's information society. MCIT launched its national plan which included several projects, like establishing Information Technology Clubs (IT Clubs), providing training in all fields related to Information and Communication Technology (ICT), promote the software and hardware industry in Egypt, etc.

One of the most successful projects presented by MCIT was the subscription-free Internet initiative. This initiative is based on a public-private partnership model between Telecom Egypt and the majority of the ISPs. Launched in 2002, this free-Internet initiative enabled Internet users to access the Internet by dialing the number of their preferred ISP without prior registration and only being charged for the price of local phone calls, which is currently about US$0.2/minute. This initiative had a direct implication on the remarkable growth of the Internet adoption rate, as the number of Internet users reached 2.8

million in January 2004 (*Ministry of Communications and Information Technology, MCIT, 2004*).

Another factor contributing to the diffusion of the Internet in Egypt is the development in the IT infrastructure. The number of total switch capacity telephone lines increased to reach 11.2 million in October 2003 compared to 6.4 million in 1999 (MCIT, November 2003). Moreover, improvements in Internet services worldwide such as DSL, fiber optics, GSM communications, WiFi technology, etc., are now applicable in Egypt.

E-Commerce in Egypt

E-commerce in Egypt is still in its early stages of evolution. It first started when the Internet Society of Egypt (ISE) formed the E-Commerce Committee, whose main objective is to bring the public and private sectors together to discuss and reveal e-commerce challenges and opportunities in Egypt. Since 1998, various seminars, workshops and conferences were held to boost e-commerce applications and exchange e-business experiences among Arab countries.

Precise estimates for e-commerce in Egypt still require more investigation and research. The only available figure for B2C in Egypt was provided in a recent study conducted by ACNielson AMER Research Group, which asserted that only 6% of Internet users make purchases online. This was based on a sample survey of 452 Internet users in Egypt (*American Chamber of Commerce in Egypt, 2002*).

Despite being in their initial phases, all levels of e-commerce (Business-to-Business, Business-to-Consumer, Government-to-Government, Government-to-Citizen, Government-to-Business) exist in Egypt.

The main barriers to developing e-commerce in Egypt are the traditional trade barriers, the lack of a robust and efficient business model, and the absence of a legal foundation that regulates all e-commerce transactions.

Only recently, in April 2004, the parliament of Egypt is discussing a new e-commerce law that addresses several e-commerce related issues like e-signatures, e-contracts, taxation, custom duties, privacy, intellectual property rights, etc. "Tailoring a law for e-commerce is not an easy task since a number of the crucial issues affecting the economy in general and e-commerce in particular still do not have a foundation in the Egyptian legal system" (*Kamel, Ghoneim & Ghoneim, 2003*).

Online Recruitment Business in Egypt

The increase in Internet awareness, e-business and e-commerce concepts in Egypt over the past years have encouraged the creation of dot.com (virtual) recruitment services. Other brick-and-mortar recruitment agencies realized the advantages of online recruitment and decided to establish their presence on the Web. Businesses mixing brick-and-mortar channels with virtual ones are classified as "click and mortar". Several sites achieved a remarkable success and attracted a large number of users to profit from the valuable services they offer. Other sites couldn't continue due to many reasons like the

Table 2. Main Egyptian recruitment sites

Site Name	Launching Date	Type
www.careermideast.com	1998 (re-launched in April 2004)	Virtual
www.skill-link.com	2000	Click and mortar
www.egyrec.com	1999	Virtual
www.thejobmasters.com	2000	Click and mortar

inability to set clear objectives, the lack of an efficient marketing strategy, the failure to recognize users' needs, the inattention to adopt Web design optimization guidelines, etc.

Four Egyptian recruitment sites are considered the most popular and successful ones. Table 2 shows the name of these sites, their launching dates, and their types.

All recruitment sites in Egypt are primarily focused on providing job-posting services through a range of subscription-based pricing models, while giving candidates free access to post their CVs, look for job vacancies or apply for available vacant jobs. They also include resume repositories and management sites offering data acquisition, conversion, maintenance and validation services that support the management of resumes online.

Revenue is derived from the provision of information management service, or through selling access to a database of candidate resumes to recruiters and customers alike. The revenue also results from advertisements and from extra services to be offered to candidates like providing help in writing CVs, offering training courses, etc.

Online recruitment charges are cheap compared to the market in Egypt and to publishing for job vacancies in other traditional media.

The success achieved by such a model attracted also other portals, government sites or even newsgroups to include a job matching section, like for example, *www.amcham.org*, *www.masrawi.com*, *www.mcit.com*, and *elshella@yahoogroups.com*.

Technology Acceptance Model

The acceptance and effective utilization of Information Systems (IS) by individuals and organizations are areas of research that have gained importance in recent years. The Technology Acceptance Model (TAM), introduced by Davis and associates (Davis, 1989), is one of the most widely used research constructs in the study of computer usage behavior. The theoretical basis of the model was Ajzen and Fishbein's (1980) Theory of Reasoned Action (TRA). The TRA and its successor, the Theory of Planned Behavior (TPB) (Ajzen, 1985) are well known, and have been widely employed in the study of specific behaviors. All three theories, TRA, TPB, and TAM, state that a Behavior (B) is determined by Intention to perform the Behavior (BI), which means that intention and actual behavior are highly correlated.

TRA explains people's action by identifying the causal connections between various components: beliefs, attitudes, intentions, and behavior. The theory reasons that BI is determined by a person's Attitude toward a behavior (A), and the Subjective Norm (SN) concerning that behavior. According to TRA, A is determined by a person's beliefs and evaluations that the behavior will lead to certain outcomes. SN results from the social pressure (normative beliefs and motivation to comply) exerted on the person to perform the behavior.

TPB identified three factors responsible for changing people's behavior: attitude to the behavior, subjective norms, and perceived behavioral control. Perceived behavioral control is defined by the perception of somebody on the difficulty or ease of the change of behavior.

TAM proposes that two particular beliefs, Perceived Usefulness (PU) and Perceived Ease of Use (PEOU), are the primary drivers for Technology Acceptance (TA) (Figure 2). PU stands for the degree to which a person believes that using a particular system would enhance his or her job performance. PEOU depicts the degree to which a person believes that using a particular system would be free of effort. PU is affected by PEOU since an easy-to-use system will allow a person to enhance his/her work performance and make him/her more productive. PU and PEOU then affect jointly a person's Attitude toward using the system (A), and thus further to Behavioral Intention (BI) to use the system, which then leads to actual system use or Behavior (B).

The measures of PU include performance increase, productivity increase, effectiveness, overall usefulness, timesavings, and increased job performance. Correspondingly, measures for the PEOU include ease of learning, ease of control, ease of understanding, ease of use, clarity, and flexibility of use.

Both PU and PEOU are directly affected by external variables. These differ depending on many issues like the type of information system used, user's culture, time and place of deployment of the information system, etc. For example, in case of systems related to monetary transactions, trust and risk are the first and forecast factors to be considered. When adopting a new system in an organization, external variables influencing PU or PEOU could be training, documentation, user support consultants, management encouragement, social pressure, etc.

The main goal of TAM is to predict user's acceptance of computer-based technologies and to provide an explanation so that researchers, information systems' investors and

Figure 2. Technology Acceptance Model (TAM)

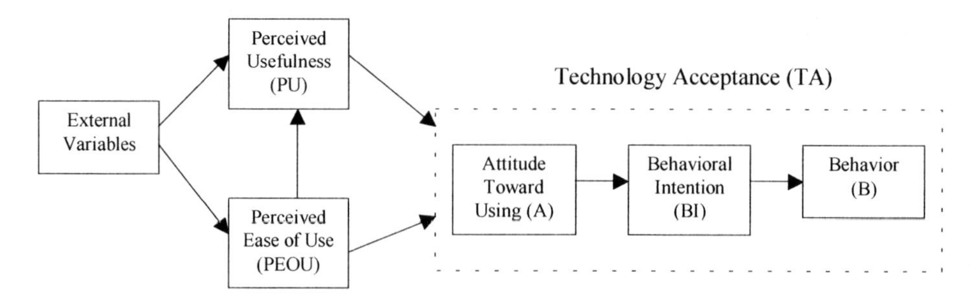

designers will be acquainted with the causes of a system's acceptance. Based on this knowledge, efforts and investments could be rectified towards developing more utilized and accepted systems by the users.

The robustness of TAM has been established through several applications and replications, which proved its efficiency (Davis, Bagozzi & Warshaw, 1989; Mathieson, 1991; Adams, Nelson & Todd, 1992; Segars & Grover, 1993; Davis & Venkatesh, 1995; Taylor & Todd, 1995; Chau, 1996; Igbaria et al., 1997; Doll et al., 1998; Fenech, 1998; Agarwal & Prasad, 1999; Kucuk & Arslan, 2000; Venkatesh et al., 2003).

Main Thrust of the Chapter

The study will start by examining the factors that contributed to the creation and dissemination of online recruitment in Egypt. Secondly, the research model and research hypotheses will be presented focusing on how they were reached. This will be followed by explaining the methodology used in the research. Finally, results obtained will be analyzed leading to proposing solutions and recommendations.

Factors Contributing to the Emergence and Diffusion of Online Recruitment in Egypt

The awareness and adoption of online recruitment technology in Egypt on part of employers and candidates is due to many reasons that can be classified into three parts: reasons driving job seekers, employers, or both to use this service (Figure 3).

Factors Related to Both Job-Seekers and Employers

- The increasing growth of the number of Internet users in Egypt (2.8 million Internet users in January 2004 compared to 2.7 million in December 2003, approximately 44.5% annual growth) (Ministry of Communications and Information Technology (MCIT), 2004).
- Privacy policies provided by recruitment sites encourage people to use them because in most sites, users can choose not to expose their names or workplaces
- Popular recruitment sites are always dynamic, which make users always log in to them to be aware of new opportunities each single day
- The fact that Internet usage has become an integral part of most desk jobs nowadays made it easy for workers to look for information and services through the Internet
- Recruitment sites give organizations and job-seekers a global search exposure

Figure 3. Factors contributing to the diffusion of online recruitment in Egypt

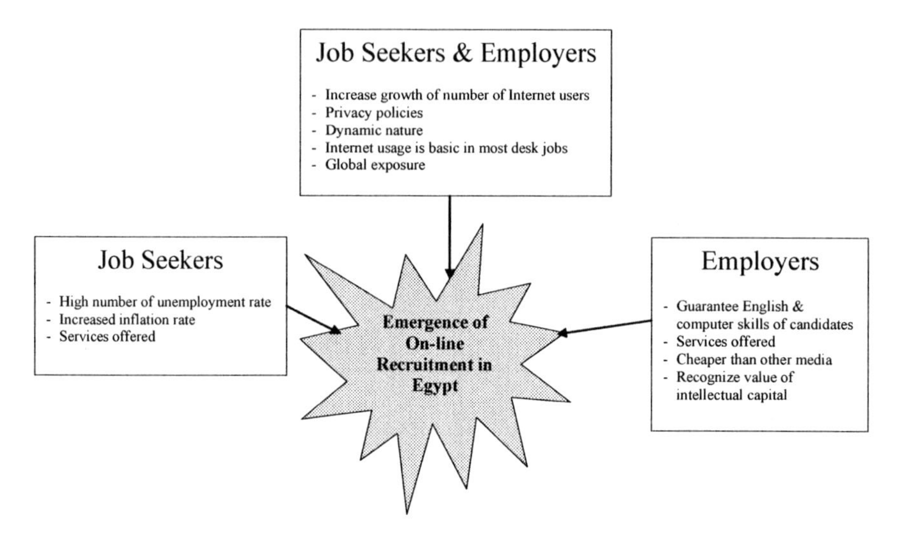

Factors Related to Job-Seekers

- The high number of unemployment rate (10.7 % - *Central Authority for Public Mobilization and Statistics, CAPMAS, January, 2004*) drives candidates to look for jobs in all channels available including online recruitment.
- The increase in inflation rate in Egypt (from 2.4% in 2001/2002 to 3.2% in 2002/2003) (*Ministry of Foreign Trade, February 2004*) during a prevailing low growth rate (or recession) period urges candidates to look for more lucrative jobs to enhance their income.
- Recruitment sites' services offered to candidates like filtering available jobs based on the criteria they select, helping them in writing their resume, etc., encouraging them to use these sites.

Factors Related to Employers

- Many companies prefer looking for employees through recruitment sites since this guarantees implicitly that candidates applying for online positions must have Internet skills and a reasonable command of English language, since English is the default language on Egyptian recruitment sites.
- Management realized the features provided by online recruitment services like filtering their needs easily according to their preferred criteria, assisting them during the whole recruitment procedure, etc.

- Publishing for required jobs on recruitment sites is cheaper than other media. For example, the highest cost of posting one job through recruitment sites is L.E. 250/ one job posting (US$38). The number of purchased job postings is valid for one year, and each job posting expires one month after posting date. Whereas, the cost of an average job vacancy advertisement (around two or three jobs) in a newspaper ranges from L.E.12,000 to L.E.19,000 (US$1,826 to US$2,891) to be published only one time.
- Organizations are recognizing the value of intellectual capital and the relationship between skill resources and workers' competitive advantage. As a result, more attention is being placed on the recruitment and management of human resources in general in order to ensure the efforts and investments made result in the attraction and retention of qualified knowledge workers.

Proposed Research Model

The research model for this chapter will focus on candidates' acceptance of online recruitment technology. Adopted from the TAM, the model presents PEOU and PU as the proposed constructs affecting TA directly. According to TAM, PEOU will also have an indirect effect as well on TA through PU. External factors indirectly influencing acceptance of this technology through PEOU or PU are Socio-Economic status (SE), (such as income or position), and Job Market (JM) (Figure 4).

Proposed Research Variables

Based on the research model, the research variables are as shown in Table 3.

Figure 4. Proposed research model

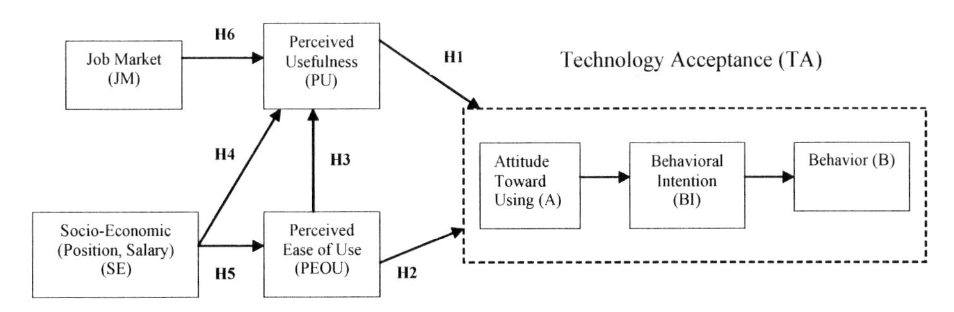

Table 3. Research variables

Variable	Description	Type	Scale
TA	Technology Acceptance	Dependent	Discrete (1-7) 1: Accept; 7: Reject
PU	Perceived Usefulness	Dependent/Independent	Discrete (1-7) 1: Accept; 7: Reject
PEOU	Perceived Ease of Use	Dependent/Independent	Discrete (1-7) 1: Accept; 7: Reject
SE	Socio-Economic Status	Independent	Discrete (1-7) 1: Accept; 7: Reject
JM	Job Market	Independent	Discrete (1-7) 1: Accept; 7: Reject

Proposed Research Hypotheses

Six hypotheses are presented in the research. These are:

Perceived Usefulness (PU)

A significant number of studies have shown that perceived usefulness is an important antecedent of technology utilization (*Davis et al., 1989; Igbaria & Livari, 1995; Keil et al., 1995; Satzinger & Olfman, 1995; Igbaria et al., 1996*). In these studies, PU has proven to be the stronger of the two TAM variables. PU is defined as "the user's subjective probability that using a specific application system will increase his or her job performance within an organizational context" (Davis, Bagozzi and Warshaw, 1989). Thus the first hypothesis is:

Hypothesis 1 (H1): Perceived Usefulness (PU) will be positively related to the candidates' acceptance of online recruitment technology (TA)

Perceived Ease of Use (PEOU)

Several researchers proved that PEOU has a direct affect on TA. PEOU is defined by "the degree to which a person believes that using a particular system would be free of effort" (*Davis, 1989*). The influence of PEOU on TA is more ensured by Rogers' (1995) introduction of another construct opposite to PEOU, which is perceived complexity. According to Rogers, perceived complexity is defined as "the degree to which an innovation is perceived as difficult to understand and use." Rogers' research suggests that the more complex a technology is perceived, the slower will be its adoption rate. Thus the second hypothesis is:

Hypothesis 2 (H2): Perceived Ease of Use (PEOU) will be positively related to the candidates' acceptance of online recruitment technology (TA)

It was also proven that PEOU has an indirect effect on TA through PU to be more significant than its direct effect on TA. According to Davis, Bagozzi and Warshaw (1989), "efforts saved due to improved EOU may be redeployed, enabling a person to accomplish more work for the same effort." Hence, the third hypothesis is:

Hypothesis 3 (H3): Perceived Ease of Use (PEOU) will be positively related to Perceived Usefulness (PU)

Socio-Economic (SE)

The general economic review of Egypt presented earlier suggests that there has been a combination of relatively high inflation and slow growth (or even recession according to some informal reports). This led to the stagnation of salaries and the scarcity of jobs. Naturally in such circumstances the workforce would be always looking for jobs that could enhance their income or position. Due to the importance of the socio-economic factor to this workforce, it could have a great impact on its belief in the usefulness of applying for jobs online. This leads to the fourth hypothesis:

Hypothesis 4 (H4): Enhancing the socio-economic status (position, income) of a candidate (SE) will be positively related to Perceived Usefulness (PU)

The urge from the part of the candidates to enhance their position or income could drive them to concentrate on learning how to use recruitment sites. Time invested in learning leads to better acquaintance with the use of such sites as well as more resourcefulness. Psychologically there would be a general feel of easiness in adopting the technology. Thus the fifth hypothesis is:

Hypothesis 5 (H5): Enhancing the socio-economic status (position, income) of a candidate (SE) will be positively related to Perceived Ease of Use (PEOU)

Job Market (JM)

During the last few years, there has been continuous change in the job market in Egypt. Fields where there had been high demand reached saturation, and on the other hand, new jobs that never existed before are now all over the job market. For example, the traditional jobs for engineers, physicians, accountants, etc., are not required as before as the supply is much more than the demand due to the large number of universities' graduates in such fields. On the contrary, the emergence of the information age, replacing the industry age in the past century worldwide, has created new concepts like Customer Relationship Management (CRM), user satisfaction, information value, etc. These concepts have created a demand for more job openings in sales and marketing everywhere. In addition, the fast developments in Information and Communication Technology (ICT) created the

Table 4. Subjects demographics

Gender/Age Group	21-30	31-40	41-50	Total
Male	34	20	10	64
Female	29	18	5	52
Total	63	38	15	116

Table 5. Subjects socio-economic status

Position/Income	L.E. <10,000	L.E. 10,000 – 20,000	L.E. 20,000 – 50,000	L.E. 50,000 – 100,000	L.E. 100,000 – 200,000	L.E. >200,000	Grand Total
Official in Charge	9	8	21	7	1		46
Team Leader	1	2	11	6	4	1	25
Head of Department		3	7	10	4	3	27
General Manager		2			3	1	6
Others	5	2	4	1			12
Grand Total	15	17	43	24	12	5	116

continuous need for new positions in that field. Thus, job market changes drive job-seekers to use recruitment sites because they appreciate their importance in keeping them updated with the job market status. As a result, the sixth hypothesis is:

Hypothesis 6 (H6): Job Market (JM) will be positively related to Perceived Usefulness (PU)

Methodology

A research questionnaire was used as a measurement instrument (**Appendix I**). The questionnaire was circulated among a sample of computer and Internet users. The purpose of the questionnaire was to test the extent to which online recruitment technology is used by job seekers in Egypt, its effect and the most used sites. The questionnaire was also intended to help identify the major factors affecting candidates' acceptance of online recruitment technology. The research variables are measured in a 7-point Likert scale, with 1 as strongly agree, and 7 as strongly disagree. Multiple regression analysis was used to test the relation between the research variables and to compare the results obtained with the research hypotheses.

The questionnaire was mainly dessiminated through electronic mail. A few versions were distributed as a hard copy since some respondents don't use the Internet intensively. The answers received were 124, of which 116 were valid (93% of total respondents).

The sample population is well balanced demographically. 55% of the respondents were males (64), and 45% females (52). The sample was divided into three age groups: 21 to 30, 31 to 40, 41 to 50 years old (Table 4).

Figure 5. Different ways of introducing online recruitment to candidates

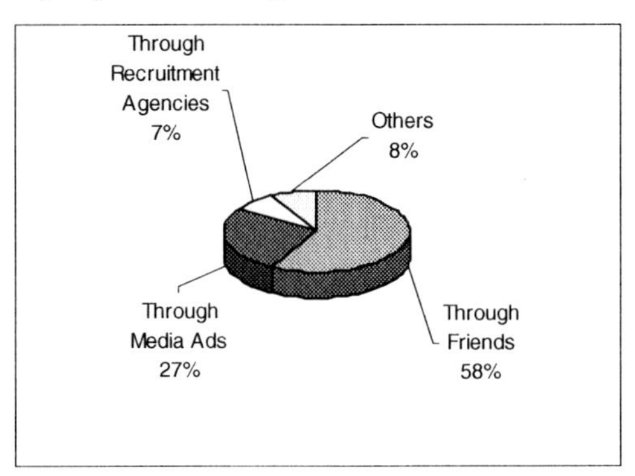

The subjects were also of varying socio-economic levels, with some working in the public sector and others in the private sector. Table 5 shows the socio-economic distribution of the respondents.

Results

The responses obtained from the subjects indicate a number of facts that will be presented in the following points:

Awareness

Ninety-four percent of the respondents, a considerably high ratio, are aware of online recruitment. This is due to the fact that a large number of them (91%) possess high computer and Internet skills and use the Internet everyday (72%).

The questionnaire results reveal also that friends are the major channel through which online recruitment sites were introduced (58%), followed by media advertisements (27%). Figure 5 shows the effect of different channels on online recruitment diffusion.

Usage

Sixty-four percent of the respondents revealed that they use recruitment sites, and 41% of the users have gotten jobs through them. This constitutes a high success rate in the efficiency of this technology in Egypt. When associating usage with the socio-economic status of the respondents, it was found that middle socio-economic class candidates are the category that uses this technology most. With respect to the respondents' ages, online recruitment is used mostly by subjects of age from 20 to 30, followed by those of age from 30 to 40. As for the sites used by the respondents, 65% of them use the four sites presented in the research (CareerMideast, EgyRec, Skill-Link and TheJobmasters). Other

Figure 6. Online recruitment usage patterns

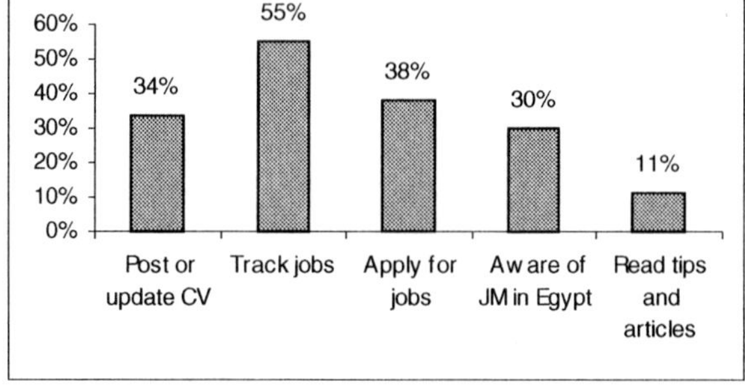

Figure 7. Respondents' evaluation of online recruitment services

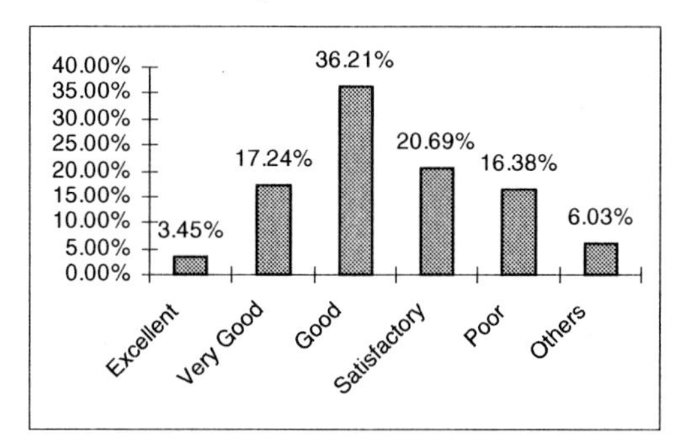

less used sites were either non-Egyptian (for example, "bayt.com," which was used by 13% of the respondents), or sections of multi-purpose sites (for example, the site of the American Chamber of Commerce in Egypt "amcham.org.eg" was used by 6% of the respondents).

With respect to usage patterns of online recruitment technology, it was found, as shown in Figure 6, that the major task the respondents do when using online recruitment sites is tracking jobs (55%), followed by applying for jobs (38%), then posting or updating their CVs (34%). A small percentage uses these sites for reading tips and articles (11%).

Such patterns can be attributed to the fact that candidates believe they could always find job opportunities through these sites. Not reading tips or articles in these sites (only 11% read tips or articles) is due to two reasons: (1) when linking to these sites, candidates focus only on performing the main tasks offered because they do not have enough time to read, or (2) they do not feel reading tips or articles can be of use to them.

Service Satisfaction

As shown in Figure 7, a large number of respondents evaluated the quality of service offered by recruitment sites as poor or satisfactory (37%). This means that there is room

Figure 8. Respondents declaration of the main reason to use recruitment sites

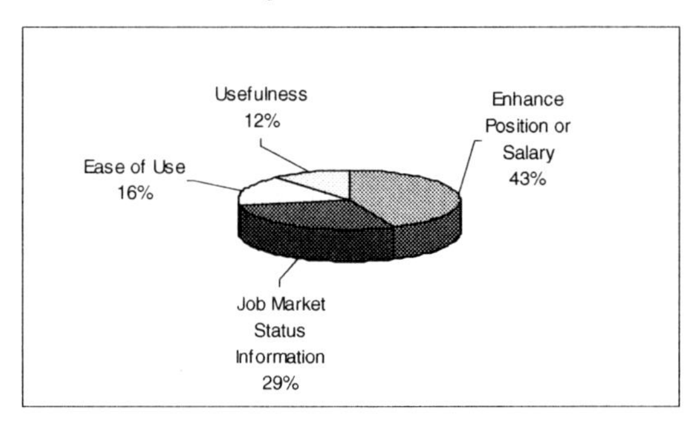

for enhancement and online recruitment agencies should exert more effort in providing better service and support for candidates, or better yet try to attract constructive feedback.

Barriers Limiting Usage

The main obstacle the respondents face is time constraints (32%), followed by their belief that this technology is inefficient and not useful for them (21%). There are other barriers like user support, security, access to the Internet, or Internet performance, but at a lower rate (ranging from 7% to 9%).

Reasons for Using Online Recruitment

A large percentage of the respondents stated that the main usage purpose was to enhance their position or salary (43%), followed by obtaining information on the status of the job market (29%). Ease of use is not a determining motive (only 16%) since most

Figure 9. Overall resulting model with correlation coefficients and coefficient of determination ($P<0.05$, ** $P<0.01$, *** $P<0.001$) (Critical Value: Correlation Coefficient R: $R>0.195$ $P<0.05$, $R>0.254$ $P<0.01$, $R>0.313$ $P<0.001$)*

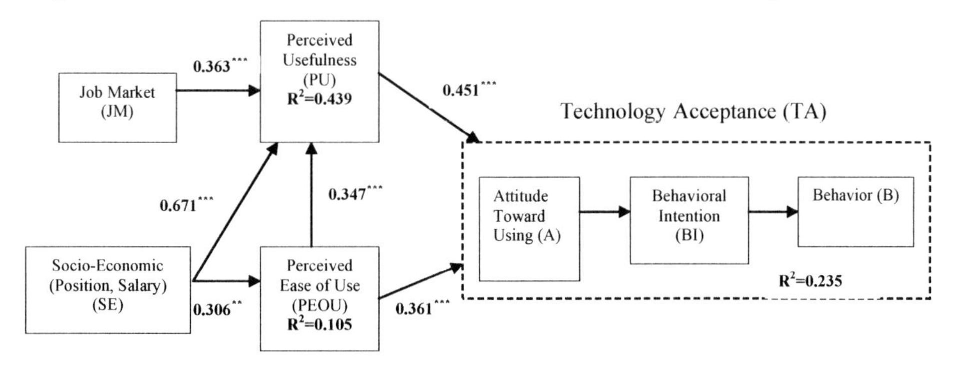

of the respondents are using the computer professionally and using the Internet frequently. Usefulness also is not a major drive (only 12%) to use this technology. This could be attributed to lack of trust in the efficiency of this technology (Figure 8).

Results Testing the Proposed Hypotheses

Validity of the Measurement Items

The descriptive statistics of the measurement items are measured with respect to mean values and standard deviations. It shows that all the items show generally positive perceptions towards technology acceptance. All mean scores are under 4 (4=Neutral), ranging from 2.70 to 3.73 while the standard deviations range from 1.23 to 1.62.

Overall Model

Multiple regression analysis is used to test the six hypotheses of the study. Figure 9 shows the resulting path coefficients for the overall model. These represent coefficient of correlation (R) between each pair of two variables. R is a number between –1.00 and 1.00 that indicates both the direction and the strength of the linear relationship between two variables.

Estimates of coefficient of determination (R^2) are shown for the variables TA, PU, and PEOU. R^2 values represent the percent of the variance of the particular dependent variable that is explained by the antecedent variables. As seen in Figure 9, PEOU, SE, and JM explain approximately 44% of the variance in PU, while SE explains approximately 11% of the variance in PEOU. PU and PEOU account for approximately 24% of the variance in TA.

Figure 9 shows that the results obtained confirm all research hypotheses. It also proves that SE has a stronger effect on PU than JM and PEOU, which means that the professional position and the income are the most predominant factors driving candidates to perceive the usefulness of online recruitment technology. Moreover, the structural model shows that PU has a more significant effect on TA than PEOU, which is consistent to what Davis (1989) originally developed. This finding shows that users are driven to adopt an

Table 6. Result of t-test analysis of gender difference

Variable	Males	Females	P
TA	MEAN=3.19 SD=1.66	MEAN=3.10 SD=1.59	0.76 Non-significant
PU	MEAN=3.33 SD=1.63	MEAN=3.17 SD=1.57	0.60 Non-significant
PEOU	MEAN=2.80 SD=1.34	MEAN=2.57 SD=1.08	0.31 Non-significant
SE	MEAN=3.73 SD=1.61	MEAN=3.73 SD=1.50	0.98 Non-significant
JM	MEAN=3.34 SD=1.34	MEAN=3.31 SD=1.53	0.92 Non-significant

*Figure 10a. Male resulting model with correlation coefficients and squared multiple correlations (*P<0.05, **P<0.01, ***P<0.001) (Critical Value: Correlation Coefficient R: R>0.250 P<0.05, R>0.325 P<0.01, R>0.400 P<0.001)*

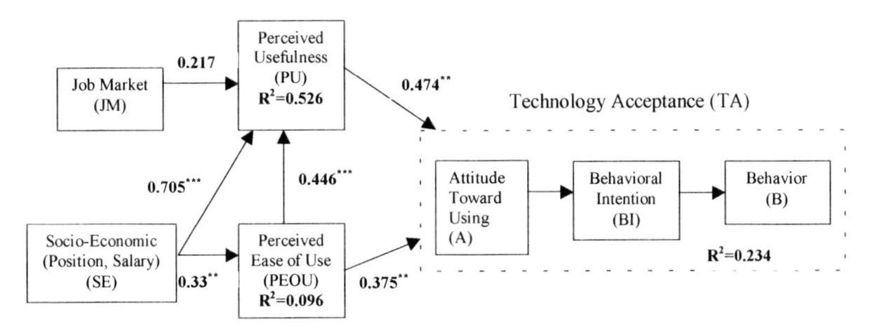

*Figure 10b. Female resulting model with correlation coefficients and squared multiple correlations (*P<0.05, **P<0.01, ***P<0.001) (Critical Value: Correlation Coefficient R: R>0.273 P<0.05, R>0.354 P<0.01, R>0.441 P<0.001)*

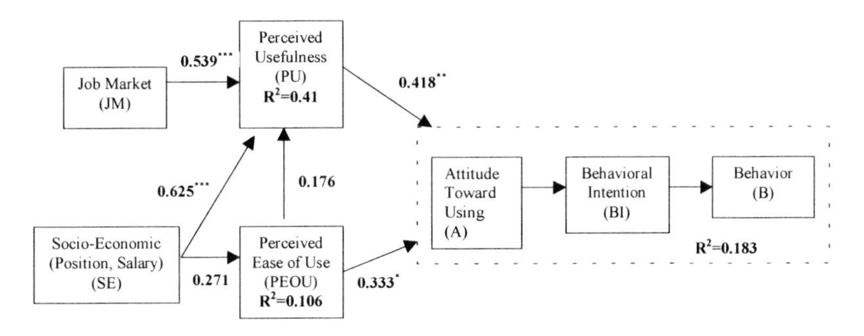

application primarily because of the functions it perform for them, and secondly by how it is easy or hard to get the system to perform those functions.

Gender Difference Effect on the Research Model

The model was analyzed on males only and females only in order to detect any differences between them and also to find out the effect of gender differences on each of the model's construct.

Comparing the two groups, males and females, using unpaired student T-Test analysis, no significant difference was found (Table 6).

Analysis of path coefficients for each gender summarized in Figure 10a and Figure 10b, indicates that for the male model, all data sets provide an overall fit of the model postulated by TAM and confirm all research hypotheses except H6, which states that PU is dependent on JM. For the female model, the resulting model is not consistent with TAM in H3 that declares that PEOU has an effect on PU.

Comparison of both models denotes that men placed a great emphasis on PEOU in determining PU. On the contrary, PEOU has no significance on PU for women (0.446 for male, 0.176 for female). On the other hand, JM has a strong effect in determining PU for women, and has no significant effect on PU for men (0.217 for male, 0.539 for female).

The resulting models show also that men's position and salary had more weight in determining PU than women (0.705 for men, 0.625 for women). These findings could help online recruitment agencies, when setting marketing and promotion strategies, to determine which construct to emphasize for each gender.

Recommendations and Outlook

The findings obtained from this study shed light on a number of important factors that influence candidates' acceptance of online recruitment technology. Online recruitment agencies should attend to such factors.

Since the study revealed that a large number of subjects are not satisfied with the service provided by recruitment sites, online recruitment agencies should focus on improving the level of service offered to job-seekers by responding regularly and repeatedly to their needs, and by providing support tools to empower them to reach their objectives. The key is to learn from the experience of off-line recruitment, but also act well armed with the value propositions offered by the Internet.

Given the fact that perceived usefulness has a great effect on the acceptance of such technology, online recruitment agencies should set strategies to raise job–seekers' awareness of the efficiency of this technology since a large number of respondents do not trust that this technology would make a difference. Moreover, some of the subjects believe that online recruitment agencies do not offer good opportunities in terms of position and income. Other candidates see that making themselves available in these sites lowers their value in the job market, and that this technology is not targeting real qualified applicants. The contents of the sites should pinpoint on the value added in terms of enhancing socio-economic status. The sites should also be updated continuously to reflect changing job market demands.

The study findings indicate that a large number of candidates do not use recruitment sites because of the lack of time. Recruitment sites should therefore contain valuable and summarized information to attract these candidates and improve their perception of the value added by this technology.

When investigating recruitment sites' usage, the study demonstrated that job seekers use recruitment sites to achieve definite tasks like tracking or applying for jobs, or to post or update CVs, but they don't find these sites attractive to spend more time in reading their contents. Based on these findings, several approaches could be done to retain users a longer time. Contests could be made to attract candidates, for example conduct online tests in order to inform them of their strengths and weaknesses and to suggest the best jobs relevant to their skills. Another contest could be to choose the best or worst reply of a question during an interview, etc. The sites could also include links to other sites that could direct users to find resources corresponding to their fields of expertise. A discussion board section could also be added to let candidates share the interviews they went through and their job application experience with others.

Some of the subjects commented that they prefer sometimes a face-to-face interaction with representatives from recruitment agencies to direct them to the appropriate opportunities. This could be attributed to the fact that fully integrated e-business models are still new to candidates in Egypt. Hence, recruitment agencies should consider allowing each candidate to choose a different mixture of online and off-line services.

The presented study highlights the effect of ease of use on accepting such technology. However more research should still be conducted to enhance the performance and usability of recruitment sites. Easier processes could be provided. For example, the sites could offer job-seekers the ability to fill their resumes according to several levels of details because many candidates do not like to post their CVs just to avoid filling very long forms.

Investment in online recruitment in Egypt is expected to boost in the near future due to many facts: the increasing number of university students graduating each year (200,000 per year approximately) along with predicted growth in the unemployment rate, will drive them to look for available jobs in the market. Since the age group of fresh graduates constitutes a large sector of users of this technology, then it is fair to expect that more will embrace it in the future.

Another factor that will help the diffusion of online recruitment in the future is the growing competition for qualified people from the part of many organizations, and the drive to maintain a competitive advantage in an increasingly knowledge-based economy.

As various factors combine to escalate the demand for Web-based recruitment, new recruitment sites will eventually emerge. Among them the successful players will be those who set a clear business model, provide a core set of customer requirements efficiently (some of candidates requirements are presented in the study), and possess a proactive vision in understanding users' needs and responding promptly.

Suggestions for Future Research

The scope of this research is limited to the study of job-seekers' acceptance and utilization of online recruitment technology. It is recommended for future research to assess these same attitudes but from the part of employers and recruiters rather than job-seekers.

Further research is also recommended to perform similar follow-up studies to investigate the effect of lapse of time (i.e., after two years and five years) on the development of perceived usefulness and on perceived ease of use which are the constructs affecting acceptance of this technology.

Similar studies could be performed on one (or more) developed country to highlight the factors affecting job seekers' acceptance of online recruitment.

Conclusion

The study provides a clear idea about candidates' behavior toward online recruitment technology and its usage in Egypt. The study also applies the TAM model to assess job-seekers' acceptance of this technology. Besides findings that are consistent with previous studies that perceived ease of use and perceived usefulness contribute to the acceptance of such technology, it is also proved that external factors like job market and socio-economic status of candidates have a great effect on perceived usefulness. The desire to enhance the socio-economic status has also an impact on perceived ease of use. Acceptance models for men and women are presented to detect significance in gender difference.

Assessing candidates' acceptance of online recruitment in a developing country like Egypt raises several questions when trying to understand the differences in job-seekers' acceptance of such technology between developing and developed countries. For example: Are the barriers preventing job-seekers from using this technology the same?, Does the target audience have the same characteristics?, How about their usage patterns? Regarding the identification of factors affecting online recruitment acceptance, are they the same in developed countries? To which extent they do affect job–seekers' perceptions toward this technology?

References

Adams, D.A., Nelson, R.R., & Todd, P.A. (1992). Perceived usefulness, ease of use and usage of information technology: A replication. *MIS Quarterly*, *16*(2), 277-247.

Agarwal, R. & Prasad, J. (1999). Are individual differences germane to the acceptance of new information technologies? *Decision Sciences, 30*(2), 361-392.

Ajzen, I. (1985). From intentions to action: A theory of planned behavior. In J. Kuhl & J. Beckman (Eds.), *Action control: From cognitions to behavior* (pp. 11-39). New York: Springer-Verlag.

Ajzen, I. (1991). The theory of planned behavior. *Organizational Behavior and Human Decision Processes*, *50*, 179-211.

Ajzen, I. & Fishbein, M. (1980). *Understanding attitudes and predicting social behavior.* Englewood Cliffs, NJ: Prentice Hall.

American Chamber of Commerce in Egypt. (2002). E-commerce and its applications in Egypt. Retrieved December 2003, from *http://www.amcham.org.eg/BSAC/StudiesSeries/Report39.asp*

American Chamber of Commerce in Egypt. (2004). The Geneva Summit: Implications for Egypt. Retrieved March 2004, from *http://www.amcham.org.eg/operation/Events/Events04/AhmedNazif.asp*

Boling, E., & Sugar, W. (1995). User-centered innovation: A model for early usability testing. *Proceedings of the 17th Annual Proceedings of Selected Research and Development Presentations at the 1995 Convention of the Association for Educational Communications and Technology* (pp. 563-570).

Central Authority for Public Mobilization and Statistics (CAPMAS). (2004, January). *AlAhram, 128*(42819).

Chau, P.Y.K. (1996). An empirical assessment of a modified technology acceptance model. *Journal of Management Information Systems, 13*(2), 185-204.

Dahlberg, T., Mallat, N., & Oorni, A. (2003). Trust enhanced technology acceptance model – Consumer acceptance of mobile evidence solutions – Tentative evidence.

Davis, F.D. (1985). *A technology acceptance model for empirically testing new end-user information systems: Theory & results.* Doctoral Dissertation, MIT Sloan School of Management, Cambridge, MA.

Davis, F.D. (1989). Perceived usefulness, perceived ease of use and user acceptance of information technology. *MIS Quarterly, 13*(3), 319-339.

Davis, F.D., Bogozzi, R.P., & Warshaw, P.R. (1989). User acceptance of computer technology: A comparison of two theoretical models. *Management Science, 35*(8), 982-1002.

Davis, F.D., & Venkatesh, V. (1995). Measuring user acceptance of emerging information technologies: An assessment of possible method biases. *Proceedings of the 28th Annual Hawaii International Conference on System Sciences* (pp. 729-736).

Doll, W.J, Hendrickson, A., & Deng, X. (1998). Using Davis's perceived usefulness and ease-of-use instruments for decision making: A confirmatory and multigroup invariance analysis. *Decision Sciences, 29*(4), 839-869.

E_Cruiter.com. (n.d.). White Paper: The business case for Web-based recruiting. Retrieved December 2003, from *http://www.ecruiter.com/corporate/wrbc.htm*

Egypt Guide for Travel, Ancient and Modern Egypt. (n.d.). Retrieved March 2004, from *http://touregypt.net/aboute.htm*

Fenech, T. (1998). *Using perceived ease of use and perceived usefulness to predict acceptance of the World Wide Web.*

Ghoneim, S., Ghoneim, A., & Kamel, S. (2002). *The role of the government in electronic commerce diffusion: The case of Egypt.*

Gretchen, S. (2001). Technology adoptability: Is it possible to herd cats. Electronic Recruiting Exchange: Information and Networking for Recruiters. Retrieved March

2004, from *http://www.erexchange.com/articles/default.asp?cid=%7B15C08604-4287-4D8F-A97A-3016BDBECE20%7D*

Igbaria, M., & Livari, J. (1995). The effects of self-efficacy on computer usage. *Omega, 23*(6), 587-605.

Igbaria, M., Parasuraman, S., & Baroudi, J.J. (1996). A motivational model of microcomputer usage. *Journal of Management Information Systems, 13*(1), 127-143.

Igbaria, M., Zinatelli, N., Cragg, P., & Cavaye, A.L.M. (1997). Personal computing acceptance factors in small firms: A structural equation model. *MIS Quarterly, 21*(3), 279-305.

Kamel, S., & Hassan, A. (2003). *Assessing the introduction of electronic banking in Egypt using the technology acceptance model.* Hershey, PA: Idea Group.

Keil, M., Beranek, P.M., & Konsynski, B.R. (1995). Usefulness and ease of use: Field study evidence regarding task considerations. *Decision Support Systems, 13*(3), 75-91.

Kucuk, S.U., & Arslan, M. (2000). A cross cultural comparison of consumers' acceptance of the Web marketing facilities. *Journal of Euro Marketing, 9*(3), 27-43.

Kwon, H.S., & Chidambaram, L. (2000). A test of the technology acceptance model – The case of cellular telephone adoption. *Proceedings of the 33rd Hawaii International Conference on System Sciences.*

Levi, M., & Conrad, M. (1996). A heuristic evaluation of a World Wide Web prototype. *ACM Interactions.*

Lightner, N.J., & Bose, I. (1996). What is wrong with the World Wide Web? A diagnosis of some problems and prescription of some remedies. *Ergonomics, 39*(8), 995-1004.

Lopez, D.A., & Manson, D.P. (1997). A study of individual computer self-efficacy and perceived usefulness of the empowered desktop information system. *Interdisciplinary Studies.*

Malhotra, Y., & Galletta, D.F. (1999). Extending the technology acceptance model to account for social influence: Theoretical bases and empirical validation. *Proceedings of the 32nd Hawaii International Conference on System Sciences.*

Mathieson, K. (1991). Predicting user intentions: Comparing the technology acceptance model with the theory of planned behavior. *Information Systems Research, 2*(3), 173-191.

Ma, W. (2003). Implementation strategies and the technology acceptance model: Is "ease of use" really useful or easy to use in implementation? *Information Technology and Organizations: Trends, Issues, Challenges and Solutions*, *Proceedings of the 2003 Information Resources Management Association International Conference (IRMA2003).*

Ministry of Communication and Information Technology. (n.d.). Retrieved March 2004, from *www.mcit.gov.eg*

Ministry of Foreign Trade. (2004, February). *Monthly Economic Digest, 5*(12).

Online Recruitment Market. (2002). US Report. Retrieved November 2003, from *http://www.marketresearch.com/map/prod/768360.html*

Onrec.com. (2001). Mr. Ted pioneers multilingual European recruitment technology market. Retrieved November 2003, from *www.onrec.com*

Pavlou, P.A. (2001). Consumers' intentions to adopt electronic commerce – Incorporating trust and risk in the technology acceptance model.

Pitkow, J., & Kehoe, C. (1996). Surveying the territory: GVU's five WWW user surveys. *The World Wide Web Journal, 1*(3), 77-84.

Rogers, E.M. (1995). *Diffusion of innovations* (4th ed.). New York: The Free Press.

Satzinger, J.W., & Olfman, L. (1995). Computer support for group work: Perceptions of the usefulness of support scenarios and end-user tools. *Journal of Management Information Systems, 11*(4), 115-148.

Segars, A.H., & Grover, V. (1993). Re-examining perceived ease of use and usefulness: A confirmatory factor analysis. *MIS Quarterly, 17*(4), 517-525.

Song, S., & Song, J. (2002). Collaborative electronic media usage for information sharing: Technology competence and social ties. *Eighth American Conference on Information Systems.*

Taylor, S., & Todd, P. (1995). Understanding information technology usage: A test of competing models. *Information Systems Research, 6*(2), 144-176.

Venkatesh, V., Morris, M. G., Davis, G.B., & Davis, F. D. (2003). User acceptance of information technology: Toward a unified view. *MIS Quarterly, 27*(3), 425-478.

Wholesome Words. (2003). Retrieved March 2004, from *http://www.wholesomewords.org/missions/greatc.html*

The World Bank Group. (2003). Retrieved February 2004, from *http://www.worldbank.org/data/countrydata/countrydata.html*

Appendix I

Thank you for taking the time to answer this questionnaire. This questionnaire is intended to measure candidates' acceptance of online recruitment in Egypt. The data collected by this questionnaire is exclusively for research purposes only, and will not be used or distributed elsewhere.

Please Put "x" in the box you choose, and write on the line dedicated for data entry.

1. Age:

☐ 20 to 30

☐ 31 to 40

☐ 41 to 50

☐ More than 50

2. Gender:

☐ Male

☐ Female

3. Current Position:

☐ Official in charge

☐ Team leader

☐ Head of department

☐ General manager

Other ______________________________

4. Average **Annual** income:

☐ Less than 10,000 EGP

☐ 10,000–20,000 EGP

☐ 20,000–50,000 EGP

☐ 50,000–100,000 EGP

☐ 100,000–200,000 EGP

☐ More than 200,000 EGP

5. Are you aware of on-line recruitment technology?

- ☐ Yes
- ☐ No

6. If you are aware of on-line recruitment, how was it introduced to you?

- ☐ Through friends
- ☐ Through media advertisements
- ☐ Through recruitment agencies
- ☐ Others

7. Do you use any on-line recruitment site?

- ☐ Yes
- ☐ No

If yes, specify which sites: ______________________________

8. What are the impediments that prevent you from using on-line recruitment technology?

- ☐ Access to a computer
- ☐ Access to the Internet
- ☐ Internet Performance
- ☐ Security
- ☐ Time constraint
- ☐ User support
- ☐ Others

Please specify: ____

9. What do you use an on-line recruitment site for:

- ☐ Post or update your CV
- ☐ Track available jobs
- ☐ Apply for jobs
- ☐ Be always aware of the job market in Egypt
- ☐ Read tips and articles about jobs in Egypt

10. Did you get a job through an on-line recruitment site?

- ☐ Yes
- ☐ No

11. What do you think of the quality of service provided by on-line recruitment sites?

- ☐ Excellent
- ☐ Very Good
- ☐ Good
- ☐ Satisfactory
- ☐ Poor

12. What is the main reason that makes you use a recruitment site?

- ☐ Enhance your position or salary
- ☐ Job market status
- ☐ Ease of use
- ☐ Usefulness

13. How do you rank the following factors in terms of their importance to you when using on-line recruitment? (1 being the most important, please choose each value once)

- ☐ Enhance your position or salary
- ☐ Job market status
- ☐ Usefulness
- ☐ Ease of use

14. How often do you use the Internet?

- ☐ Everyday
- ☐ Several times a week
- ☐ Once or twice a week
- ☐ Occasionally

15. How would you classify your computer/Internet skills?

- ☐ Professional
- ☐ Good
- ☐ Neutral
- ☐ Know the basics
- ☐ Do not know how to use a computer

For the following questions, please give to which extent you agree/disagree with the argument:

a) I would prefer using on-line recruitment rather than traditional one?

Strongly Agree	Agree	Somewhat Agree	Neutral	Somewhat Disagree	Disagree	Strongly Disagree

b) I find using on-line recruitment useful and convenient?

Strongly Agree	Agree	Somewhat Agree	Neutral	Somewhat Disagree	Disagree	Strongly Disagree

c) My interaction with on-line recruitment sites is easy and understandable?

Strongly Agree	Agree	Somewhat Agree	Neutral	Somewhat Disagree	Disagree	Strongly Disagree

d) Using on-line recruitment would enhance my position and/or salary?

Strongly Agree	Agree	Somewhat Agree	Neutral	Somewhat Disagree	Disagree	Strongly Disagree

e) Changes in the job market are reflected on on-line recruitment sites?

Strongly Agree	Agree	Somewhat Agree	Neutral	Somewhat Disagree	Disagree	Strongly Disagree

Thank you very much for taking some time to answer this questionnaire.

Chapter VIII

Evolution of Telecommunications and Mobile Communications in India: A Synthesis in the Transition from Electronic to Mobile Business

Chandana Unnithan, Deakin University, Australia

Bardo Fraunholz, Deakin University, Australia

Abstract

Electronic business is a concept that has been adopted by businesses all over the world. The developing world takes it as a viable economic opportunity to catch up with other economies. A significant underlying factor in this development is the evolution of telecommunication infrastructure, especially in developing economies. In this chapter, we have synthesized this critical evolution in India. In the process, we found that there is a second layer of evolution into mobile communications and subsequently mobile business, which is gaining momentum in India. We conclude with an outlook for the future for these developments.

Introduction

The electronic business revolution, an idea that caught the imagination of many businesses, governments and individuals across the world during the second millennium has now become a reality (Chen, 2001). Many economies across the globe have transitioned their business processes and service delivery into the electronic mode, ushering in the digital or rather electronic business era. It is now widely accepted by policy makers, enterprises and societies that information communication technologies are at the centre of an economic and social transformation that is affecting all countries (UNCTAD, 2003). Lately, the technological advances in mobile communications, which form part of the information communications infrastructure, has caught the attention of many an economy. While it is a natural transition for developing nations, to use mobile technologies to facilitate electronic businesses and progress from electronic to mobile business, developing nations find potential in the low-cost, convenient infrastructure it offers.

Among the world's population, more than 80 percent live in developing countries where socio-economic progress continues to be slow, due to a variety of reasons such as poor infrastructure, low education, etc. (Splettstoesser, 2002). In this context, India is a geographically disparate developing economy, with a population of over 1.2 billion people spread over 35 states, speaking different languages, relatively unequal in the distribution of wealth, education and progress, is finding the transition of its domestic economy into the digital world rather challenging. To begin with, there was regulation not permitting foreign direct investment in the country for decades until the 1990s. However, during the latter half of 1990s, with the deregulated telecommunications industry opening up to private competition, combined with the federal regulations permitting foreign direct investment into the economy, the domestic economy began its transition. This transition has been enabled by the government and the growing middle-class information technology professionals, who found potential for progress through information communications technologies. Lately, it has been realized that mobile communications provide a low-cost infrastructure which provides for economic progress within the economy by enabling electronic business processes beginning with electronic government delivery to revenue generating models for network operators.

In this chapter, we aim at capturing the transition of India into the digital world by closely examining the key influencers, i.e., the evolution of telecommunications and mobile communications. In the process, we have also touched upon other influencers in the evolution, such as the effect of information technology and some government initiatives. This chapter is meant to inform academia, policy makers as well as all concerned forums involved in developing nations. This chapter offers an example for other developing nations who wish to exploit the power of telecommunications and especially mobile communications, in their economic progress as well as their transition into electronic business processes.

Background

Examining the impact of globalization and the Internet on developing nations, Kshetri (2001) contends that the main factors that lead to the explosion of e-commerce include the development of better and faster computer technology, the creation of more user-friendly software, people's trust in electronic transactions, and low costs. However, some factors such as cultural beliefs, lack of computer literacy, technological infrastructure, and government policies are a major deterrent for the spread of e-commerce in developing nations. The UNCTAD (2003) report, which analyzed developing nations, confers with this view, suggesting that developing nations have manifold challenges such as coping with new technologies as well as exploiting their full potential, and managing embedded logical relationships with the developing world in their transformation to knowledge societies. This section leads into various subsections that briefly examines and traces the evolution of India in this context.

India, being a developing nation, has many interesting challenges. The main issues have been the availability of bandwidth and power across the nation (Unnithan, 2002). Although the situation has improved over the past five years, as of March 2003, there were only 13 million Internet users in India out of a population count of over 1.2 billion (ITU, 2003). It is commonly believed that these issues impact computer literacy and obstruct technology absorption into the society. Paradoxically, India is one of the largest exporters of software, human capital and information technology enabled products in the world, which sets it apart from other developing nations (NASSCOM, 2002). Therefore, it can be established that computer literacy and availability of technologies does not have a direct correlation in India.

If the impact of government policy is examined, over the fragmented population, there is a visibly growing middle-class layer of information technology professionals, or knowledge workers, who seem to be driving the technology-enabled processes within the economy by influencing government policies (Unnithan, 2002). A small example is that these professionals, who now seem to be the growing working population, are impatient to stand in a queue to pay something as mundane as a utility bill, which usually requires a day's leave. Over the past decade, they have influenced the government to facilitate technology-enabled processes. The economy seems to be in the midst of a subtle social revolution which is pushing it into the digital world (Unnithan, 2002).

As the UNCTAD (2003) report rightly suggests, government intervention is necessary for universal access to facilities and to provide a climate for foreign direct investment within the country, which in turn helps e-commerce growth. The form of political system and legal laws adopted by people around the world may also contribute to a low enrollment in global e-commerce (Kshetri, 2001). Countries with authoritarian governments may interfere with their constituents' freedom of speech and association. Such governments can also impose great trade tariffs on business conducted via the Internet to discourage and control global e-trade. In the following subsection we briefly trace the impact and evolution of information technology sector and policies that impacted on telecommunications reform, which the government had a significant role to play, as a preamble to our major thrust in the chapter.

Transition and Impact of Information Communication Technology Revolution

The Information Communication Technology revolution in India had its beginnings in 1975, when the government of India strategically decided to take effective steps for the development of information systems and the utilization of information resources (Moni & Vijayaditya, 2002). The federal government of India, with a view to informatics-led development, decided to introduce decision-support systems within government ministries and departments, mainly to facilitate planning/implementation of socio-economic programs during the fifth planning period. The National Informatics Centre under the Electronic Commission/ Department of Electronics was the outcome of this view and was assisted by the United Nations Development Program (UNDP).

The 1990s were a period of rapid development in the technology-based industries, and de-regulation of markets following the removal of protection by the government lead to the growth of entrepreneurial activity. These developments were supported by the growing levels of expertise in information technology; venture capitalism and increasing amounts of foreign investment (Reddy, 2000). Toward the end of 1990s, with the opening up of the economy, deregulation and privatization, India became a favored destination for software development due to cheap labor and highly skilled manpower. Investment in knowledge-based industries was expected to boost India's dominance in the next millennium (Ministry of Technology, 1999).

The formation of the National Association of Software Service Companies in the early 1990s reflects on India's strength in this sector (Nasscom, 2002). This association along with the government has been a catalyst in forming the "software backbone" of India. The Indian IT industry has grown from USD 1.73 billion in 1995 to USD 13.5 billion in 2002. In terms of GDP, the figures have risen from 0.59 percent to 2.87 percent. The export orientation of the information technology industry in India is very evident from these figures. In 1999, taking in the all around socio-economic growth with the emerging digital economy, the government created a new Ministry of Technology by merging the Department of Electronics and the National Informatics Centre with the Electronics and Software Export Promotion Council (Nasscom, 2002). The Ministry of Technology envisioned Internet-based information facilitation for the common public by various government agencies at all levels to be made available by 2005, the establishment of 100 million Internet connections and one million Information kiosks (i.e., one to two connections per village) by 2008 with private sector and unorganized sector participation, promotion of Indian language content over the Internet, re-engineering of government processes leading to e-governance and launching of a mass campaign on IT awareness (Ministry of Technology, 2000).

The Centre for Development of Advanced Computing (C-DAC) was one of the pioneering ventures of the Government of India in the early 1990s to promote research in the area of advanced computing. Towards the end of 1990s, this organization became a catalyst in developing multilingual software that facilitated e-governance India-wide (C-DAC, 2002). India's software and services industry grossed annual revenue of USD 8.26 billion during 2000-01, with its export market taking USD 6.2 billion. One out of every four global software giants outsourced their mission critical software requirements to India in 2000-01. The sector also accounted for almost 2 percent of the country's GDP (Nasscom, 2002). Parallel to these developments was the Internet growth in terms of subscribers projected to touch 1.5 million in 2002. Within India, the Central/State administrations, insurance companies, financial institutions, defence segment, the public tax system, ports, customs, telecommunications and educational institutions rapidly adopted e-governance, thus boosting domestic software revenues (Nasscom, 2002).

The Central Board of Direct Taxes, Ministry of Finance, Government of India issued a notification in September 2000 listing IT-enabled services exempted from income tax including, back office operations, call centres, data processing, engineering and design, geographic information system services, human resource services, insurance claim processing, legal databases, medical transcription, payroll, remote maintenance, revenue accounting, support centres and Web site services. The provision of including other services in this list is progressively being implemented (Nasscom, 2002). The Information Technology Act of 2000 heralded a new Cyber Law regime in the country. Nasscom is committed to catalyze Internet proliferation in the country, the ultimate goal to get 2Mbps of bandwidth for every adult citizen by 2005.

According to Nasscom (2003), India currently offers a strong value proposition of all IT-enabled services due to its abundance of skilled English-speaking manpower, which rates high in the area of qualifications, capabilities, quality of work and ethics. This places India ahead of competitors such as Singapore, Hong Kong, China, Philippines, Mexico, Ireland, Australia and Holland, among others (Nasscom, 2003). Nasscom is working with international certification agencies to set standards and India has been found to be uniquely capable for setting, measuring and monitoring quality targets. When compared to their western counterparts, the number of transactions per hour for back office processing has achieved higher productivity levels. With its unique geographic positioning that makes it possible to offer 24x7 services and reduction in turnaround times by leveraging time differences is yet another strong point. The regulatory environment, specially relating to ICTs is highly progressive and most of the policy recommendations made to the government have been accepted and acted upon. Incentives such as income tax holiday until 2010 have been provided for the export of IT-enabled services. The Government of India has announced a special policy for call centers. Many state governments in India are offering incentives and infrastructure for setting up IT-enabled services (Nasscom, 2003).

In the next section, we examine the evolution of the telecommunications sector, which is considered as one of the key influencers in the transition of India into the digital world.

Evolution and Impact of Telecommunications

Jain (2001) argues that many developing countries have noted the constraint of a state monopoly in telecommunications as standing in the way of spurring internal growth and competing in an increasingly global economy. Even though it is a century since telecommunications emerged, developing nations such as India do not share the benefits of a universally distributed telecommunications service (Maxwell, 2000). Historically, until the mid-1990s, India was still struggling with electronic business activity, especially for the domestic market. However, several developments constituted a sudden, although modest, surge in this activity after 1996 (Nasscom, 2003). Liberalization of the telecommunications sector and the opening up of the economy to foreign direct investment constituted the major facilitators. Before the early 1990s, telecommunications in India was a state-owned and bureaucracy laden. For example, to obtain a telephone connection, there was a high fee and a few years of waiting period. In a geographically disparate terrain, laying down infrastructure across the nation was and is still a challenging dream (Unnithan, 2002).

Indian telecommunications sector was wholly under government ownership until 1984 (Jain, 2001; India Infoline, 2001; Sinha, 1997, Dhar, 2000). The Post and Telegraph was separated from the sector in 1985 to form the Department of Telecommunications or DOT. Subsequently, DOT set up two public sector corporations, Mahanagar Telephone Nigam Limited (MTNL) and Videsh Sanchar Nigam Limited (VSNL), to allow greater autonomy and flexibility. While MTNL took over the operation, maintenance and development of telecommunications services in the metropolitan areas of Mumbai and New Delhi, VSNL was set up to plan, operate, develop and accelerate international telecommunications services in India (Jain, 2001; India Infoline, 2001; Sinha, 1997). MTNL enjoyed a monopoly position in the two metropolitan cities, until recently, but VSNL was given a monopoly over all international access to India through its gateways (India Infoline, 2001).

The Telecommunications Commission with representatives from many government departments, including electronics and finance, was a result of government realization that regulation of the Telecommunications sector remained with the DOT (Jain, 2001; Dhar, 2001). In 1997, a separate regulatory body, the Telecommunications Regulatory Authority of India (TRAI) was formed by an act of Parliament, with the main function of finalizing toll rates and settling disputes between the main players (India Infoline, 2001; Bagchi, 2001). Following the National Telecommunications Policy of 1994, the government announced private participation in basic and cellular services. The country was divided into 20 "circles" and one private operator was allowed to compete with the DOT in each of these circles. However, DOT was to give the licenses to operators with a fee, driving out competition with heavy license fees and tariffs (Bagchi, 2001). This led to the announcement of the National Telecommunications Policy of 1999, taking into account the convergence and existing anomalies in the sector (Bagchi, 2001). As a result of this policy, 70 ISPs became operational in India. The government also encouraged several ISPs to set up international gateways to the Internet, bypassing the VSNL monopolized gateway (Bagchi, 2001). In addition to opening up international telephony, the government also decided to end VSNL's monopoly, two years before the WTO-set deadline of 2004.

The Centre for Development of Telematics (CDOT) in India was perhaps one of the earliest government initiatives to research and develop technology suited for the Indian climate (Jain, 2001). CDOT was able to champion the idea of technology for the masses, with rural automatic exchanges designed specifically for Indian climatic conditions. Many regional areas including villages, small towns and B class cities were connected and public telephone booths became part of Indian society. An indigenously developed technology, adaptable for Indian conditions, was successful and, by end of the year 2000, 10 million of the 20 million lines installed in India were using CDOT exchanges (Jhunjunwala, 2001).

Several technological changes made it imperative for the government to view IT, telecommunications and broadcasting legislation in a coherent and convergent manner, which led to the drafting of the Information, Communications and Entertainment Bill (Jain, 2001). The Communication Bill 2000 had the objective of facilitating the development of a national infrastructure for an informed society, establishing a licensing framework for carriage and content of information in the converging areas of television, broadcasting, data communications, multimedia and other technologies (Bagchi, 2001). This convergence bill along with the Information Technology Act of 2000 clearly indicate that India's government is moving towards a single communication network catering to all types of technologies including the Internet, Datacom, Telecommunications, Wireless, Fixed, Mobile, Cellular, Satellite Communications and e-commerce (Jain, 2001).

India Infoline (2001) pointed out that the country had an approximate tele-density of only 2 fixed lines per 100 persons (India Infoline, 2001). However, telephone penetration is not dependant on phone ownership. As in many developed countries, private space in houses is not abundant, and phones tend to be shared (Unnithan, 2002). In many interior areas, public call offices or telephone booths tend to be used (Jain, 2001). The socio-economic changes within the country spurred by the Internet have seen the emergence of cyber cafes and computer institutes all over the country. Interestingly, this development has been accentuated by the growing need for technology education within the country, essentially facilitated by the software industry. Although telephones may not have reached every home, cyber cafes are in great demand (Unnithan, 2002).

Convergence of ICTs, telecommunications, broadcasting and entertainment toppled most of the old value chains, bringing forth yet another revolution within India (Moni & Vijayaditya, 2002). Studies have shown that a large populace of television users would embrace the Internet, video-on-demand and greater interaction with content, but may be diffident about buying or using a personal computer. India has the highest cable penetration percentage of 46.8 percent among low telephone penetration countries (Nagaraj, 2001). There is a drive for cable modems, particularly by work-at-home households and Internet users. Satellite is another broadband access technology, which has immense potential for a country as large as India. Technological developments now permit the network used to carry broadcasting signals to the customer premises to be used for carrying telecommunications and data. India already has had a critical mass addicted to television and cable channels and this has fuelled the growth of cable Internet (Unnithan, 2002).

Indian society has had limited resources to absorb unprofitable innovations and the majority of the population has historically responded only with caution and economic necessity. Interestingly, the growing upper-middle class, characterized by the computer professionals, is increasingly driving the diffusion of technological innovation. Many of these professionals are non-residents willing to invest in innovative telecommunications ventures, as they are seen as progressive icons for the economy. The impact of global trends, technological innovations, and a growing generation of technically skilled youth who are driven by rational views, moving away from the older generation with their nationalist attitudes, making further modification of attitudes and actions inevitable (Jeevan, 2000).

Together with Nasscom, the government of India is now committed to push electronic commerce in India, as reflected in the announcement of the Information Technology Act of 2000, the announcement of ISP policy for the entry of private Internet service providers in 1998, permission grants to private ISPs to set up international gateways, permission of Internet access through cable TV infrastructure, initiation of a National Internet backbone, announcements of national long distance service beyond the service area to private operators, complete non-monopolization of undersea fibre connectivity for ISPs in 2000, free right of way facility with no charge to access providers to lay fibre optic networks along national, state highways; interconnectivity of government and closed user networks, and establishment of public tele-info centres (PTIC) with multimedia capabilities (Nasscom, 2002, 2003).

After having examined the telecommunications sector, we now propose to examine the mobile communications sector, which is fast becoming a key enabler in facilitating electronic business process delivery.

Evolution and Impact of Mobile Communications

Towards the turn of the century, the government of India recognized the key role of telecommunications in its developing economy and decided to invest significantly in the mobile communications sector (COAI, 2002, 2003). Mobile services were introduced initially as a duopoly under a fixed-license regime for a period of 10 years. With liberalisation in the telecommunications sector, the country was divided into four metropolitan cities and 19 telecommunications circles which were then roughly analogous with the states of India, and licenses awarded to private operators, bringing in competition. Cellular licenses were awarded to the private sector, first in the metropolitan cities of Delhi, Mumbai, Kolkata and Chennai in 1994 and then in the 19 telecommunications circles in 1995 (COAI, 2002, 2003).

The initial response of the private sector was very encouraging with the attractiveness of the Indian market — the low teledensity, the high latent demand and a burgeoning middle class — brought in some of the largest global telecommunications players, foreign institutional investors and the major Indian industrial houses to invest. Annual foreign investment in telecommunications increased steadily from an insignificant INR 20.6 Million in 1993 to INR 17,756.4 Million in 1998. However, the attractiveness of the Indian market did not last for very long, as by 1997-98 the private cellular operators were confronted with a series of problems that threatened their very viability and survival. As a result of this, Foreign Direct Investment inflow into telecommunications dropped sharply, declining by almost 90 percent to INR 2126.7 Million in 1999 (COAI, 2002, 2003). As private-sector participation preceded the set up of regulatory authority and tariff rebalancing, licenses were auctioned at exorbitant amounts, leading to high cost structure and unaffordable tariffs. Therefore, for the common public, although mobile telephony was a convenient faster option, as against a fixed phone, the unaffordable tariffs did not help the situation.

According to the COAI (2003), one of the key factors of this critical state was the manner in which liberalization was undertaken. Usually, deregulation is preceded by tariff rebalancing, institution of a strong and independent regulator and then private sector participation is invited. In India, private sector participation was invited in 1992, the Regulatory Authority was set up in 1997 and the tariff rebalancing exercise commenced in 1999 and is still far from complete. The regulatory authority had considerable ambiguity on its powers, which resulted in virtually each and every order of the authority being challenged by the licensor. In addition, consumer benefit was the least priority by the government and the sector was a key revenue generator for the government. Although the National Telecommunications Policy of 1994 identified the primary objective as affordable cellular services, this was almost disregarded during implementation. Licenses were granted through an auction process to an enthusiastic private sector deluded by the huge potential of the Indian market and lured into bidding exorbitant sums of money for cellular licenses. These huge license fees resulted in a high cost structure leading to unaffordable tariffs and lower growth of the market. Subsequently, the cellular industry was on the verge of bankruptcy by the end of 1998 (COAI, 2002, 2003).

Under these circumstances, the government introduced a new policy called NTP 99 and the amendment of the Telecommunications Regulatory Authority Act in January 2000. The policy replaced the high–cost, fixed licensing regime with a lower cost licensing structure through revenue sharing, providing a greater degree of competition and flexibility in the choice of technologies. Existing private cellular operators migrated to the new telecommunications policy regime beginning in August of 1999. Cellular tariffs have dropped by over 90 percent since May of 1999. The average airtime tariff in 2001 was prevailing around INR2 per minute as against the peak ceiling tariff of INR16.80 per minute when NTP 99 was announced. There was also a significant drop in the mobile phone costs with cellular handsets costing around INR 30,000 or US$645 in the initial days to INR2000 or US$42 (COAI, 2002, 2003).

More specifically, as the government rationalized levies, resulting in high turnover, and cellular operators were able to venture more into cities and towns. Parallel to this development, the operators are able to offer services to consumers on a contract or plan basis, subsidizing the cost of phones. Consumers may be offered a certain tariff for buying the phone over a period of 12 months on a contract, where they also are bound to the operator for that period for services provision. Thus, on a plan the consumer may be charged as low as USD42 for buying the phone, over a period of 12 months, which may be the term of the contract. The government is promoting the mobile sector as it generates revenue for the exchequer, but also reduces the costs of infrastructure rollout especially when connecting remote villages. Low tariffs, along with price wars by cellular operators are supported by massive consumer demand, especially the youth in metropolitan cities (Fraunholz & Unnithan, 2004a).

By the end of 2002, the mobile subscriptions surpassed fixed-line networks. As against owning a PC and getting an Internet connection at home, or using an inconvenient option of a public Internet booth without privacy, ownership of a mobile phone constituted to this surge. Another enabler of mobile telephony success is touted to be the Short

Messaging Service, which has facilitated not only the growth of a new culture but also many business models that support electronic business. Although there is modest growth in the area of electronic commerce as such, there seems to be an interesting trend toward developing mobile commerce. Many vertical industry segments, especially the banking industry, have introduced mobile commerce successfully. The transition into mobile commerce is supported by the introduction of 3G networks in 92 cities in 2003 (COAI, 2003).

On the technological frontier, the Indian Government, when considering the introduction of cellular services into the country, made a landmark decision to introduce the GSM standard, thus avoiding adolescent technologies and standards. Although cellular licenses were made technology neutral in September 1999, all the private operators are presently offering only GSM-based mobile services. In July 2001, cellular licenses were awarded and all of the new licensees have opted for the GSM standard to offer their mobile services (COAI, 2002). According to Gartner (Indiantelecomnews, 2003a). CDMA technology is particularly attractive to India, as the point-to-point concept of communication within specific circles is an important factor for India. The Indian market had clearly defined points of usage within a telecommunications circle where CDMA is likely to work better as opposed to GSM or unlimited mobility. IDC forecasts (Indiantelecomnews, 2003b) that the cheap CDMA connections are likely to affect GSM operators, with up to 20 percent of subscribers willing to try cheaper CDMA services. Beginning with 3G advanced wireless services, the Reliance India Mobile service marked the first CDMA2000 1X nationwide commercial launch in India, bringing advanced wireless data and voice services to 92 cities (3g, 2003) in May 2003. However, in India, the CDMA is being adopted as a platform for launching 3G but it still is expected to co-exist with current GSM and future 3GSM services.

On another note, the unique nature of multiple network operators licensed within each circle, leading to co-operation and competition — or co-opetition (Xu et al., 2003) is the way SMS operates within India. The term co-opetition describes competing businesses cooperating to create and enlarge the market rather than competing to divide (Brandenburger & Nalebuff, 1996). The success of SMS in India has been a stimulant for network operators who seem to be optimistic about the forthcoming MMS to go along the same success route. The vast geographic terrain also offers opportunities in niche segments such as farming or agro sectors in central and north western India, trade sector in Gujarat, IT professionals in the southern region, industrialists in the central region, literary communities of eastern region and so forth. Each of these communities along with the growing youth population in metros offer significantly different opportunities for the providers as they increasingly demand value-added services tailored to their needs, whether they be different languages, script, content and so forth (Fraunholz & Unnithan, 2004a).

The infrastructure problems associated with the geographic region motivates the population — due to the convenience it offers — as well as the government authorities to promote the mobile sector. It not only generates revenue for the exchequer, but also reduces the costs of infrastructure rollout especially when connecting remote villages. With stimulation from the government, and population demand, cellular network providers seem optimistic about their future growth in India. However, the licenses issued clearly indicate some significant players who hold the major market shares, whether it is through subsidiaries or sister organizations. The industry itself is showing the signs of becoming an oligopoly in the future (Fraunholz & Unnithan, 2004a).

Figures from the Cellular Operators Association of India or COAI showed that the industry had 5.725 million subscribers, up from 3.27 million at the end of January 2001 and 5.48 million subscribers at the end of year 2001. The data showed that the industry added 246,281 users in January 2002 alone, led by the four main city markets of Bombay, New Delhi, Madras and Calcutta, which together added 93,070 customers. The overall Indian mobile subscribers jumped to more than 12 million in first quarter of 2003 (Fraunholz and Unnithan, 2004a,b).

A recent Reuter (2002) report claimed that the number of mobile phone subscribers in India is likely to rise to 120 million by 2008 because it has the cheapest call rates in the world. India's US$5.0-billion mobile phone sector, billed as one of the fastest growing markets globally in this decade, has eight million users spread across some 1,500 cities and more than 60,000 villages. The main driver for a more than 100 percent growth each year in the past six years has been falling tariffs in a sector where a dozen money-losing firms have launched a fierce price war to grab market share. The eight-year-old sector has the lowest rate of USD16 a month for a 300-minute talk time plan compared with other developing nations such as USD21 in China and USD77 in Brazil (Reuters, 2002; Rediff.com, 2002). However, mobile operators pay between 8.0 to 12 percent of their revenues as license fee compared with no license fee in China. Data released by the Cellular Operators Association of India showed that the industry is expected to be one of the world's fastest growing markets this decade (Rediff.com, 2002).

As of late 2002, there were 24 companies and 42 networks on air all over India (Ramachandran, 2002). For the first time, over a three-month period from April to June 2002, the cellular subscriptions went up by 960,000 as against the fixed lines increase of 300,000 over the same period (COAI, 2003). Low tariffs, along with price wars by cellular operators are supported by massive consumer demand, i.e., the youth and businessmen in metropolitan cities as well as the relatively new and upcoming households in rural areas (Reuters, 2003). The tight monopoly control over telecommunications and aggressive efforts to curtail competition had led into slow growth of the Internet. This in turn led to the boom of entrepreneurs in India.

India has a low teledensity of 4.5 percent compared with a global average of more than 15 percent (Reuters, 2003). The number of households in the rural areas is expected to grow to 360 million by 2010, making them an attractive audience. On the other hand, the thickly populated urban city areas are less motivated to get a fixed-line network. To explain the cause of this de-motivation, an example would be the thickly populated metropolitan Mumbai, where every suburb is connected with metro railway lines lined by illegal slums (Fraunholz and Unnithan, 2004a). To get a fixed-line cable network that runs across these slums into the 12th level apartment, in itself is a significant feat, as it would require cutting through many bureaucratic angles including the people employed for installation (Fraunholz & Unnithan, 2004a).

On the other hand, once a fixed line is in place, the fear of a slum dweller cutting into the line, resulting in massive bills every month — which cannot be traced — often deters households from connecting fixed lines. On yet another note, due to the small limited spaces available within households and the relative lack of privacy often drives people into public booths — which are perceived to have more privacy — especially for the youth who would like to be away from the earshot of family members (Fraunholz and Unnithan, 2004a). The mobile phone, although starting off as a high cost affair, is now a lure for those who seek privacy, relative safety from slum dwellers and freedom from bureaucratic tangles. Most firms expect the market for mobile services to grow by between 10-14 million new subscribers in 2003 (CellularOnline, 2002).

The affordability of mobile phones has become more and more possible in India, with the government cutting down on the levies. In addition, the network operators have opened up the avenue of subsidizing the phone, through call plans, which has almost brought the cost of the phone down to INR2000 or USD43. In spite of a heady growth in the cellular services market, following the subsidies offered through call plans, the legal market for cellular handsets has remained very small (Indiainfoline, 2002). A large percentage of the handsets sold in the country are through the unauthorized or the grey channel, which includes smuggled handsets, parallel import and handsets brought by people travelling abroad.

The share of the unauthorized market in the overall market has shot up from 74 percent in the year 2000 to 86 percent in the first half of 2001 and an estimated 89 percent in second half of 2001, according to an IDC report on the Cellular Handsets Market in India (Fraunholz and Unnithan, 2004a). This increase can be attributed to the price differential between handsets bought from legal and grey channels. The difference in price is at least 25 percent to 35 percent and arises due to the high level of duties like customs or sales tax paid by the vendors selling in the legal market. In a metropolitan city like Mumbai, there is also additional taxes, such as Octroi (a levy on the basis of getting into the metro area). Yet another reason for this flourishing grey market is that the handset vendors do not provide extensive after sales support because of the small size of the actual legal cellular market and therefore absence of economies of scale. Without any supporting infrastructure, buyers do not feel the need to go on a call plan — to buy a phone — when the same phone is available with much cheaper rates from the grey market. There is no incentive for buyers in real terms (Indiainfoline, 2002).

Mobile communications has re-invented the role of fishing captains into logistics and supply chain managers (Karkera, 2002). For example, the fishing industry in Kerala, a southwestern state of India, generates USD600 million in a year in revenues. During the day prices vary throughout the day at 17 landing ports around the main port of Cochin. Currently, 8,000 fishing boats carry mobile phones, to locate the best offers before landing in the port, saving expensive fuel by calling in carrier boats that take the catch to the shore. In addition, the agents, handlers and middlemen also carry mobiles to get their best deals. Two competitive firms are offering services to these "communities of interest" (Keen and Mackintosh, 2001). The boom of young IT professionals carrying PDAs and mobiles and also the growing concept of mobile workers in densely populated metropolitan cities of Mumbai, where commuting otherwise takes hours, are becoming increasingly commonplace.

Short messaging services has brought mobile communications to Indian life, whether it is student or executive and urban or rural life (Thomas, 2002). India is an economy widespread geographically, but prefers the closeness in society. The cost-effectiveness and convenience of the mobile combined by this new SMS and lately multimedia messaging services or MMS, is becoming increasingly common. The growth of mobile workers, especially in the IT area, within metropolitan cities and with increasing demands on their time have further added to the increasing popularity of SMS. A fair example of an office executive, stuck in a traffic jam before a presentation, sending an SMS or even connecting through a PDA to send the agenda or presentation through is becoming part of daily life (Fraunholz & Unnithan, 2004b).

Mobile Youth (2002) claimed that SMS is creating a revolution in India with an estimated 60 messages sent per phone per day from India's 8 million mobile phones owners in early 2002. On special days such as national festivals, SMS traffic increases to clog most networks. SMS was reported to be four times more than normal as people sent festival greetings on Diwali, as this is the most economical, convenient and instant mode of communication (Chatterjee, 2002). There seems to be a 500 percent jump over normal usage during this major festival season equivalent to New Year's Eve (MobileYouth, 2002). In a joint initiative, Ericsson Mobile and Bharti Telecom — the network provider who holds major market share within India — worked together to develop an SMS-based service for school children (Ericsson, 2001). Four to five million school children obtained their test results by sending their identification numbers and receiving their results in SMS form. Not only did they receive their overall percentage, but also individual subject marks. Following this project, Bharti's (which has 25 percent market share) SMS traffic grew more than 100 percent (Ericsson, 2001).

Pereira (2002) reported that on the 13th of May 2002, the capital city of New Delhi was introduced to traffic police SMS. The service was aimed at providing aid in answering the average queries of a motorist as well as to help the traffic police operating the field. Commuters can get information on traffic blockages and diversions while investigative journalists can acquire information on accidents or prosecutions immediately through this service. A vehicle being towed away is immediately notified to the owner. News sites in India such as Mid-day and the India Times have expanded their SMS alert options. Major portals such as Yahoo and Rediff have launched SMS services for instant messaging via gateways on their Web sites. In the southern state of Kerala, fishermen use their mobiles to send SMS messages to their partners on the shore about their catch, so that the price can be fixed and faster transactions can be made. For many families living apart in the large geographic region, SMS is a lifeline to keep in touch (Thomas, 2002).

The southern Indian city of Chennai outstripped all the other metropolitan cities in SMS usage with 80 percent of the mobile owners sending SMSs (Times, 2003). In an interesting IDC survey, women were found to be SMS'ing more than their male counterparts — averaging 4.2 messages per day within the same period, in Chennai, which reported maximum SMS usage. Interestingly, human rights activists in India have condemned the diffusion of SMS especially among the youth as a cause of breaking up relationships (Mobileyouth, 2002). For example, a typical 'U4Me' message was cited to have sparked marital discord ending in a divorce. However, the growth of SMS seems undeterred with operators clocking a staggering nine million short messages in one single festival day, in the capital city of New Delhi alone. Many celebrities now provide mobile numbers to fans as SMSs do not intrude their privacy (Thomas, 2002). A vital aspect behind the success of SMS is that the costs range from INR 2 or USD 0.042 in some circles to 50 Paisa or USD 0.010 and free in others. The income from SMS is currently 10 percent of the total revenue for many network providers (Thomas, 2002). The affordability and the trendy, cool aura that it provides to the youth seem to be the key factor in SMS success in India.

From the corporate point of view, at least in the metro cities, SMS is becoming vital for communications amongst traffic congestions and traveling (De, 2001).

Evidently, government support and stimulation has progressed the telecommunications reform. Over the 1990s, growth of professional youth has supported the massive demand of mobile communication services. Combined with these factors were the after effects of deregulation that stimulated the previously stagnant or rather inflexible telecommunication sector. With digital transmission becoming increasingly popular in India during late 1980s, the next evolution into mobile phones was slow but gradual over the period. However, initially the non-affordability of mobile handsets itself was impeding the growth of this sector. With the new telecom policies that subsidized the levy on handsets, and parallel growth of the grey market, once the influx of foreign goods became open after 1995, handsets became more affordable (Fraunholz & Unnithan, 2004a).

Businesses as well as individuals have now become common users of mobile phones, due to the convenience and cost-effectiveness that it offers in the Indian context, especially compared to the hassles of obtaining a fixed-line phone. With the blessing of government subsidies, increased demand and subsidized handset levies, network service providers (or cellular network providers as they are known in India), have launched into a fierce price war, especially targeting the youth market (Fraunholz and Unnithan, 2004b). In the end, the network providers who will offer the most affordable as well as innovative services will drive the mobile communications market. The uptake of SMS/MMS as a value-added service seems to be driving a mini social revolution within India. It has in a way provided the youth with "affordable freedom" within a restrictive society. It has to be noted that a fixed-line telephone is not considered "private" as it exists within a closely condensed household, where people can eavesdrop. The administrative hassles of obtaining a fixed phone, as compared to a mobile phone, are making individuals as well as businesses opt for mobile communications (Fraunholz & Unnithan, 2004b).

In turn, SMS is an attractive business model for network providers to extend into mobile commerce. While a combination of SMS and voice transmission is able to help governance, service provisioning such as traffic information, it will be a while before mobile commerce can take over from electronic business over the Internet. India still has a long way to go to reach the critical mass or saturation regarding mobile phones. 3G with CDMA standard is becoming a beacon of hope for rural and geographically spread out areas, as well as affordability (3G, 2003). However, the CDMA standard will co-exist with the established GSM networks for a long time to come. If the 3G is absorbed as fast as it is perceived to be, and holds up to its promises within India, it may be the perfect launching platform for many mobile commerce models such as location-based services, which would be very lucrative in the disparate terrain (Fraunholz and Unnithan, 2004a).

As it can be seen from this section, mobile communications have not only facilitated a low cost infrastructure, which has overtaken fixed-line subscriptions, but also have facilitated revenue generating models within the domestic economy. With rapid absorption of mobile technologies, the day is not too distant when electronic business in India may transition into mobile business. The next section is a brief appraisal of the electronic business processes within the economy.

Indian government had recognized the potential savings for the exchequer by introducing digital service delivery (Nasscom, 2003). It would cut down transaction costs of governance, thereby stretching the taxes paid by the average citizen to provide more services across the economy. Therefore, a major effort was made to introduce digital governance into the country with the Central and State administrations, customs, ports, the public tax system and education system pioneering the venture. A number of state governments implemented e-commerce initiatives aimed at cost effectively taking various facilities to citizens. Innovations in the area of land records, taxation, procurement, etc., were witnessed in the sector, with the Internet pervading significant government transactions (Nasscom, 2002). The government of India issued guidelines that 2-3 percent of every ministry or department plan budget was to be utilized in achieving digital governance using IT (Raje, 1999). As pointed out, many state governments have taken initiatives to provide "one-stop shops" to deliver a host of services to citizens such as domicile certificates, driving licenses, property tax payments, electricity and water bills, etc. In parallel, to achieve mass customization, the government of India decided to set up a National Institute of Smart Government as a tripartite venture between government, business and community (Raje, 1999).

With the increasing recognition that information technology is catalyzing economic activity and efficient governance, Indian government has made significant investments in the sector. One of the interesting challenges for the government was to implement a common e-governance thread in a geographically dispersed, demographically multilingual India. Out of the 1.2 billion population, 95 percent (950 million) speak or practice 18 officially recognized languages. For the Centre for Development of Advanced Computing (C-DAC) this presented an opportunity. An initiative in developing Indian language tools with natural language processing, in evolving script and font standards through GIST technology was pioneered by C-DAC, with directives from the government (C-DAC, 2002). Some of the successfully commissioned initiatives include:

- **Hospital Management System:** implemented to improve healthcare services for patients in speciality and government hospitals across India.
- **State of Maharashtra:** Public Works Department with 250 state-wide offices was networked, the GIS-based land management was implemented providing Web-based access to land data covering allotment, transfer, mortgage, surrender, etc., of the industrial development units. Archives Computerization was deployed for the Department of Archives and Octroi (a type of tax) collection was computerised and networked.
- **Stamp registration in Maharashtra and UP States:** Online property registration, valuation and report generation across 366 offices at various state administrative units, reducing time and increasing revenue.
- **Karnataka State:** Major functions of property tax valuation/collection, issue and record of death/birth certificates, water supply billing, consumer complaints and internal MIS functions were computerised to provide improved citizen services.

- **Andhra Pradesh:** Implemented a data warehouse of land and population data of 60 million people to enable well informed, timely policy decisions by government officials across various departments.

Compaq India established a memorandum of understanding with Electronics Development Centre of India (ER & DCI) in the year 2000 to initiate e-governance in NOIDA city and extend it subsequently to various states. The project was to smart link/interface between citizens and to develop a system that automates rural development, arms and licenses, regional transport offices, land records, citizen databases, electricity board payments, and set up GIS (Compaq, 2000). This project brought Internet/intranet infrastructure up to section officers level, IT empowerment of officers and officials through training, IT-enabled services including government G2G, G2B, G2C portals and development of BPR methodology for electronic services delivery, among other initiatives (Moni & Vijayaditya, 2002).

In December 2001, the entire state of Gujarat was networked up to the small taluk level of government. Everything from collecting the posts to disposal of files is computerised and the GSWAN has come into existence. The Concept Centre of Electronic Governance set up by the Indian Institute of Management in Ahmedabad was able to identify worthwhile applications and disseminate knowledge for successful implementation of e-governance applications amongst bureaucracy and other stakeholders. E-governance has been able to bridge the digital divide in this state with a 50 percent literacy level and more than half of the population living in rural areas. The priority services include pension processing and ration cards (CIOL, 2002). As Zdnetindia (2002) reports, Tamil Nadu State has a comprehensive state government information site with application forms in English, Tamil, comprehensive land records, a pilot project of utility bill payment over the Internet, tele-medicine projected proposal, application software for regional transport offices, registrar office and major intended IT projects for high court and police departments. It is evident that similar ventures in other states are ongoing here as well.

The most important initiative of the Karnataka government is the lodging of taxes online. Computerised land records — the Bhoomi Scheme — and registered transfer certificates are on the anvil. In addition, the policing system, forestry, agriculture and regional transport system is being computerised. The state also focuses on education with YUVA, a program aimed at underprivileged youth, women and families with low income (Banerjee, 2001). In Madhya Pradesh, 5,500 centres for computer literacy were announced with a program called Headstart. The state is advocating e-governance through education. In addition, the commercial tax department, registration department, treasuries and agriculture marketing departments were computerised. The state is promoting a "build-operate-transfer" system for smart cards in the transport department for registering vehicles and issuing driving licenses (Singh, 2002).

The Indian government has been taking key initiatives over the past few years to create an environment conducive for e-commerce activity and some of them include the Information Technology Act 2000, which brought in the cyber-law regime in the country, entry of private Internet service providers in 1998, granting permission of Internet access through Cable Television networks, the establishment of Public TeleInfo Centers (PTIC) with multimedia capabilities and allowing 100 percent foreign direct investment in B2B ventures. The nationwide Internet backbone has also been initiated. Whilst the federal government has laid out several such initiatives, there has also been support from state-level ministries initiating various developmental initiatives for public welfare and for promoting business, especially Small and Medium Size businesses (Ministry of Technology, 2003).

Despite these, there is a modest e-commerce activity estimated to be around USD 300 million in the year 2002 (Nasscom, 2002). The Business-to-Commerce spending in India is estimated to have grown in the year 2002, with the travel sector accounting for 23 percent of transactions. Business-to-Business e-commerce implementation was low except in certain vertical sectors such as automobiles, banking and finance (Nasscom, 2003). Experts have argued that the low cost of personal computers, a growing installed base for Internet use, and an increasingly competitive Internet Service Provider (ISP) market will help fuel e-commerce growth. Dataquest, an Indian computer journal, has found that the rise of Indian Internet subscribers will ultimately depend on the proliferation of network computers and Internet cable (Gartner, 2003). Cyber cafes will also continue to provide low-cost access.

Currently, the lion's share of current e-commerce revenue is generated from an ever-expanding business-to-consumer (B2C) rather than business-to-business (B2B) market. As in the United States, B2C transactions have taken the form of online purchases of music, books, discounted airline tickets, and educational resources. In a recent McKinsey-Naccson report, it was estimated that some 80 percent of e-commerce in India over the next few years could be B2B if the correct environment were developed. The B2B market is expected to increase following greater investment in the Indian telecommunications infrastructure, once intellectual property rights and legal protections for commerce over the Internet are addressed. There are still enormous challenges facing e-commerce sites in India. The relatively small credit card population and lack of uniform credit agencies create a variety of payment challenges unknown in the United States. Increased distribution of online purchases could be complicated by India's complex postal system and an uncertain regulatory environment. Nonetheless, everyone from Yahoo, Microsoft, and IBM to local carpet vendors, hotels, and some 300 Indian ISP's are trying to claim a slice of the rapidly emerging Indian e-commerce market (Nasscom, 2003).

As Kripalani and Clifford (2000) have rightly commented, India has always had enormous potential, but a difficult time living up to it. Corrupt governments, outbursts of Xenophobia or communal violence have affected its confident progress, not to mention the disparate socio-economic, cultural and geographic spread. Successive governments have made great strides to reduce stifling, socialist-era regulations, but the shadow still exists as many politicians still are reluctant to relinquish power. For e-commerce to surge, there is a need for ample telecommunications capacity, computers, and electricity. Even with progressive reforms, India still needs to attain the critical mass with telephony — wired or wireless. Power shortages are chronic while access to PC/Internet continues to be low. However, on the positive side, India has free media and democracy which many developing economies lack. And the country is proactive to solve its shortcomings. In telecommunications, the government is dismantling curbs on foreign investment and competition. Consortiums are building fiber-optic networks. Satellite communications and TV set-top boxes are expected to help bring the Internet to households that still lack phone lines. Wireless telephony has gained momentum (Kripalani & Clifford, 2000).

Mobile Business

According to Nasscom's Strategic Review 2003, in the year 2002, the m-commerce market was in the region of USD50 billion worldwide (Nasscom, 2003). While the US and European markets are expected to dominate forecasted revenues until 2005, Asia-Pacific and the rest of the world are expected to account for 40 percent of the estimated $225 billion m-commerce market by 2005. However, the m-commerce market in India has not seen as much growth as was expected. Experts opine that it is still in a very nascent stage and will take time to reach the maturity level to match EU and the US standards.

According to Anil Lekhi, VP–IT for Spice Telecom in Punjab (Expresscomputer, 2003), the company that witnessed hardly any m-commerce transactions a year ago, is now doing business of around INR 25,000 to 30,000 daily through m-commerce. The average value of these transactions varies from INR 1.50 to 10. Presently almost 9 percent of Spice's overall revenues come from value-added transactions of which m-commerce is a part. The optimistic country operations manager of Motorola PC division commented on m-commerce potential in India based on the ability to address nearly 50 million subscribers by 2005 (Expresscomputer, 2003).

Indian cellular operators today are under tremendous pressure to sustain and grow their Average Revenue Per User or ARPU. Fierce competition among operators has consistently driven down tariffs, reducing revenue from the voice-based operations of the wireless networks. Operators today are providing value-added services to sustain and grow their ARPU (Fraunholz & Unnithan, 2004a). Optimism prevails that with subscribers already checking movie schedules and airline/railway booking status among other things using the mobile handset, it is only a matter of time before they start booking tickets as well. The advent of wireless Internet in the form of GPRS and CDMA 1x is expected to further boost m-commerce in India (Expresscomputer, 2003).

The two types of m-commerce transactions are low-value and high-value. Low-value transactions usually imply music downloads, logo downloads, picture downloads, ring tone downloads, etc., some banking, value-added services like news, stock alerts and services like m-coupons and wallets. On the other hand, there are high-value transactions, which involve credit and debit card transactions, point-of-sale terminals, going to the merchant location and paying through the handset. There are only low-value transactions in India at present. Major advances in m-commerce are not going to happen until higher-bandwidth networks are deployed and wireless service providers cooperate with each other instead of pushing competing standards (Expresscomputer, 2003).

Yet another major reason is the lack of commerce-capable cellular networks, which can route real-time transactions over the cellular network to a remote payment gateway and guarantee security over the transactions (Alexander, 2003). Operators need to instill confidence among mobile users to start taking buying decisions on deals initiated from mobile phones and turning the mobile into a wireless debit card. The different parties involved in the entire m-commerce value chain are wireless infrastructure providers, wireless service providers, certifying authorities, applications/software providers, equipment manufacturers, credit card companies and banks. And the absence of proper coordination between them will hamper growth (Alexander, 2003). Also all players in the value chain, from biometrics to SIM card providers, cellular operators, network providers, application developers, banking to semiconductor companies have to coordinate.

However, on the progressive side, in their rollouts the operators are actively deploying next generation wireless Internet technologies to facilitate data services. The cellular subscriber base is growing phenomenally and reaching respectable levels of adoption. So, the stage seems to be set for m-commerce and the industry is already seeing the first signs of evolution in value-added services that operators have started providing. Quick adoption is imminent, keeping in mind the ARPU decline from voice (Expresscomputer, 2003). If statistics are anything to go by, the SMS rage will drive m-commerce in India. According to Merrill Lynch, SMS could bring in as much as $75.6 million of revenues for Indian GSM operators by the year 2005. The stock-brokering firm predicts that by the end of 2003, close to 700 billion application-driven SMS would be sent from mobile phones, which would be almost half of the total SMS traffic. During 2000-2003, while peer-to-peer messaging has been growing at a the annual growth rate of 46 percent, the application-driven SMS traffic has been growing by 204 percent during the same period (Expresscomputer, 2003).

On the other hand, the software services sector, which has been a catalyst in Indian electronic business is aiming at competing in emerging markets of wireless communications and mobile commerce through client software development and embedded systems design (Shankar, 2004). As the growth rate of mobile phones has already outnumbered the growth of fixed-line phones in India, with the development of a secure, easy-to-use method for paying over a mobile is devised, m-commerce will become a reality in India.

Conclusion

Over the past decade, progress is visible from the Indian perspective as regards government initiatives in promoting Information Communication Technologies infrastructure, telecommunications and especially mobile communications that supports electronic/mobile business. Despite this, in a geographically diverse nation, with different grades of socio-economic progress, density of population and attitudes, it is still a challenge to catch up with developed nations as regards electronic business. Although mobile communications seem to be a boon for an economy with infrastructure issues for telecommunications, it will be a while before the critical mass is achieved for mobile communications, Internet and overall technology access.

The overall attitude in the economy is now driven by growing middle-class professionals, who form the majority of the population. Policies seem to be based on the needs of these professionals and a proactive government, which seem to realize the potential of ICTs and electronic/mobile business for the socio-economic progress within the economy. The use of wireless technologies has become increasingly common especially in densely populated urban markets. This in turn offers potential for lucrative business models such as location-based services which drives mobile business.

Outlook

India's ruling government has made information technology the cornerstone of its political agenda of generating high economic growth while surrendering little sovereignty to multinationals (Kripalani and Clifford, 2000). The hope is to spread information communication access to unify a nation otherwise divided by cultural and economic disparities. Many remote villages in India are now connected to the Internet. From craftspeople to daily farmers, rural Indians have begun using mobile communications or mobile Internet facilities to sell goods and monitor prices. It seems to be a novel, but very effective, approach for a progressive resurgent India. This chapter has provided a synthesis of many interesting factors, mainly in the telecommunications area that influenced the growth of electronic business and its transition into mobile business within India. It is expected to inform researchers, academia, policy makers, and all players who are involved in electronic business development in developing nations.

References

3g. (2003). 3G Launches in India. *3G Newsletter*. Retrieved June 30, 2003, from *http://www.3g.co.uk/PR/May2003/5369.htm*

Alexander, G.C. (2003). It is time to shop through cellphones. Retrieved from *http://www.rediff.com/money/2003/dec/22betterlife.htm*

Bagchi. (2000, December). Telecommunications reform and the state in India: The contradiction of private control and government corporation. Center for Advanced Study of India, CASI Occasional Paper #13. Retrieved from *http://www.sas.upenn.edu/casi/reports/Bagchipaper120000.pdf.*

Brandenburger, M. A., & Nalebuff, J. B. (1996). *Co-opetition.* New York: Doubleday.

CellularOnline. (2002). India's mobile industry grows 75% in January. Retrieved from *http://www.cellular.co.za/news_2002/021502-india_growth_75%25.htm*

Chatterjee, S. (2002). 'Shrt n swt. C u 2nite,' with SMS, love goes mobile at touch of a button. *News India times Online*. Retrieved from *http://www.nesindia-times.com/2002/02/22/sp-valentinesday-sms.html*

Chen, S. (2001). *Strategic management of e-Business.* London: John Wiley.

COAI. (2002, August 23). COAI News Bulletin 20. Retrieved from *http://www.coai.com/docs/nb20.pdf*

COAI. (2003). Cellular operators association India statistics. Retrieved from *http://www.coai.com/stats.2003.q1.htm*

De, R. (2001). SMS from Yahoo: Net profits not a myth. *Express Computer.* Retrieved from *http://www.expresscomputeronline.com/20011231/cover1.shtml*

Dhar, S. (2001). *Indian telecommunications liberalisation and development* (a report). Essar Comvision Limited, India.

Ericsson. (2001). Ericsson mobility world India and Bharti India's largest private communication services provider launch SMS project for students. *Ericsson Online.* Retrieved from *http://www.ericsson.com/mobility world/sub/articles/success_stories/india_bharti_launch_sms_project_ for_students?PU*

Expresscomputer. (2003). Dial M for m-commerce. Retrieved from *http://www.express computeronline.com/20030818/indtrend1.shtml*

Fraunholz, B., & Unnithan, C. (2004a). Critical success factors in mobile communications: A comparative roadmap for India and Germany. *International Journal of Mobile Communications, 2*, N1.

Fraunholz, B., & Unnithan, C. (2004b). SMS growth and diffusion: A preliminary investigation of three economies. *Proceedings of ISoneworld 2004*, Las Vegas.

Gartner. (2003). Gartner Press Room, Quick Statistics, Mobile Phones. Retrieved June 30, 2003, from *http://www.dataquest.com/press_gartner/quickstats/phone.html*

India Infoline (2001). Telecommunications. Indian Telecommunications Industries Sector Report. Retrieved from *http://www.indiainfoline.com/sect/tesp/ch01.html*

India Infoline. (2002, August 29). 85% of the cellular phones come from the grey market. Retrieved from *http://www.indiainfoline.com/nevi/cell.html*

Indiatelecommunicationsnews. (2003a). CDMA is better for India – Gartner, India Telecommunications News. Retrieved from *http://www.indiatelecommunications news.com/technology.htm*

Indiatelecommunicationsnews. (2003b). GSM way ahead-IDC, India Telecommunications News. Retrieved from *http://www.indiatelecommunicationsnews.com/technology.htm*

Jain, R. (2001). The telecommunications sector. India Infrastructure Report 2001, IIM Ahmedabad. Retrieved from *http://www.iimahd.ernet.in/ctps/iir8.pdf*

Jhunjunwala, & Ramamurthy, B. (2001). *Enabling telecommunications and Internet connectivity in small towns and rural India.* India Infrastructure Report 2001.

Karkera. (2002). Of fishermen, mobile phones and changing lifestyles. Rediff.com special report/George Type. Retrieved from *http://www.rediff.com/news/2002/aug/13spec.htm*

Keen, P.G.W., & Mackintosh, R. (2001). *Freedom economy: Gaining the mCommerce edge in the era of the wireless Internet.* CA: Osborne/McGRawHill.

Kripalani, M., & Clifford, M.L. (2000, February 21). Information technology is lifting the economy, and the politicians are backing it. *Business Week, Asian Edition.* Retrieved from *http://egov.mit.gov.in/*

Kshetri, N. B. (2001). Determinants of the locus of global e-commerce. *ElectronicMarkets, 11*(4), 250-257.

Ministry of Technology. (1999). *Annual Report of 1999-2000.* Ministry of Information Technology, Government of India.

Mobile Youth. (2002). Indian youth love affair with SMS condemned by cultural activists. Mobile Youth Online. Retrieved from *http://www.mobileyouth.org/news/mobileyouth629.html*

Moni, M., & Vijayaditya, N. (2002). Convergence and eGovernance: National informatics centre – An active catalyst and facilitator in India. Retrieved from *http://waterinfo.nic.in/news/egover_convergence.html*

Nair, S. (2002). *Governance and public management.* Strategy Paper.

Nasscom. (2002). IT industry. Retrieved from *http://www.nasscom.org/it_industry.asp*

Pereira, M. (2002). Traffic cops and SMS. *Online NIC.* Retrieved from *http://www.delhitrafficpolice.nic.in/art3.htm*

Ramachandran, T.V. (2002). The Indian cellular mobile sector - Activities and concerns. *Proceedings of the June 13th Conference in Bangkok.* Retrieved from *http://www.itu.int/ITU-D/pdf/4597-11.1-en.pdf*

Rediff.com. (2002). Indian mobile sector grows 4.8% in July. *Rediff.com report.* Retrieved from *http://www.rediff.com/money/2002/aug/09cell.htm*

Reuters. (2002). Indian mobile users to touch 120 million by 2008. *Telephony.* Retrieved from *http://031102.coverstory.telephonyonline.com/microsites/newsarticle.asp?mode=print&newsarticleid=2652994&releaseid=&srid=10750&magazineid=7&siteid=3*

Reuters. (2003). Indian rural market has huge potential – Telecommunications sector. *Reuters Online.* Retrieved from *URL:http://in.tech.yahoo.com/030306/137/21u37.html*

Rheingold, H. (1993). *The virtual community: Homesteading on the electronic frontier.* UK: Addison-Wesley.

Shankar, J. (2004). mcommerce new mantra for Indian software firms. Retrieved from *http://www.islamonline.net/iol-english/dowalia/techng-2000-june-21/techng3.asp*

Singh, D. (2002). Keynote address by Chief Minister of Madhya Pradesh. *Proceedings of the Second Roundtable on IT in Governance,* March 12, Hyatt Regency, New Delhi.

Sinha, S. (1997, July-September). The risks of financing telecommunications projects. Indian Institute of Management, Ahmedabad, Vikalpa. *The Journal of Decision Makers, 22*(3), 1- 15.

Splettstoesser, D., & Kimaro, F. (2000). Benefits of IT-based decision-making in developing countries. *The Electronic Journal on Information Systems in Developing Countries, 3,* 1-12.

Thomas, K. S. (2002). r u hooked? Communications. The Week. Retrieved from *http://www.the-week.com/22feb03/life9.htm*

Times. (2003). Chennai beats other cities in SMS usage. *The Economic Times.* Retrieved from *http://economictimes.indiatimes.com/cms.dll/html/comp/articleshow?artid=14872448*

Unnithan, C. (2002). eGovernance in India – Initiatives and drivers – A preliminary investigation. *Proceedings of 2nd European Conference on eGovernment,* October 1-2, Oxford University, UK

Xu, H., Teo, H.H., & Wang, H. (2003). Foundations of SMS commerce success: Lessons from SMS messaging and co-opetition. *Proceedings of the 36th Hawaii International Conference on System Sciences (HICSS 36),* January.

Chapter IX

Critical Success Factors For E-Brokerage: An Exploratory Study in the Brazilian Market

Allan Marcello De Campos Costa, Getulio Vargas Foundation, Brazil

Luiz Antonio Joia, Getulio Vargas Foundation, Brazil and
Rio de Janeiro State University, Brazil

Abstract

The scope of this chapter is to investigate the critical factors for the success of stock brokerage processes via the Web, using financial portals on the Brazilian Internet, from the perspective of the investor. The framework of the online stock trading system is presented in order to compare the traditional form of stock brokerage with that made possible by the Internet and discuss some of the issues regarding intermediation and disintermediation that has occurred in the process. Conclusions are drawn based on a survey conducted with Brazilian investors who operate using Internet stockbrokers for the intermediation of their stock operations. The findings were analyzed from the perspective of the theoretical reference framework created and outlined here. Critical aspects for the success of stock brokerage over the Internet, from the investor's standpoint, were assessed and presented, while areas for further research were also identified.

Introduction

The objective of this chapter is to develop an exploratory study where some of the Critical Success Factors (CSF) for stock brokerage over the Internet in the Brazilian market will be singled out. The findings described are based on a survey that was conducted with a group of investors who use Brazilian financial portals for trading stocks and are analyzed from the perspective of these investors, i.e., based on their perception of what they consider the critical factors for the success of the stock trading process over the Web.

After this introduction, the background for this work is presented by developing a contextual assessment of the situation where relevant factors in the Brazilian and worldwide stock markets in the context of this study are discussed. This is followed by the elaboration of a theoretical framework where the differences between traditional and online stock trading are defined and where intermediation and disintermediation concepts are discussed. The methodology used in the research is briefly discussed and the findings of the survey are presented. Finally, possible future trends in this area are listed , followed by the conclusions drawn from this study.

Background

Contextual Appraisal of the Situation

After the stabilization of the Brazilian economy in the mid-1990s, a new reality was presented to local investors. Before this, high inflation rates were a determining factor for giving preference to investment strategies involving low-risk assets, and the main strategy was to protect funds from being whittled away by inflation. However, a stable economy forced investors to seek out new investment alternatives and since then Brazilian investors began migrating to the stock market looking for higher returns (Fortuna, 1999). Halfeld (2000) illustrates this change emphatically in the investment strategies of the Brazilian investor, presenting a study developed by the Brazilian financial institute *Economática*. Using various timeframes, the study compares the gains obtained by a hypothetical Brazilian investor who had invested US$1.00 in the Brazilian stock market in a portfolio similar to that used in the IBOVESPA index (the index of the largest Brazilian stock exchange), and another person who hypothetically invested the same amount in a traditional savings account. The comparison can be seen in Graph 1 below and the results of the study clearly reveal the higher returns afforded by the stock market, notably in the long term.

During the same period, use of the Internet grew exponentially in Brazil and around the world and, with the fastest growing number of people connected to the network, one of the sectors which saw the most marked improvements in Brazil was the financial market. Halfeld (2000) states that due to progressive access to better services and lower costs

Graph 1. Gains on US$1 invested in the Brazilian stock market and in a savings account in different timeframes

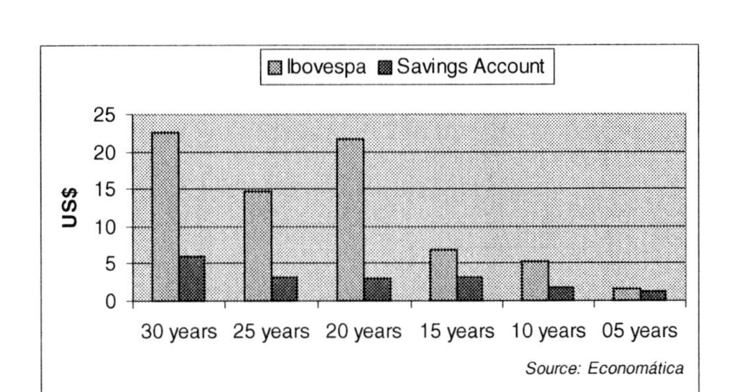

for transactions made over the Internet, in addition to swift IT implementation by the major consumer banks, a significant percentage of the population started operating bank accounts using the new channel. Indeed, as pointed out by Albertin (2001), the Internet is now part of the everyday life of a large section of the population.

Finally, an important event in the context of this research was the creation of the home-brokerage system in the major Brazilian stock exchange, the *Bolsa de Valores do Estado de São Paulo – BOVESPA*. Before the advent of home-brokerage, stock trading used to be conducted through brokerage houses that acted as the bridge between the investor and the stock market, and most of the transactions were made over a conventional phone line. After March 1999, the home-brokerage system enabled the investor to place trading orders directly over the Internet to a broker and follow it up on a real-time basis using the network (BOVESPA, 2001). These three factors together, namely the stabilization of the Brazilian economy, the rapid expansion of the Internet and the launch of the home-brokerage system on the Brazilian stock market were determining factors for the emergence of financial portals on the Brazilian Internet.

At the same time, the process of trading stocks over the Internet took off around the world. By way of example, Prabhudev et al. (2000) state that stockbrokers operating over the Internet in the North American market attracted a contingent of 12 million investors since the launch of this service in 1994. These investors were also responsible for 33% of the stocks traded in the US market in the year 2000. Barber and Odean (2001) present a study developed by Cerulli Associates according to which it is estimated that 42 million investment accounts will be opened by private investors with online stockbrokers in the USA. Fan, Stallaert and Whinston (2000) also discuss the expansion in online investment in the US based on projections made by Jupiter Communications, predicting that the online brokerage market will top 3 trillion US dollars by 2003.

The Social and Economic Context of Internet Users in Brazil

Brazil is a federation with three levels of government:

- **Central:** Federal Government
- **Intermediary:** 27 states, plus the Federal District of Brasília (the capital of Brazil)
- **Local:** more than 5,500 municipalities with the constitutional status of autonomous members

The Federation was not created on a base-upward premise, rather it was based on a perceived awareness of the need to divide the state. It is important to remember that Brazil is a country with no significant cultural frictions generated by differences of language, religion or race.

Most of the bibliography about Internet users addresses the OECD countries (OECD, 1999). Some developing countries, such as India, Brazil, Mexico, etc., use Information Technology in a very intensive way. This fact in itself may represent a highly optimistic perception of e-government initiatives. However, as these countries have large populations, absolute figures can lead the reader to misconceptions. If, for instance, one compares Canada and Brazil, it can be seen that while Brazil has almost the same number of Internet users as Canada, nearly 50% of the population of the latter is digitally included, whereas little more than 5% of the population of the former has access to the Internet.

The number of Internet users in Brazil is estimated at around 10 million. So, as already said, it represents a very small percentage of the 170 million total population of a country whose GDP was close to US$588 billion in 2000 (Afonso, 2001; Neri, 2003). According to Afonso (2001), fixed-line telephones are the predominant option for Internet access (62.5 million lines, of which 38.8 million are hard-wire and 23.7 are cellular phones). Individual services still serve only 39% of the population. Hence, reduced line access remains a limiting factor for Internet expansion, both for individual users and service providers. The high price of conventional telephone services represents another barrier in a country where the income per capita is around US$3,500 per year. Connections between the local and international backbones are still very expensive. Furthermore, ISPs do not offer local-line Internet connections in many small towns.

As may be imagined, the price of hardware is another obstacle to providing nationwide Internet access, as per-capita income in Brazil remains very low. Besides, the predominance of English-based content on the Web sites limits accessibility to a very small percentage of the Brazilian population, namely that which has the highest educational level. The lack of training, not only in accessing the Internet, but also in providing support to infrastructure as well as in developing new services and software is another constraint in Brazil.

Naturally, all of these issues create an environment that fosters digital inequality. Most of these inequalities are merely the consequence of longstanding social and economic disparities in Brazil. Nonetheless, according to the United Nations, Brazil was ranked 18th among the 190 UN members with respect to Electronic Government benchmarking. In 2001, Brazil achieved an index of 2.24 (higher than Italy – 2.21) within a maximum index score of 3.25. USA was ranked in first place with an index of 3.11 (Ronaghan, 2002, p. 7).

However, even with such difficulties in Internet availability and access, electronic commerce in Brazil shows healthy signals of constant and steady growth. The work of Albertin (2001) has shown that the degree of electronic commerce adoption within Brazilian companies can already be considered advanced, mainly with respect to the use of these technologies in relationship processes with clients.

An additional source for understanding the present stage of electronic commerce in the Brazilian market is the "Internet POP" report conducted regularly by the Brazilian IBOPE Research Institute. According to the latest report, published in the second quarter of 2004, the number of Brazilians connected to the Internet is still growing. However, what is even more important for the purposes of this chapter is the fact that 61% of the population with Internet access has been connected for more than two years (IBOPE, 2004), showing a high percentage of users that cannot be considered newcomers and therefore tend to be able to use more sophisticated services, such as stock brokerage over the Internet.

Intermediation and Disintermediation in Electronic Commerce

According to Evans and Wurster (2000), intermediaries usually make their profit based on a combination of excellence of service and client outreach. In the first aspect, profit is obtained on the basis of a differentiated service with a high level of excellence, usually superior to the level that could traditionally be delivered straight to the final consumer without intermediary participation. In the case of value enhancement based on outreach, the intermediaries exploit their capacity of reaching a higher number of customers to reap a profit.

With the growing use of new technologies by companies targeting the practice of electronic commerce, many argued that this would herald the end of intermediation (Carr, 2000). However, according to Evans and Wurster (2000), it is wiser to see these changes as leading to the emergence of a new kind of intermediary that can take advantage of a new form of disintermediation as shown in Figure 1 (based on Evans & Wurster, 2000, p. 70).

Evans and Wurster (2000) go further in explaining these ideas, arguing that in the history of disintermediation new players usually compete with the intermediaries already in the market by offering a wider reach to the detriment of excellence. Typically, different versions of the same product or service are offered, characterizing a proposition with different, though not necessarily higher, value.

In the new form of disintermediation proposed by the authors, a more radical event has occurred where technology makes it possible to offer both advantages, i.e., where competitors offer both wider reach and more facilities and resources for the customer.

Figure 1. New forms of disintermediation

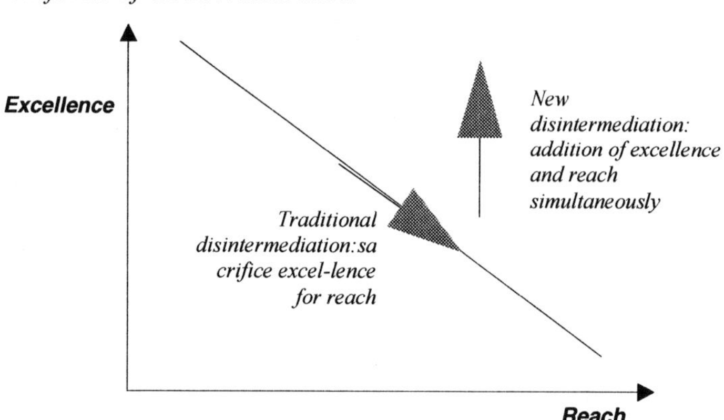

According to Evans and Wurster (2000), this is exactly what happened in the North American stock trading market. New players like Charles Schwab took advantage of the possibilities offered by the Internet and increased their reach while simultaneously delivering a full range of products and services to customers relating to stock trading processes that had not previously been available.

The Stock Market in Brazil

The stock trading process in the Brazilian market requires intermediation by brokerage houses that are responsible for taking orders from their clients and executing them in the market. These orders are cleared in what is known as the *Pregão*, which is the Brazilian expression to denominate the place where the agents of the brokerage houses congregate to execute the orders received from their clients (Pinheiro, 2001). Only registered agents have access to the *Pregão* and tables and cabins owned by the brokerage houses are distributed around it. Each of the houses transmits the orders to be executed to their agents in the *Pregão* and after receiving the orders the agents seek out the posts corresponding to each specific stock and try to execute them. This mechanism is represented in Figure 2.

Stock Trading in the Real World *vis-à-vis* via the Internet

Every stock investment process involves four steps: (i) trading order placed by the investor; (ii) routing of this order for execution; (iii) fixing of the price; and finally, (iv) execution of the order (Dasgupta & Dickinson, 1999). In the traditional brokerage model, the investor places a telephone call to the broker. The broker receives the order and

Figure 2. Example of a Pregão layout

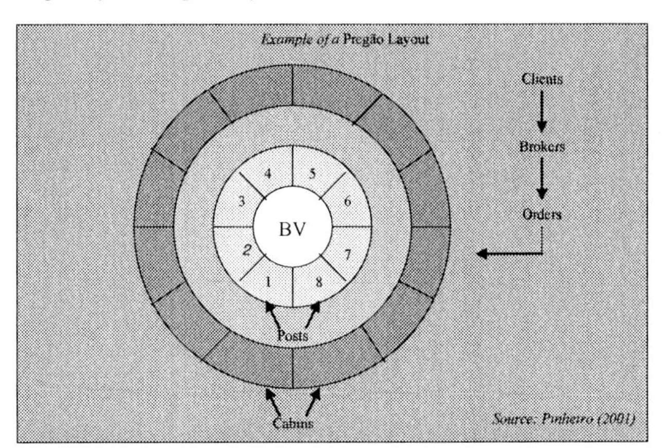

Figure 3. Trading process: Traditional brokerage model

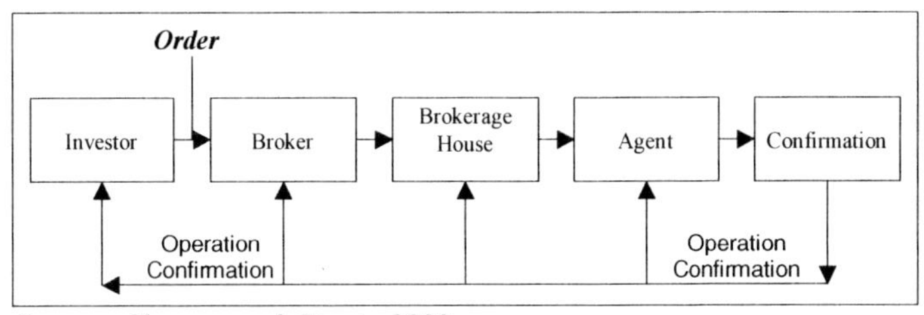

Source: Sharma and Bingi, 2000

transmits it to the brokerage house, which then routes the order through to the agent who will try to find the best offer for the execution of the order (Sharma & Bingi, 2000). Figure 3 illustrates the process.

The same transaction via the electronic system involves the following sequence described by Sharma and Bingi (2000) and presented in Figure 4 below: the investor places the trading order directly through his computer, the order is routed by the brokerage house's system directly to the agent who establishes the price and executes the order by finding the best possible offer.

The analysis and comparison of the two models presented shows that the operation conducted using the traditional model involves at least two additional intermediaries. The elimination of these two steps helps decrease the costs involved in the transaction and this reduction is usually passed on to the investor (Sharma & Bingi, 2000).

Besides the cost reduction, the online process also has other advantages for the investor. Cotsakos (2001) describes some of the things that can be done by the investor using online operations:

i. To analyze preferred stocks in any market, by any index;

ii. To analyze stock performance using graphic tools;

Figure 4. Trading process: Web-based brokerage model

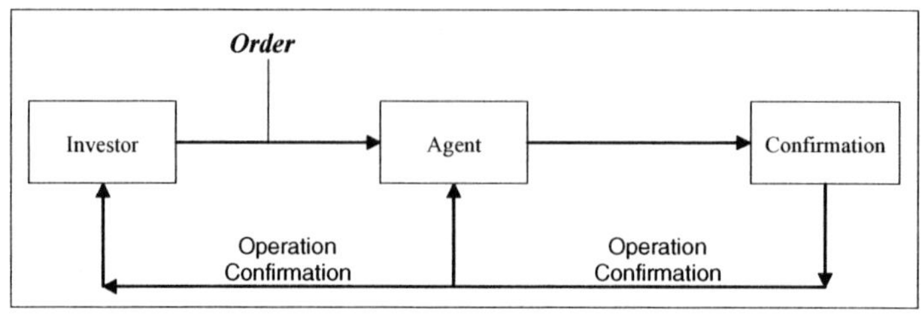

Source: Sharma and Bingi, 2000

iii. To compare prices and information of two or more stocks at the same time;

iv. To research the financial situation and historical profitability of a specific company or sector;

v. To access up-to-date reports generated by specialists.

Research Methodology

This work can be characterized as an exploratory study due to the scarcity of research and published material on the issue in question. The methodology was based on a survey conducted among investors on the Brazilian Internet applied through an online questionnaire published on the Web at the time of data collection. According to Blaxter, Hughes and Tight (2001, p. 77), "surveys are usually associated as a research approach with the idea of asking questions to groups of people." This would appear to be the most valid approach for this work, since its findings will be based on investors' opinions about Brazilian online stockbrokers.

The population parameters for the research can be seen as the group of private individuals who have used Internet portals for trading stock online. Due to the difficulty of identifying these individuals, a non-probabilistic approach was used to select a sample, based on definitions by Rea and Parker (2000). This approach can be seen as having a lower degree of scientific rigor, however, it is also a very useful tool for the researcher. This statement is based on the fact that this kind of non-probabilistic approach can quickly generate a preliminary comprehension of the key questions in a research venture, besides allowing researchers to refine their research instruments (Rea & Parker, 2000).

The non-probabilistic sampling strategy used was that defined by Rea and Parker (2000) as snowball sampling. They describe this technique as being particularly useful when it is hard for the researcher to identify potential targets. After the identification of some targets, these people are then asked to indicate other individuals who might answer the questionnaire and this process is used until the sample is complete. Another useful definition of this technique is given by Kumar (1999, p. 162), according to whom "snowball sampling is the process of selecting a sample using networks."

The sample size was defined based on statistical criteria, with a 95% confidence level and a tolerated error of 5%. Based on Rea and Parker (2000, p. 128), the minimum sample size for a survey with these characteristics is 97 individuals, therefore, this is the minimum number of individuals who were interviewed during the data collection phase.

As stated before, the objective of this research is to investigate the Critical Success Factors (CSF) for the stock trading process over the Internet from the perspective of the investor. The main questions this research intends to answer are:

- What are the CSFs in the selection of an electronic stockbroker by the investor?
- What are the CSFs with respect to the quality and quantity of information available on electronic stockbrokers for the investor?
- What are the CSFs regarding the resources available for investors on electronic stockbrokers?
- What are the CSFs relating to the user-friendliness of the tools available on the electronic stockbrokers?
- What are the CSFs with respect to the alternatives offered to investors to establish and maintain their relationship with electronic stockbrokers?
- What are the CSFs regarding investor concerns about security?
- What are the CSFs regarding the costs involved in the trading process for the investor?

Data Collection and Findings

The survey was conducted with a sample of 100 individuals in the first semester of 2002. Data collected was treated with version 2.09 Sphinx Lexica statistical software, and analysis was conducted leading to the conclusions presented in this section.

The findings listed are presented and based entirely on the survey conducted with the sample described above. In a general way, these findings did not possess any relationship with other variables relating to investor profile (e.g., prior experience in the stock market, gender or time operating through the home-brokerage system), since the potential links were investigated through the use of cross tabulations and statistical tests and no relationships could be identified. Therefore, it is feasible to conclude that the findings presented here can be attributed to the whole sample independent of any specific attributes.

The questionnaire was designed with three objectives in mind. Firstly, to identify the profile of the investors included in the sample. Secondly, based on a single multiple-answer questionnaire, to ask the investors directly about the critical factors for selecting an online stockbroker in their opinion. Thirdly, to conduct further investigation based on the use of a summated rating scale, or Likert scale, to ascertain investor perception of different variables and aspects. The scale used was a five-point, three-directional categorical scale, where the variables were rated as “unimportant,” “low importance,” “indifferent,” “important,” and “very important.” For the selection of the Critical Success Factors, the criteria adopted was to consider as critical the factors which were classified as important or very important by more than 70% of the respondents.

Investor Profile

The investor profile analysis produced the following characteristics:

- Forty-five percent of the sample invested on the stock market for over a year, with the other 55% having less experience with this kind of investment.
- The home-brokerage system would seem to be an important factor in the attractiveness of the stock market to new investors, since 61% of the investors in the survey started their operations with this type of investment after creation of this system.
- The sample can be divided almost in the same proportion between small and large investors, as shown by the percentage of 49% investing amounts lower than R$10,000 (approximately 4,000 Euro at the date of the survey).
- With respect to their objectives, 54% of the investors in the sample had a medium/long term horizon for their investments against 46% who could be considered short-term investors and speculators.
- Sixty-seven percent of the individuals had a portfolio with no more than five different companies, with this proportion rising to 95% if we include individuals whose portfolio had up to 15 companies. This revealed a tendency for low diversification in the Brazilian investor’s portfolio.
- A very significant percentile (74%) of the sample relied on information provided by stockbrokers to obtain information about the market and investment opportunities. However, other online stockbrokers of which they were not clients (66%), financial newspapers (55%) and general sites on the Internet (41%) also appeared with marked frequency.

Critical Success Factors in Selection of an Electronic Stockbroker

In order to determine the critical factors in the selection process of an online stockbroker, a multiple choice questionnaire was included where the following factors could be rated as relevant or otherwise by the investors:

1. Image and credibility of the online broker
2. Transaction security

Graph 2. Critical success factors in selection of an electronic stockbroker

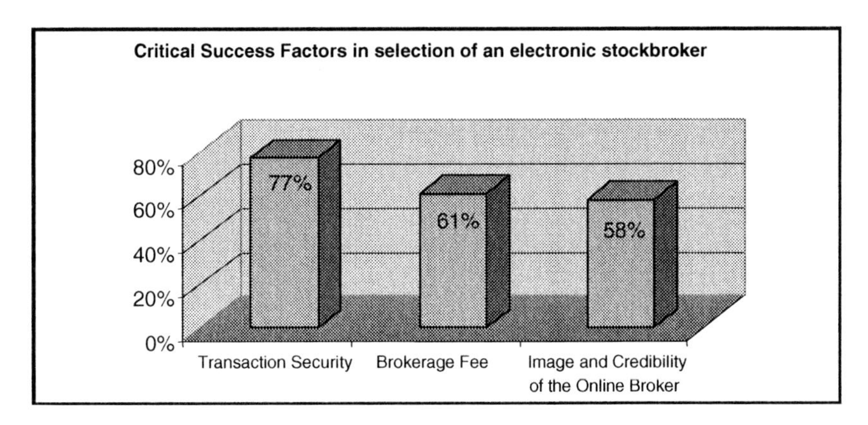

3. Level of information available on the online broker
4. Response time of the online brokerage system
5. Brokerage fees
6. Protection of investor privacy
7. Relationship channels available to contact the online broker
8. Offer of free Internet access for clients
9. Association of the online broker with a large retailing bank

According to the survey, three items were considered more relevant, as can be seen in Graph 2.

Critical Success Factors with Respect to Information Available on the Electronic Stockbroker

The survey demonstrated that the Brazilian investor tends not to attribute a high level of importance to the quantity or quality of information available on Brazilian online stockbrokers. Among the variables investigated shown in Table 1 none of them was ranked as being critical from the perspective of the investors.

The only aspect that scored highly was that regarding the quality of information available, albeit the score was not sufficient for it to be considered a critical factor.

Table 1. Evaluation of the critical success factors for the online brokerage process with respect to information available on the electronic stockbroker

Variable	% of positive evaluation
Quality of information available about investments through online stockbrokers	65%
Quantity of information available about investments through online stockbrokers	47%
Availability of online courses and tutorials for self-tuition	45%
Availability of games and simulations for practice and learning about the home-brokerage mechanism	42%
Availability of online courses with attendance by a consultant	28%

Table 2. Evaluation of the critical success factors for the online brokerage process with respect to resources available from the electronic stockbroker for the investor

Variable	% of positive evaluation
Active monitoring of stock prices with warnings sent to the investor according to pre-defined parameters	79%
Availability of reports analyzing the stock market generated by independent consultants or consulting firms	73%
Availability of graphic information about the stock market generated by the online broker	71%
Availability of graphic information about the stock market generated by independent consultants or consulting firms	69%
Availability of reports analyzing the stock market generated by the online broker	64%
Periodic newsletters sent by e-mail with stock prices and market information	62%
Recommendations about specific stocks and strategies elaborated by independent consultants or consulting firms	57%
Recommendations about specific stocks and strategies elaborated on by the online broker	51%
Availability of reports about the macro-economic scenario generated by the online broker	49%
Availability of reports about the macro-economic scenario generated by independent consultants or consulting firms	49%
Availability of tools to elaborate simulations and projections about return on investment	44%

Critical Success Factors with Respect to Resources Available from the Electronic Stockbroker for the Investor

This was the group where more variables were investigated and, consequently, the section where more critical factors could be identified. The variables investigated with respect to resources available from the online broker are shown in Table 2.

Graphs 3, 4 and 5 show the level of importance attributed to the variables considered critical.

Graph 3. Importance of the availability of an active monitoring service

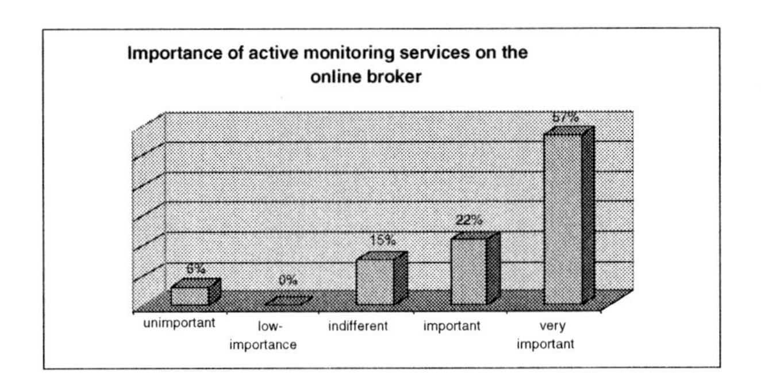

Graph 4. Importance of the availability of reports analyzing the stock market generated by independent consultants or consulting firms

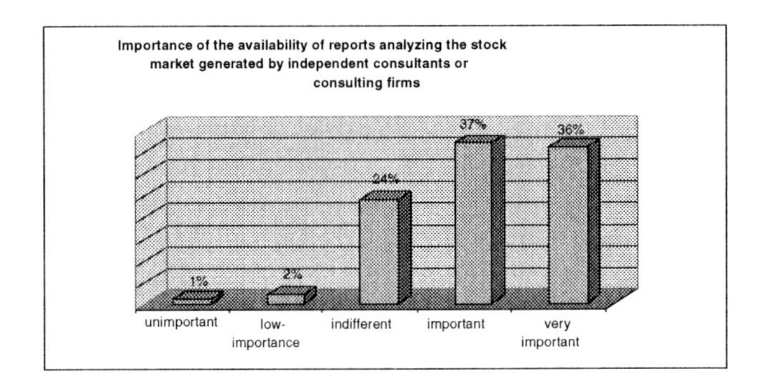

Graph 5. Importance of the availability of graphical information about the stock market generated by the electronic stockbroker

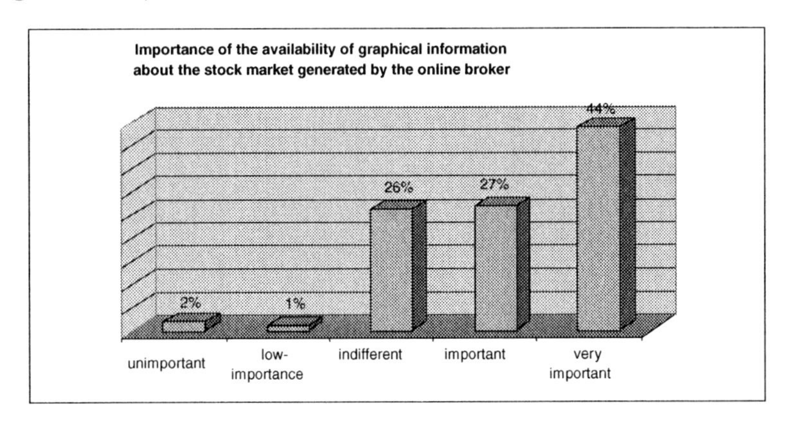

Table 3. Evaluation of the critical success factors for the online brokerage process with respect to the user-friendliness of the tools on the electronic broker's site

Variable	% of positive evaluation
Response time of the electronic broker's Web site	89%
Availability of an interactive and user-friendly layout on the electronic broker's Web site	82%
Availability of a simple and intuitive interface on the electronic broker's Web site	82%

Critical Success Factors with Respect to the User-Friendliness of the Tools Available at the Electronic Stockbrokers

In this section, all the variables were classified as critical by the investors who took part in the survey, as can be seen in Table 3.

Besides the fact that all the variables were marked as critical, this section was also the factor that received the highest evaluation by the respondents. The response time of the electronic broker's Web site was considered the single most critical factor. What is important to note is that user-friendliness is not related to the experience of the user, as one might suppose (more experienced users could rank this aspect as less important than less experienced ones) since, as explained before, there are no dependences between any of these variables.

The individual values for each of the variables in this section can be seen in Graphs 6, 7 and 8.

Graph 6 . Importance of the response time of the electronic broker's Web site

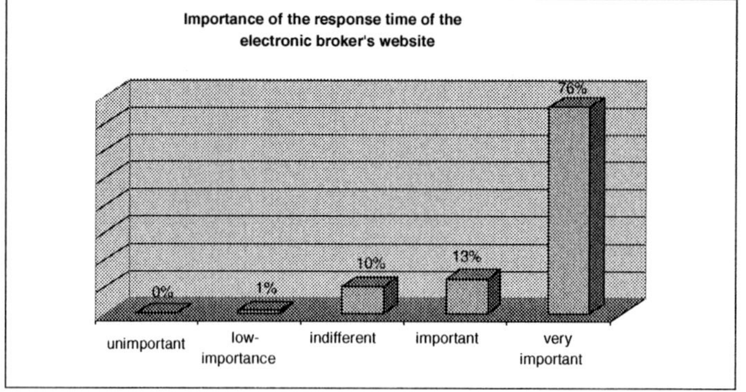

Graph 7. Importance of the availability of an interactive and user-friendly layout on the electronic broker's Web site

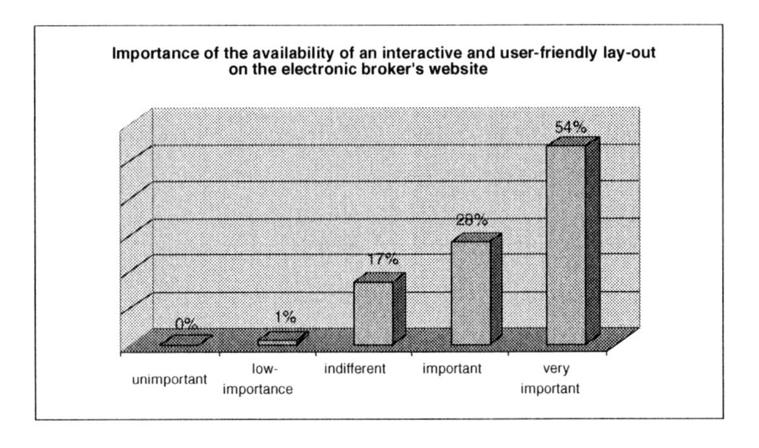

Graph 8. Importance of the availability of a simple and intuitive interface on the electronic broker's Web site

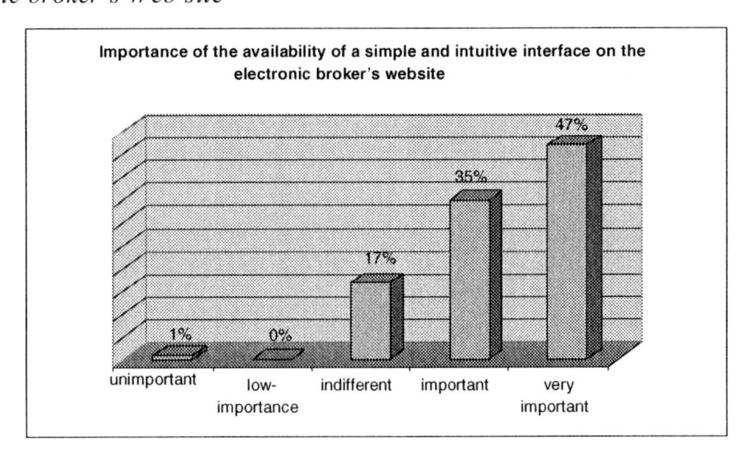

Critical Success Factors with Respect to the Alternatives for Establishing a Relationship with the Electronic Stockbroker

Four variables were investigated in this section, as presented in Table 4. The only variable ranked critical was the "availability of services on a 24-7 basis." This fact tallies with one of the most important benefits for the consumer made possible by the Internet, namely affording the consumer the option of buying and making transactions 24 hours a day, from virtually anywhere (Turban et al., 2000). The specific values for this critical factor are shown in Graph 9.

Table 4. Evaluation of the critical success factors for the online brokerage process with respect to the relationship possibilities with the electronic stockbroker

Variable	% of positive evaluation
Importance of the availability of services on a 24-7 basis	79%
Availability of consultants for interaction by telephone	55%
Availability of consultants for interaction via "chat service"	42%
Existence of investor communities supported by the online broker for exchange of ideas and opinions about investments and the market	41%

Graph 9. Relative importance of the availability of services on a 24-7 basis

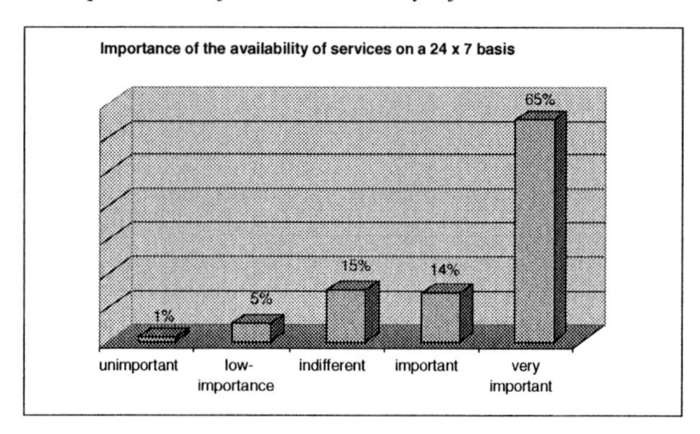

Critical Success Factors with Respect to the Security Provided by the Online Stockbroker

This group of factors was ranked as the single most important factor in the process of choosing an online stockbroker and a further investigation was conducted with the variables related to this issue confirms this investor perception. As presented in Table 5, 85% of the respondents classified the security certification of the electronic broker's Web site as an important or very important aspect, while 63% of these observations ranked it very important (Graph 10). This highlights the importance of taking security issues as a fundamental concern on the part of the electronic brokers.

Critical Success Factors with Respect to Brokerage Fees

Brokerage fees also appeared at the top of the list of critical factors for choosing online brokers, where "brokerage fees" received a very high ranking as "important" or "very important," as shown in Table 6 and in Graph 11.

Table 5. Evaluation of the critical success factors for the online Brokerage Process with respect to the security provided by the electronic stockbroker

Variable	% of positive evaluation
Certification of the electronic broker's Web site, with respect to its security infrastructure, by some independent institution (e.g., Verisign)	85%
Association of the online broker with a large retailing bank	60%

Graph 10. Relative importance of the availability of certification of the electronic broker's Web site, with respect to its security infrastructure, by some independent institution (e.g., Verisign)

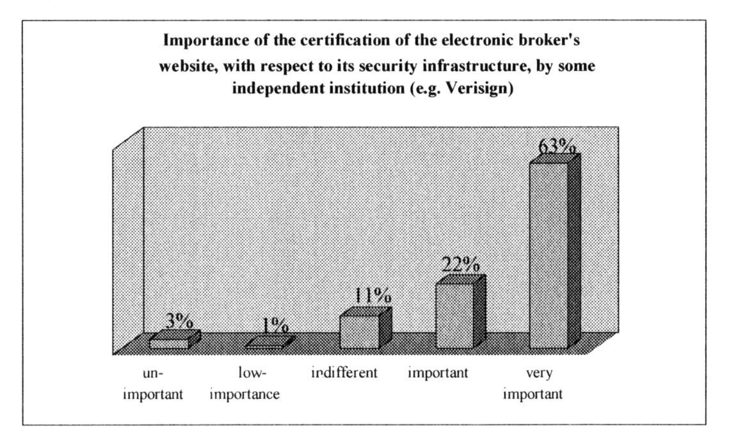

Table 6. Evaluation of the critical success factors for the online brokerage process with respect to brokerage fees

Variable	% of positive evaluation
Brokerage fees	**87%**
Availability of a free-tool line for contacting the online broker	66%
Offer of free Internet access for clients of the online broker	37%

Graph 11. Importance of brokerage fees

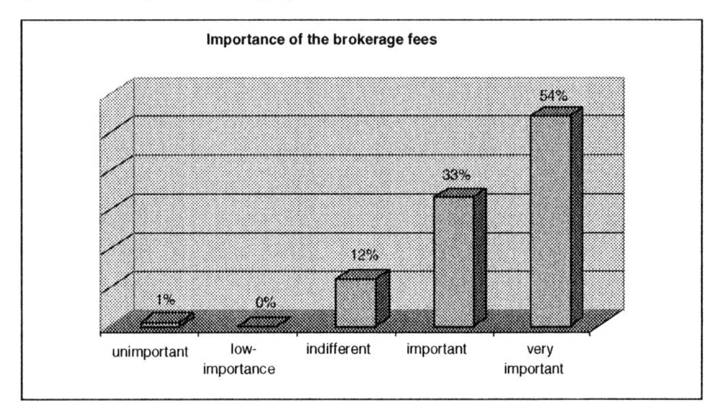

Future Trends

The research conducted in conjunction with the analysis of the data collected reveals some interesting reflections regarding future trends for the e-brokerage process, especially in developing countries like Brazil. Maybe the most relevant trend observed relates to the need for the investor to become better acquainted with this new form of investment. The factors perceived as more critical by the investors reinforce the perception that they are still getting accustomed to operating through online brokers on the Brazilian market. This statement is based on the fact that most of the Critical Success Factors identified in the survey reflected concern with fundamental questions (e.g., security and price), to the detriment of more sophisticated services. If on the one hand we may establish that the success of an electronic stockbroker may depend on how professionally and efficiently the simplest issues are handled, on the other hand it should be remembered that this is still a market at an evolutionary stage. This tendency can be confirmed by the information in Table 2, where the "active monitoring service of the stock prices with warnings being sent to the investor according to pre-determined parameters" was considered important or very important by 79% of the respondents. Therefore, as the market keeps evolving, the relative importance of more sophisticated services also tends to increase. This will certainly put some pressure on e-brokers to find ways to differentiate their service from that of their competitors.

Furthermore, this study also seeks to contribute by giving directions for additional research that will be required to continue exploring the issues discussed here. Further research is recommended in other countries and markets, especially in developing countries, where online stockbrokers are operating in order to generate the compilation of broader knowledge about online investor profiles and their needs. Similarly, segmented investigation should be conducted to investigate additional links between the factors considered critical and variables not considered in this study, such as geographical location or investor income.

Conclusion

As stated before, this work should be seen as an exploratory study. Therefore, its findings seek to raise some issues related to the topic of "online stock brokerage on the Internet" rather than claim to be conclusive. However, some relevant observations can be seen from the results of the survey. The groups of factors where some of the critical factors investigated were considered as being more critical were those related to user-friendliness of the broker Web site, security concerns and brokerage fees, as presented in Table 7. These findings corroborate the findings of the second section of the survey, where the investors were directly asked about the critical factors in the process of choosing an online stockbroker, presented in Graph 2.

The findings of this research also reveal that investors are still getting accustomed to the e-brokerage process and, consequently, their concerns at this juncture are mainly

Table 7. Most important critical factors for the success of the stock brokerage process via the Internet from the investor's perspective

Group	Critical Success Factors	Positive Evaluation
User-friendliness	Response time of the broker Web site	89%
User-friendliness	Use of an interactive and user-friendly layout on the electronic broker's Web site	82%
User-friendliness	Use of a simple and intuitive interface on the electronic broker's Web site	82%
Security	Certification of the electronic broker's Web site by independent organizations	85%
Brokerage fees	Brokerage fees	87%

related to essential services usually made available by the e-brokers. However, this should not be interpreted by e-brokers as being a consolidated situation, since it is possible to detect that some investors have already started to focus on more sophisticated services. Therefore, it is essential for the continued success of electronic brokers to monitor perceived investor needs continuously, since the behavior of this sector tends to develop as the e-brokerage process becomes more mature in the environment of developing countries.

References

Afonso, J.R. (2001). e-Government in Brazil: Experiences and perspectives. *Forum of Federations*, April, Montreal, Canada. Retrieved October 9, 2003, from *http://federativo.bndes.gov.br/destaques/egov/egov_estudos.htm*

Albertin, A.L. (2000). *Evolução do comércio eletrônico no mercado Brasileiro*. EAESP-FGV.

Albertin, A.L. (2001). *Pesquisa FGV comércio eletrônico no mercado Brasileiro*. EAESP-FGV.

Barber, B., & Odean, T. (2001). The Internet and the investor. *Journal of Economic Perspectives*, *43*(1), 41-54.

Barber, B., & Odean, T. (2001, February). Boys will be boys: Gender, overconfidence, and common stock investment. *The Quarterly Journal of Economics*.

Barber, B., & Odean, T. (2000). *Online investors: Do the slow die first?* Working paper, University of California, Davis.

Blaxter, L., Hughes, C., & Tight, M. (2001). *How to research* (2nd ed.). Buckingham: Open University.

BOVESPA. (2001). Porque seus sonhos precisam de ação. Retrieved 08/05/2002, from *http://www.bovespa.com.br*

Byron, C. (2001). *Deleteyourbroker.com: Using the Internet to beat the pros on Wall Street.* New York: Simon & Schustler.

Carr, N. (2000, January). The future of commerce. *Harvard Business Review,* 63-78.

Cotsakos, C.M. (2000). *It's your money: The E*TRADE step-by-step guide to online investing.* New York: HarperCollins.

Credit Suisse First Boston. (2000, June Quarter). On-line brokerage quarterly roundup.

Dasgupta, S., & Dickinson, F. (1998). Electronic contracting in online stock trading. *Electronic Markets, 8*(3), 20-22.

Evans, P., & Wurster, T. (2000). *Blown to bits: How the new economics of information transforms strategy.* Boston: Harvard Business School.

Fan, M., Stallaert, J., & Whinston, A.B. (2000). The Internet and the future of financial markets. Association for Computing Machinery. *Communications of the ACM, 43*(11), 82-88.

Fortuna, E. (1999). *Mercado financeiro: Produtos e serviços.* Rio de Janeiro, RJ: Qualitymark.

Halfeld, M. (2001). *Investimentos: Como administrar melhor seu dinheiro.* São Paulo, SP: Editora Fundamento.

IBOPE. (2004). 16o. Internet pop – Consolidado Nacional. Retrieved September 27, 2004, from *http://www.netstore.com.br/ibope/prodvar.asp?codigo_produto=0002*

Kumar, R. (1999). *Research methodology: A step-by-step guide for beginners.* London: Sage.

Lameira, V. (2000). *Mercado de capitais.* Rio de Janeiro, RJ: Forense Universitária.

Modahl, M. (2000). *Now or never: How companies must change today to win the battle for Internet consumers.* New York: Harper Business.

Neri, M. (2003). Mapa da exclusão digital, Centro de Políticas Sociais, EPGE/FGV, Rio de Janeiro. Retrieved October 9, 2003, from *http://epge.fgv.br/portal/pesquisa/livros/2003.html*

OECD. (1999). Report on impact of the emerging information society on the policy development process and democratic quality. Retrieved October 9, 2003, from *http://www.olis.oecd.org/olis/1998doc.nsf/LinkTo/PUMA(98)15*

Pinheiro, J. (2001). *Mercado de capitais: Fundamentos e técnicas.* São Paulo, SP: Editora Atlas.

Porter, M. (2001, March). Strategy and the Internet. *Harvard Business Review,* 63-78.

Prabhudev, K., Menon, N., & Balasubramanian, S. (2000, January). The implications of online investing. Association for Computing Machinery. *Communications of the ACM, 43*(1), 35-41.

Rea, L.M., & Parker, R.A. (2000). *Metodologia de Pesquisa: do Planejamento à Execução.* São Paulo, SP: Pioneira.

Ronaghan, S.A. (2002). Benchmarking e-government: A global perspective – Assessing the UN member states, United Nations – Division for Public Economics and Public

Administration & ASPA – American Society for Public Administration. Retrieved October 9, 2003, from *http://www.unpan.org/e-government/Benchmarking%20E-gov%202001.pdf*

Sharma, M., & Bingi, P. (2000, Spring). The growth of Web-based investment. *Information Systems Management, 17*(2), 58-64.

Siegel, J. (1998). *Stocks for the long run.* New York: McGraw Hill.

Turban, E., Lee, J., King, D., & Chung, H.M. (2000). *Electronic commerce: A managerial perspective.* Upper Saddle River, NJ: Prentice Hall.

Chapter X

E-Readiness and Successful E-Commerce Diffusion in Developing Countries: Results from a Cluster Analysis

Alemayehu Molla, The University of Manchester, UK

Abstract

This chapter explores the impact of organizational and environmental e-readiness factors on the successful diffusion of e-commerce. It presents a conceptual framework of e-readiness and e-commerce success, identifies the constructs and variables to operationalize the framework and reports the e-readiness and e-commerce success assessments of a sample of business organizations from South Africa. Business managers and policy makers, by understanding the e-readiness factors affecting e-commerce diffusion can make effective decisions to exploit specific e-commerce opportunities. Furthermore, researchers can use the developed framework to study the level of e-commerce adoption, the success of the process that led to it and the benefits to users, organizations and the economy.

Introduction

The volume and value of transactions on the Internet are still on the rise and are predicted to continue unabated for the foreseeable future. Likewise, national and international institutions are commissioning strategies to facilitate the conduct of e-commerce. However, global e-commerce diffusion is uneven. Particularly, the participation level of developing countries is still insignificant. The relatively slow e-commerce diffusion in developing countries is mainly attributed to their lack of e-readiness to transcend technological, legal, financial, business and social obstacles. Hence, understanding the relationship between e-readiness and successful e-commerce diffusion is highly important.

Although existing e-readiness literature does help to identify macro level inhibitors of e-commerce diffusion, the drivers of successful e-commerce diffusion at a firm level have not been well studied. In particular, what is missing from the existing literature is: (1) a relevant framework underlining the relationship between e-readiness and successful e-commerce diffusion in developing countries, and (2) an empirical assessment that explicates such relationships. This chapter aims to address these gaps.

Background

Of late, the notion of e-readiness has gained popularity in the e-commerce in developing countries literature. However, the concept of readiness can be traced to prior literature in organizational change, information systems (IS), business process reengineering (BPR) and innovation (Todd, 1999; Raymond et al., 1998; Jay & Smith, 1996; Guha et al., 1997; Clark & Cavanaugh, 1997; Grover et al., 1995; Stoddard & Jarvenpaa, 1995).

Most consider readiness as a necessary precondition (or set of pre-conditions) for the implementation of change or information systems or BPR projects. A few (Raymond et al., 1998; Guha et al., 1997) extend this to relate readiness to the success of such projects, but stop short of explicitly defining the nature of the relationship. Others (Clark & Cavanaugh, 1997) consider readiness as an indicator of the agility of a business and a goal an organization must strive to achieve.

From the literature, it can be surmised that having resources such as skilled human resources, leadership, technology (Grover et al., 1995; Stoddard and Jarvenpaa, 1995), appropriate organizational culture, capabilities and learning (Todd, 1999; Guha et al., 1997) and overall organizational commitment in the form of management and administrative support, staff involvement and championship (Jay and Smith, 1996) are some of the variables used to assess readiness at a firm level. Despite these indicators, what the literature does not provide us is a consolidated and systematic framework that explicitly defines the concept of readiness, sufficiently addresses its different constructs and clearly explains its relationship to success.

On the other hand, a number of studies have emerged discussing e-readiness, however defined. In particular, the e-readiness of developing countries to exploit the potential of e-commerce has attracted a number of initiatives and studies. Depending on the level of analysis the sources referred to—national vs. organizational—the existing literature on e-readiness can likewise be classified into two main categories.

The majority of e-readiness studies have a national focus. For instance, in one of the pioneering works, OECD (1999) considered "e-commerce readiness" as the first stage in its three-stage model of e-commerce maturity. McConnell and WITSA (2000, 2001), on the other hand, considered the e-readiness of a nation as a source of national economic growth and a prerequisite for successful e-business. Others (APEC, 2000; CID/HU 2000) highlighted e-readiness assessment as a mechanism for determining a nation's (and particularly that of developing nations) capacity for e-commerce and as a tool for guiding strategic planning processes for developing e-commerce. Table 1 provides an indicative, but not necessarily exhaustive, summary of previous e-readiness studies with a national focus.

Evaluation of studies summarized in Table 1 indicates duplication of efforts and proliferation of tools (Bridges, 2002a, 2002b; Choucri et al., 2003). For instance, within a period of two years, developing countries such as Egypt, South Africa, India, have been

Table 1. Summary of national level e-readiness studies

Author and Source♦	**Variables**
APEC (Asian Pacific Economic Cooperation) www.ecommerce.gov/apec	Basic infrastructure and technology, access to network services; use of the Internet, promotion and facilitation, skills and human resources; positioning for the digital economy.
CID/HU (The Center for International Development at Harvard and IBM) http://www.cid.harvard.edu/ciditg	This guide measures 19 different categories covering network access, network learning, network society, network economy, and network policy.
CSPP (Computer Systems Policy Project) http://206.183.2.91/projects/readiness/	Rate communities on infrastructure access, applications and services, economy; and "enablers" (policy, privacy, security, ubiquity).
The Economist Intelligence Unit http://www.ebusinessforum.com/index.asp	Gauges countries "e-business environment" and network "connectivity" based on 70 different indicators such as the strength of the economy, the outlook for political stability, the regulatory climate, taxation policies and openness to trade and investment.
KAM (World Bank, Knowledge Assessment Matrix http://www1.worldbank.org/gdln/kam.htm	Uses 61 metrics to assess the economic and institutional regime, educated and skilled population, dynamic information infrastructure, and an efficient innovation system of firms, research centers, universities, and consultants.
McConnell and WITSA http://www.mcconnellinternational.com/ereadiness/default.cfm	It assesses connectivity, e-leadership, information security, human capital, and e-business climate.
M-N (Metric-Net E-Economy Index) www.metricnet.com	Statistics on country's technological sophistication and strength using metrics of knowledge jobs, globalization, economic dynamism and competition, transformation to a digital economy and technological innovation capacity.
MQ (Mosaic) http://som.csudh.edu/fac/lpress/gdiff/	Assesses the diffusion of Internet in terms of, pervasiveness, geographic dispersion, usage within the economy, technology infrastructure, the Internet service market, and sophistication of use.
CIDCM (University of Maryland, Center for International Development and Conflict Management) http://www.bsos.umd.edu/cidcm/projects/leland.htm	The method gauges the background and history, key players in Internet development, Internet development and ICT policy over time, and negotiations of a country.

assessed for their e-readiness more than seven times. While at the surface each e-readiness assessment appears to measure the extent to which a developing nation's environment is conducive and nurturing for the information economy in general and e-commerce in particular, there is a clear lack of consensus among the studies. Practically, each assessment differs from the other in terms of its conceptualization of e-readiness, its goal, the e-readiness being measured, results, standards and beneficiaries (Bridges, 2002a). As a result, Choucri et al. (2003) characterize these studies as "first-generation" e-readiness studies. They also argue that first-generation studies lack the rigor and focus to guide governments and businesses to exploit specific e-commerce opportunities. In general, from the existing studies of e-readiness with a national focus, it is difficult to obtain a clear and theory-based understanding of e-readiness and its impact on various facets of e-commerce at a firm level.

At the organizational level, a number of studies have suggested their own conceptualization of e-readiness and suffused it with a list of indicators (see Table 2 for a summary). Of the studies summarized in Table 2, a few have also attempted to empirically investigate *organizational e-readiness* as a construct affecting various facets of e-commerce but mostly that of e-commerce adoption (McKay et al., 2001; Tabor, 2000; Van Akkeren & Cavaye, 1999). Although these studies addressed a limited repertoire of variables, interestingly, their findings indicate that what each of the studies constructed as *organizational e-readiness* significantly affects intention to adopt e-commerce (McKay et al., 2001). It is also identified as both facilitator (when present) and inhibitor (when missing) of e-commerce adoption (Van Akkeren and Cavaye, 1999) and differentiates adopters from non-adopters (Tabor, 2000).

To summarize, previous literature provides us with some background information on the concept of readiness in general and e-readiness in particular. It also provides support to the hypothesis that e-readiness has a significant impact on e-commerce success. From the review, it can also be learned that e-commerce readiness can be assessed at different levels such as national and organizational. Some of the variables that need to be considered in investigating e-readiness and e-commerce success are also highlighted. The review also indicates that e-commerce readiness is considered mostly as a necessary condition to e-commerce and its success.

However, some limitations that motivated our study should also be noted. First, most of the e-readiness studies in developing countries focus on macro-level indicators. These studies have limited power in explaining how a country's level of e-readiness affects individual businesses' decision to adopt e-commerce. Second, almost all of the studies that investigated e-readiness at organizational level (Table 2) are based on Western business organizations. These studies don't consider environmental factors and hence their findings cannot be generalized to businesses in developing countries. Third, existing e-readiness studies lack clear theoretical foundation.

Table 2. Summary of the literature on organizational e-readiness

Reference	Methodology	Constructs	Remark/key finding
Hartman et al. (2000)	Non-empirical	Leadership, governance, competencies and technology	Net readiness is a critical quality required to execute in e-commerce successfully. Successful companies demonstrate the four "determinants" of readiness, which jointly determine the success of any company in the e-conomy.
Pricewaterhouse Coopers (2000)	Non-empirical	Strategy, competence, processes, performance, systems, delivery operations, tax, legal, security	E-readiness assessment is used to help companies identify their e-business capabilities and map it to e-commerce maturity.
Morath (2000)	Non-empirical	Beliefs, values, strengths and offerings, partnership, diversification, staff	Organizations (both start ups and established ones) have to "know themselves" and conduct an assessment of their capability in terms of the checklists. Knowing oneself is one of the three strategic considerations to successfully run e-commerce.
Wilson (1999)	Non-empirical	Commitment, content, systems *and* developers readiness	Readiness assessment is used to uncover information to pinpoint hurdles the organization might face in its e-commerce engagement. Readiness is considered as a continuous process and not as a single stage.
Mckay et al. (2001)	Survey	Management support, IT & financial resources, perceived risk	Management support has the greatest influence on SMEs' intention to adopt e-commerce.
Van Akkeren & Cavaye (1999)	Multiple case study	Currently used level of technology	Organizational readiness and lack of it are rated as chief facilitating and inhibiting factors respectively.
Tabor (2000)	Survey	IT, Strategy	Aggressive strategy is a stronger driver for electronic commerce.

E-Readiness and E-Commerce Success Framework

For businesses in the developing world, migrating to e-commerce represents an undertaking of innovation. Success in e-commerce innovation in part depends on these businesses making necessary changes in terms of their organizational structure and product characteristics. It also entails changes in business practices and culture related to how transactions are identified and fulfilled and relationships are formed and consolidated (Odedra-Straub, 2003).

Because business transformation to e-commerce constitutes an innovation, innovation research can be used as a theoretical framework to underpin e-readiness and its impact on successful e-commerce diffusion (Wigand, 1997). Generally, innovation literature highlights that successful diffusion of an innovation is affected by the dynamic interaction within and between organizational and environmental forces (Rogers, 1995). It then follows that understanding successful e-commerce diffusion in developing countries requires careful attention to both organizational and environmental contexts. Hence, we argue that mangers' assessment of these contexts of e-commerce can give us constructs indicating the *organizational and environmental profile of e-readiness* (Molla & Licker, 2002; Molla, 2002).

On the basis of our emphasis on established businesses in developing countries, the findings of the literature review, various discussions we have undertaken with academic and practitioner experts on IT and e-commerce in developing countries, we identify six attributes to capture the organizational e-readiness profile and three attributes to capture the environmental e-readiness profile. The six attributes of the organizational e-readiness profile are: *awareness, commitment, human resources, technological resources, business resources and governance.* Under environmental e-readiness, we identify: *market forces e-readiness, supporting industries e-readiness and government e-readiness*

On the other hand, assessing how successfully e-commerce is diffused should cover the complete cycle of the innovation diffusion process. There are several models that capture this process (Zaltman et al., 1973; Rogers, 1995). These pioneer works do not address post-implementation evaluation of the innovation. Straub (1997) extended previous models and argued that innovation diffusion passes through several phases that evolve from evaluation through adoption to use and finally productivity and benefits. Therefore, successful e-commerce diffusion can be assessed as a four-facet construct evaluating the execution of the four phases of innovation diffusion. An assessment of these facets captures an *e-commerce success profile* of a given organization. Likewise, we suggest four facets of e-commerce success covering the e-commerce diffusion process. These include: *success of adoption, development, deployment and benefit.*

Figure 1 captures a visual impression of the framework while the variables under the three constructs are defined in Table 3. The main research propositions to be explicated are:

- Organizational e-readiness explains a significant part of the differences in e-commerce success
- Environmental e-readiness explains a significant part of the differences in e-commerce success

Research Methods and Data Analysis Procedure

The data reported in this study is extracted from a survey conducted in South Africa. Respondents were asked to express their degree of agreement for each of the items used

Figure 1. E-readiness and e-commerce success framework

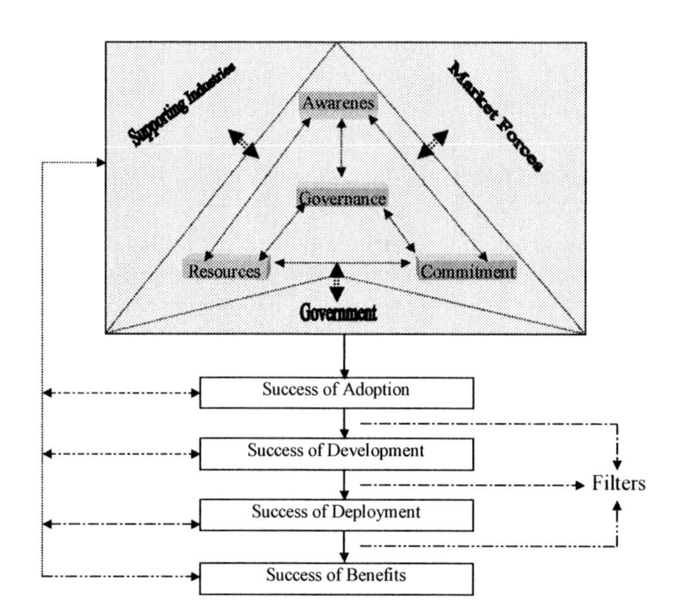

to measure the research variables on a five-point Likert scale (1=strongly agree to 5=strongly disagree). The instrument used for data collection has been rigorously tested for validity and reliability (Molla, 2002). The measures for the success of deployment didn't pass the psychometric tests and, therefore, this dimension has been excluded from further analysis. The survey covered 150 businesses. Sixty-four percent of the responses were obtained from managing directors, or their equivalent, and the rest were from directors of e-commerce, finance and information technology. Further details of the survey are discussed in Molla and Licker (2004).

Data are analyzed mainly using a cluster analysis technique. Cluster analysis represents a set of classification algorithms that are normally applied to organize observed data into meaningful structures (Hair et al., 1995). It allows classifying cases/variables into homogenous groups and developing taxonomies. A number of clustering algorithms have been applied in a wide variety of research problems ranging from natural sciences to social sciences. In general, though, the clusters identified by cluster analysis demonstrate two properties - *internal cohesion and external isolation* (Hair et al., 1995). While internal cohesion implies that cases/variables in the same cluster are similar to one another, external isolation implies that the cases/variables in one cluster are distinct from the cases/variables in another cluster.

We applied the K-means clustering algorithm to construct and compare the e-readiness and e-commerce success profiles of the businesses. K-means was used because it is consistent to our hypothesis concerning the number of clusters. That is, we were

Table 3. Description of research variables

Variables	Description
Awareness	Represents an organization's assessment of e-commerce, e-commerce initiatives of partners, competitors and government; comprehension of their meaning through an understanding of e-commerce technologies, business models, benefits and threats and projection of future trends of e-commerce and its impact on business.
Commitment	Reflects enough energy and support for e-commerce from all corners of an organization and especially from the strategic apex. It refers to having a clear-cut e-commerce vision and strategy championed by top management, e-commerce leadership and organization wide support for e-commerce ideas and projects.
Human Resources	Refers to the availability (accessibility) of employees with adequate experience in information and communications technology (ICT) and other skills (such as marketing, business strategy) that are needed to adequately staff e-commerce initiatives and projects.
Technological Resources	Refers to the ICT base of an organization at the time of assessment. It assesses the extent of computerization, the flexibility of existing systems and experiences with network-based applications.
Business Resources	This covers a wide range of capabilities and most of the intangible assets of the organization. It includes the openness of organizational communication; risk taking behaviour, existing business relationships, and funding to finance e-commerce projects.
Governance	The strategic, tactical and operational model that organizations in developing countries put in place to govern their business activities and e-commerce initiatives.
Government e-readiness	An organisation's assessment of the preparation of a nation state in terms of government commitment and the legal infrastructure to promote, support, facilitate and regulate e-commerce.
Market forces e-readiness	The assessment that an organization's business partners such as customers and suppliers allow an electronic conduct of business.
Supporting Industries e-readiness	Refers to the assessment of the presence, development, service level and cost structure of support giving institutions such as telecommunications, financial, trust enablers, and the IT industry, whose activities might affect the e-commerce initiatives of businesses.
Success of Adoption	This shows the level of sophistication of e-commerce in a given organization on a six-phase, e-commerce status indicator, i.e., no e-commerce, connected e-commerce, static e-commerce, interactive e-commerce, transactive e-commerce and integrated e-commerce.
Success of Development	This is an assessment of whether or not e-commerce projects have been completed within budget and on time.
Success of Deployment	Deployment refers to the level of e-commerce use by the intended users and for the intended purpose.
Success of Benefits	Measures the extent of perceived e-commerce benefits in terms of costs reduction, communication improvement, market reach, and improved relationship with suppliers and partners.

Source: Molla (2002)

interested to construct the clusters on the basis of the sophistication level of e-commerce adoption. The data preparation process for submitting to the STATISTICA cluster analysis statistical package involved developing a matrix consisting of cases (respondents) and variables (responses captured on the basis of the Likert scale). To assess how distinct the clusters are, we examined the magnitude of the F value and the between-group and within-group variability values. In addition to the above, the impact of ecological variables such as business size and sector on the e-readiness and e-commerce success profiles are explicated. Where applicable, ANOVA tests are used to test the significance of the observed patterns.

E-Readiness and E-Commerce Success Profile Analysis

To facilitate the cluster analysis, the success of adoption (alias e-commerce status) indicators of the businesses was used. Based on the K-means clustering algorithm, five clusters are produced (*Between SS 146; Within SS 26; F 169.88*). The members of each of the five clusters are crosschecked with the original data and there is a 100% match between the members of each cluster and their observed current e-commerce status. This classification can be accepted as "good" as the within cluster variability (within SS) is smaller than the between cluster variability. Figures 2, 3 and 4 show the organizational e-readiness, external e-readiness and the e-commerce success profiles (success of development and benefits) of each of the five clusters respectively.

Examination of Figures 2, 3 and 4 reveals that, in general, there appears to be a direct correspondence among organizational e-readiness, external e-readiness and e-commerce

Figure 2. Organizational e-readiness profile

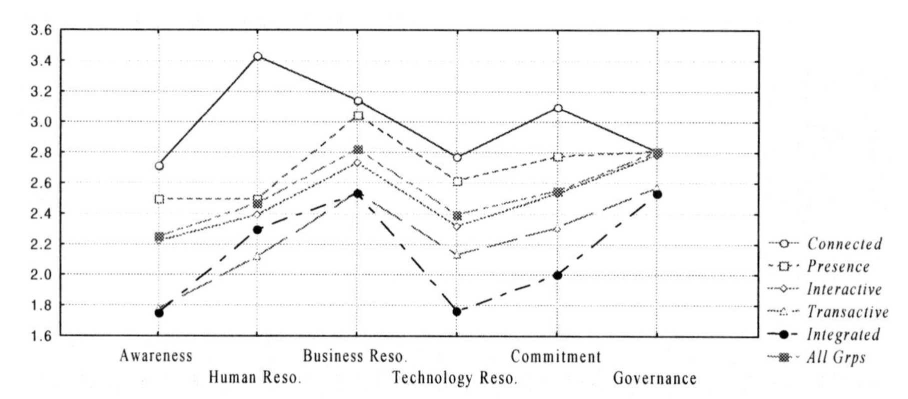

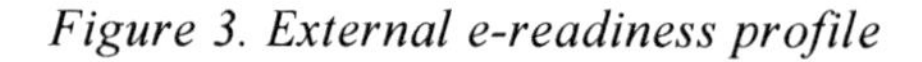
Figure 3. External e-readiness profile

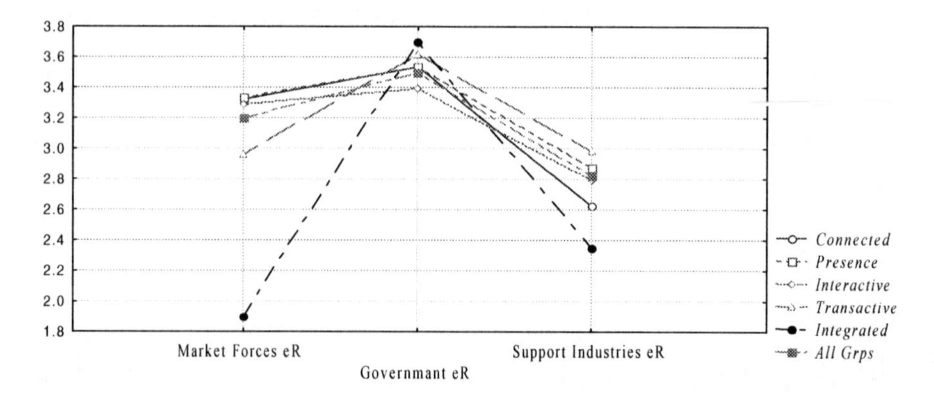

Figure 4. Development and benefits success profile

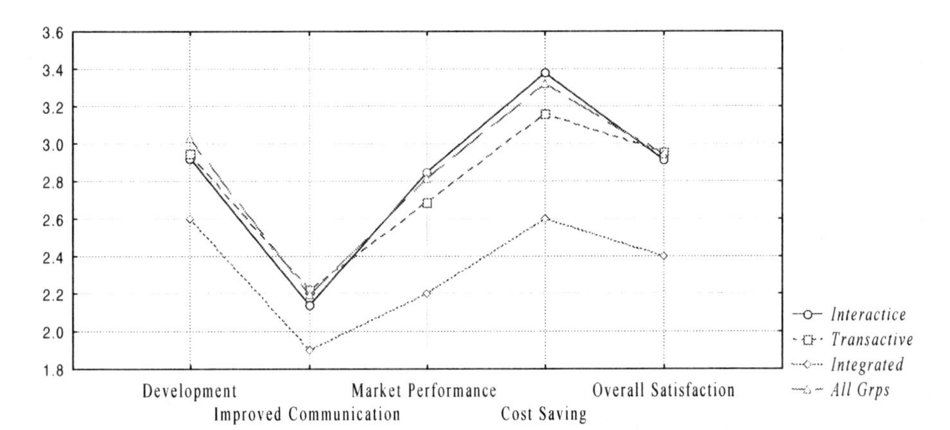

success. Businesses that demonstrate a relatively better profile of e-readiness (both organizational and external) have achieved either a transactive or integrated e-commerce status and also reported to have obtained better benefits from their e-commerce engagements.

In addition, a uniform pattern has emerged among the five clusters in terms of organizational e-readiness. That is, while organizations in each cluster appear to have a better profile of awareness and possess (have access to) technological resources, they appear to lack business resources and governance models to manage their e-commerce activities. In terms of external e-readiness, businesses that perceived that their market forces and supporting industries demonstrate better e-readiness appear to have implemented more sophisticated e-commerce applications than is the case otherwise. Overall, across all clusters, the government's e-readiness received the lowest assessment. Comparison of the organizational and external e-readiness profile shows that businesses tend to rate (perceive) their own e-readiness better than that of their environment.

The statistical significance of the above observation is further tested through ANOVA and multifactor ANOVA. The result (Table 4) indicates that e-commerce e-readiness (composite score of organizational and external e-readiness) explains a significant proportion of the variation in e-commerce success profile (Table 4, rows 1 & 2). A further test reveals that organizational rather than external e-readiness is a significant predictor of the observed differences in the success of e-commerce adoption *(current effect: $F(3,142)=8.78$, $p=0.00002$)*, development and market performance (Table 4, rows 3 & 4). However, among the external e-readiness variables, the difference in the market forces e-readiness (*current effect: $F(4, 135)=3.04$, $p=.02$)* appears to have more effect on the difference in the sophistication level of e-commerce adoption.

We further tested the impact of e-commerce status on success of benefits using ANOVA. The result (Table 4, rows 5 and 6) indicates that e-commerce status appears to have some impact on market performance and cost saving advantages. In general, the analysis hitherto outlines that rather than environmental, organizational factors appear to have

Table 4. Results of ANOVA tests

Predictor	Dependent variable	Wilks Lambda	R^2	SS	DF	F	P
E-readiness	Success of e-commerce adoption		0.18	24.48	7	4.399	0.001
	Success of development & benefits	0.63			15	2.281	0.006
Organizational e-readiness	Success of development		0.14	10.27	5	2.318	0.050
	Market performance		0.19	10.15	5	3.477	0.007
Success of adoption	Market performance		0.12	8.58	5	3.222	0.027
	Cost savings		0.10	5.35	3	2.766	0.047
Size	E-readiness	0.78			9	2.170	0.035
	E-readiness w/o human resources	0.83			8	2.070	0.057
	E-commerce success	0.87			5	2.150	0.069
	Success of development			5.43		6.140	0.002
Sector	E-readiness	0.27			90	1.024	0.428
	E-commerce success	0.49			50	1.029	0.426

greater influence on the success of adoption, development and benefits. The following few pages provide the impact of ecological factors such as business size and sector on the e-readiness and e-commerce success of the organizations.

Impact of Business Size and Sector on E-Readiness and E-Commerce Success

Statistics South Africa and the National Small Business Act 102 of 1996 classify businesses with 50 or less full time employee size (FTES) as small, 50 to 100 FTES (maximum of 200 in mining, manufacturing and construction) as medium and the rest as large. On the basis of this classification, 60% of the respondents were large while the remaining 40% were small and medium. In order to assess if business size has any effect on e-readiness, the e-readiness profile of the businesses by size is plotted (Figure 5). Examination of Figure 5 indicates a more or less comparable pattern of e-readiness except for human and business resources.

The one-way ANOVA with size as an independent predictor over e-readiness variables (Table 4, row 7) confirm the above observation that human and business resources appear to be sensitive to differences in business size. Inspection of figure 5 indicates that in both cases, small and medium enterprises (SMEs) appear to show a relatively better profile of human and business resources. Considering the measures that operationalize these two variables, this shouldn't be surprising. Because of their size, small businesses can have the wherewithal to give access to computers and the Internet to most of their employees. In addition, SMEs tend to be less bureaucratic and this might facilitate the

Figure 5. E-readiness profile of businesses by size

initiation and assimilation of changes provided that they manage to address the challenges of attracting and retaining essential ICT and e-commerce specialists. After removing the effects of the human resources variable, the ANOVA test was repeated and the result (Table 4, row 8) indicates no statistically significant impact of size on the e-readiness profile of the organizations. Therefore, it can be concluded that, overall, both SMEs and large businesses demonstrate a comparable profile of e-readiness when controlling for human resources variable.

The e-commerce success profile of the SMEs and large enterprises (Figure 6) indicates that the SMEs report a slightly better achievement from their e-commerce endeavors compared to their larger counterparts. The ANOVA test (Table 4, row 9) shows no

Figure 6. E-commerce success profile by size

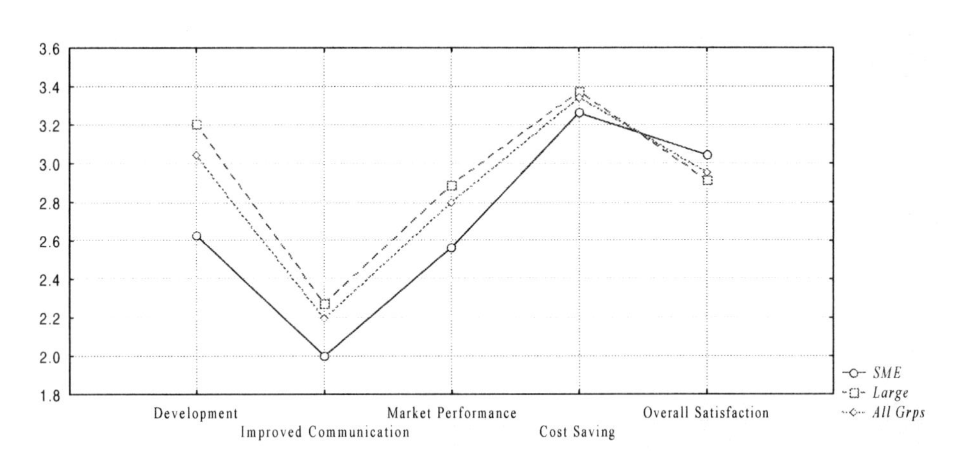

Figure 7. E-readiness profile by sector

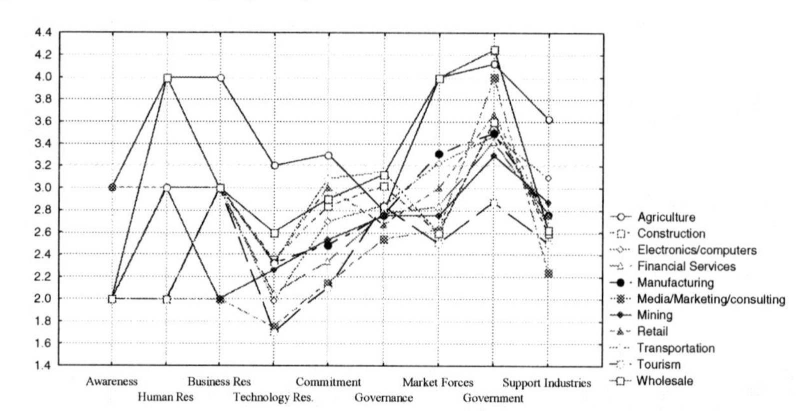

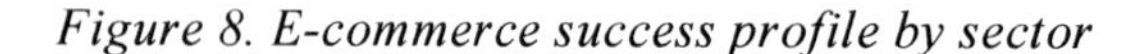

Figure 8. E-commerce success profile by sector

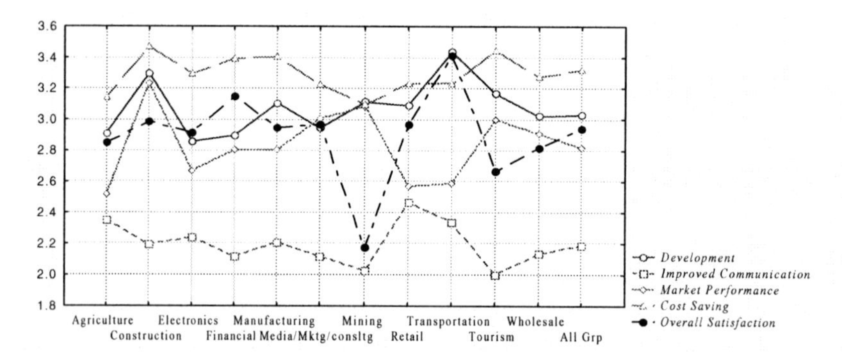

statistically significant impact of size on the e-commerce success profile of the businesses. However, size seems to have some effect on the success of e-commerce projects' delivery (Table 4, row 10) such that SMEs have reported a relatively within-budget and within-time execution of e-commerce developments. This result has to be interpreted with due consideration to the unequal distribution of the businesses sampled in the two groups.

In addition to size, the impact of sector on e-readiness and e-commerce success was assessed (Figures 7 and 8). The agriculture, retail and wholesale sectors appear to demonstrate a relatively weak profile of e-readiness, but tourism, manufacturing, electronics and computing show a stronger profile. In addition, there doesn't appear to be much difference in terms of e-commerce success because of sector. Likewise, the ANOVA test reveals no statistically significant impact of sector on e-readiness and success (Table 4, rows 11 and 12).

Discussion and Trend

Our findings demonstrate that while most of the sampled organizations appear to have developed better e-commerce awareness in terms of an understanding of the opportunities, perils, benefits, and impacts of e-commerce and monitoring the e-commerce initiatives of their competitors and partners, they do not appear to have developed an organizational structure with clear roles, responsibilities, accountabilities as well as the authority to manage e-commerce activities and change issues, that is, governance.

Better e-commerce awareness can be expected to lead to advantages in entry-level e-commerce adoption (Ang et al., 2003). Nonetheless, the lack of continuous commitment and governance model might affect how fast and how far an organization can successfully progress towards developing advanced e-commerce capability. This is essential because real cost and market benefits from e-commerce appear to go to those that have succeeded in integrating e-commerce to their other operations (Marshall et al., 2000). Another observation is that whilst the surveyed organizations reveal better profiles in terms of technological resources (such as connectivity, interactivity and flexibility), they tend to be cautious in their approach to e-commerce, perhaps preferring a wait-and-see strategy. They also appear to lack the degree of openness and aggressiveness that most in the e-commerce literature recommend. The relatively weak business resources profile attests to this and might explain why e-commerce use in most of the businesses is mainly limited to communicational rather than transactional activities. This finding is consistent with the findings of Le and Koh (2002) and Palacios (2003).

In terms of the external e-readiness, South Africa's relatively affluent IT and financial industries appear to have a bearing on the better assessment that the e-readiness of the supporting industries received compared to the rest. For instance, successive yearly ranking of countries in terms of their network readiness index (NRI) by the Global IT Report (GITR) has placed South Africa, with above the median GITR's ranking, among the top of the developing world's league (Dutta et al., 2004). This might also provide part of the explanation for the better technological capabilities the organizations demonstrated. On the other hand, contrary to visible government initiatives such as the e-commerce discussion, white and green papers, infocom 2020, etc., the legal and regulatory frameworks and the government's commitment received the least rating. One would expect this to materially affect the e-commerce activities of the organizations (Oxley & Bernard, 2001). A somewhat surprising finding related to the government's e-readiness is that it is ranked the least by respondents with e-commerce status higher up in the maturity ladder. A possible explanation could be that legal and regulatory frameworks have become more important as organizations embark on more sophisticated e-commerce operations.

The empirical results are generally supportive of the research propositions that e-readiness (both organizational and external) has a significant impact on e-commerce success. In terms of the success of e-commerce adoption, the findings indicate that initial adoption of e-commerce is significantly influenced by awareness and resources, as well as market forces readiness. However, once the organization makes the initial decision to adopt e-commerce, the institutionalization of e-commerce (i.e., what business functions and activities are to be supported and how far along the path of e-commerce the organization progresses) depends on the commitment and governance model the organization puts in place and the external e-readiness factors, namely, market forces, government and support industries. This finding is consistent to the findings of Tarafdar and Vaidya (2004) where they have discovered a combination of organizational and environmental imperatives affecting e-commerce adoption among Indian organizations.

Mansell (2001) argues that developing countries' businesses that managed to address the organizational, institutional and technological challenges of e-commerce are likely to achieve the perceived benefits of e-commerce. Pare (2002) also stresses that organizational core capabilities are critical for realizing efficiency gains promised by e-commerce. Our findings appear to reinforce the above. Organizational e-readiness factors, most notably governance and commitment (but also resources), have emerged as major contributors to the success of e-commerce development and market place benefits. The analysis also shows that while the e-readiness profile of businesses in developing countries affects the success of adoption, the sophistication of adoption in turn has a significant impact on e-commerce benefits. This tallies with Zhu and Kramer's (2002) work which correlates e-commerce capability in terms of information, transaction and interaction richness to better e-commerce value.

E-commerce benefits are, however, largely contained to improvements to internal and external communications followed by some gains in the market performance. Although cost reduction (of information gathering, transaction) is touted to represent perhaps the most obvious and immediate benefit of e-commerce to developing countries, our findings don't produce evidence to support such claims. To explore this further, we have looked at the nature of e-commerce technologies and the business functions performed electronically. The results indicate that the majority of the organizations in the survey have reported to be using Internet-based networks to conduct communications internally (79%) and externally with customers (57%) and partners (57%) and not for transactional purposes as such. Another explanation lies perhaps with the current e-commerce maturity level of the organizations, suggesting that cost-saving benefits might not appear with entry-level e-commerce activities. The e-commerce experiences of developed countries (NNI, 1999; Poon & Swatman, 1999; Young & Benamati, 2000) support this explanation.

An interesting finding is also the lack of materially significant variation of e-readiness and e-commerce success profiles because of differences in business size and sector. While some (Eze & Seong, 2001; Mckay et al., 2000; Van Akkeren & Cavaye, 1999) argue that small businesses and businesses in non-electronic sectors might lack e-readiness and tend to come lower on the maturity stage, the data here does not appear to support such an hypothesis. In addition, our results contradict Selhofer and Mentrap's (2004) conclusion, which is based on a survey of European Union businesses, that size and sector are main determinants of how businesses use ICTs. This might be because of regional differences. It might also be due to the use of employee numbers as a measure of size and other measures such as revenue might produce a different result. Therefore, further studies with larger sample size are essential before making any generalizations. Nevertheless, this finding suggests an exciting direction for future theoretical and empirical research, that is, examining the conditions under which business size and sector play important roles in affecting e-readiness and e-commerce success.

Overall, as developing countries improve their technological and other infrastructure, e-commerce diffusion is increasingly likely to be affected by organization specific variables such as commitment, business resources and governance (Ang et al., 2003; Tigre, 2003; Humphrey et al., 2003; Hempel & Kwong, 2001). For instance, some (if not many) countries have now idle telecommunications capacity. As a result, the negative weight of telecommunications development on e-commerce diffusion has been lifted. This, however, should not be interpreted as undermining the relevance of the environmental context. Rather, it highlights the importance of firm-specific issues and calls for departing from the environmental determinism view that appears to dominate most studies of e-commerce in developing countries.

Conclusion

This study identified nine key variables, classified into two constructs - organizational and external e-readiness - that are hypothesized to explain an organization's ability to succeed in e-commerce. Taken together the nine attributes underpin the concept of e-readiness. The study also defined three key facets of e-commerce success: adoption, development and benefits. Taken together these three facets also underpin the concept of e-commerce success. Different organizations might demonstrate different combinations of the nine attributes of e-readiness and the three facets of e-commerce Success. These define their e-readiness and e-commerce success profiles respectively.

There are several implications of this current work and its findings. In terms of managerial contribution, business and IT mangers can apply the framework to analyze the competency and capabilities of their businesses to exploit specific e-commerce opportunities within their macro-economic, social and technological environment. Most organizations in developing countries are often faced with "organizational inertia" that deters their innovativeness. Assessment of their awareness, commitment, resources and governance can enable one to identify the sources of such inertia and can facilitate ways of dealing with them. In addition, the framework helps to reasonably and meaningfully assess the business environment by the decision makers themselves rather than the "one-size-fits-all" assessment of existing e-readiness studies. Further, the framework helps to evaluate the level of e-commerce adoption, the success of the process that led to such a level of adoption and the nature and extent of benefits that organizations obtained from their e-commerce projects and initiatives.

Some preliminary conclusions and interpretations that need to be tested in the future can also be made from the findings. Businesses that demonstrate a relatively better profile of e-readiness in terms of resources and governance have progressed well on the e-commerce maturity continuum and have a relatively better profile of development and benefits success. In particular, the difference in the organizational e-readiness profiles is materially significant between businesses that are just connected to the Internet and those that have developed sophisticated e-commerce capabilities. Likewise, small and entry-level e-commerce activities (and investments) do not appear to be enough for realizing e-commerce benefits. Therefore, businesses need to realize that e-commerce may not materialize immediately or even in the short term and require a long-term commitment and investment.

The result also suggests possibilities of significant early and "fast–mover" advantages. Insomuch as the e-commerce success profile and e-commerce maturity of the organizations are related to e-readiness, this will give businesses with better e-readiness a competitive advantage. Persaud (*Economist,* 2000) cautions that early mover advantage in e-commerce might have a "freezing" effect on the late entrants. Of course this will materialize only if first-mover advantages outweigh first-mover disadvantages. The lesson for business mangers in developing countries, especially those that belong to established value chains, is that lack of e-commerce capabilities might put them at a competitive disadvantage and expose them to the dangers of being frozen out of global commodity chains.

In summary, the findings here represent an example of the status of e-commerce and the organizational and environmental factors that affect its successful diffusion. We expect interested researchers to replicate the current study in other developing countries' contexts and/or improve the conceptual and theoretical foundation of the framework and/ or the evidence base.

References

Ang, C., Tahar, R. M., & Murat, R. (2003). An empirical study on electronic commerce diffusion in the Malaysian shipping industry. *Electronic Journal of Information Systems in Developing Countries*, 14(1), 1-9.

APEC-Asia Pacific Economic Cooperation. (1999). *SME Electronic Commerce Study: Final Report*. Retrieved April 26, 2004, from *http://apec.pwcglobal.com.*

Bridges. (2002a). *Comparison of e-Readiness assessment models*. Retrieved April 1, 2004, from *www.bridges.org*

Bridges. (2002b). *e-Readiness assessment: Who is doing what and where?* Retrieved April 1, 2004, from *www.bridges.org*

Burgelmen, R. A., & Maidique, M. A. (1988). *Strategic management of technology and innovation.* IL: IRWIN.

Choucri, N., Maugis, V., Madnick, S., & Siegel, M. (2003). Global e-Readiness: For what? Retrieved April 1, 2004, from *http://ebusiness.mit.edu*

CID/HU (Center for International Development at Harvard University). (2000). *Readiness for the networked world: A guide for developing countries*. Retrieved September 21, 2000, from *www.cid.harvard.edu/ciditg*

Clark, C. E., & Cavanaugh, N. C. (1997). Building change-readiness capabilities in the IS organization: Insights from the Bell Atlantic experience. *MIS Quarterly, 21*(4), 425-456.

Dutta, S., Lanvin, B., & Paua, F. (Eds.) (2004). *The global information technology report 2003-2004: Towards an equitable information society*. World Economic Forum.

The Economist. (2000). The new economy: Untangling e-conomics. September 23, 5-44.

Evans, P., & Wurster, T. S. (2000). *Blown to bits: How the new economics of information transforms strategy*. Boston: Harvard Business School.

Eze, U. C., & Seong, K. T. (2001). The influence of enterprise structure and strategy on the level of e-commerce deployment in Singapore enterprises. *Journal of Global Information Technology Management*, *4*(2), 38-54.

Grover, V., Jeong, S. R., Kettinger, W. J., & Teng, J. T. C. (1995). The implementation of business process reengineering. *Journal of Management Information Systems*, *12*(1), 109-144.

Guha, S., Varun, G., Kettinger, W. J., & Teng, J. T. C. (1997). Business process change and organizational performance: Exploring an antecedent model. *Journal of Management Information Systems*, *14*(1), 119-154.

Hair, J. F., Anderson, R. E., Tatham, R. L., & Black, W. C. (1995). *Multivariate data analysis with readings* (4th ed.). Englewood Cliffs, NJ: Prentice-Hall.

Hartman, A., Sifonis, J., & Kador, J. (2000). *Net ready: Strategies for success in the e-conomy*. New York: Mcgraw-Hill.

Hempel, P.S., & Kwong, Y., K., (2001). B2B e-commerce in emerging economies: i-metal.com's non-ferrous metals exchange in China. *Journal of Strategic Information Systems*, *10*, 335-355.

Humphrey, J., Mansell, R., Pare, D., & Schmitz, H. (2003). *The reality of e-commerce with developing countries*. London: Media @LSE.

Jay, K. C., & Smith, D. C. (1996). A generic change model for the effective implementation of information systems. *South African Journal of Business Management*, *27*(3), 65-70.

Le, T. T., & Koh, A. C. (2002). A managerial perspective on electronic commerce development in Malaysia. *Electronic Commerce Research*, *2*(1-2), 7-29.

Mansell, R. (2001). Digital opportunities and the missing link for developing countries. *Oxford Review of Economic Policy*, *17*, 282–295.

Marshall, P., Sor, R., & Mckay, J. (2000). An industry case study of the impacts of electronic commerce on car dealership in western Australia. *Journal of Electronic Commerce Research*, *1*(1), 1-16.

McConnell & WITSA. (2000). *Risk e-business: Seizing the opportunity of global e-readiness*. August. Retrieved April 1, 2004, from *www.witsa.org/papers/*

McConnell & WITSA. (2001). *Ready? Net. Go! Partnerships leading the global economy*. Retrieved April 1, 2004, from *www.witsa.org/papers/*

Mckay, J., Prananto, A., & Marshall, P. (2000). E-business maturity: The SOG-e model. In *Proceedings of the Australian Conference on Information System*. [CD-ROM].

Mckay, N., Gemino, A., Igbaria, M., & Reich, B. (2001). Empirical test of an electronic commerce adoption model. In *Proceedings of the Annual Conference of the*

Administrative Sciences, Association of Canada Information Systems Division, 22(4). London: Ontarion.

Molla, A. (2002). *E-readiness and e-commerce success: Developing and exploring an antecedent model in developing countries context.* PhD thesis, University of Cape Town.

Molla, A., & Licker, P. S. (2002). PERM: A model of eCommerce adoption in developing countries. In M. Khosrowpour (Ed.), *Issues and trends of information technology management in contemporary organizations, Proceedings of 2002 Information Resources Management Association International Conference*, Seattle, USA May 19-22, 2002 (pp. 527-530).

Molla, A., & Licker, P. S. (2004). Maturation stage of eCommerce in developing countries: A survey of South African companies. *Journal of IT and International Development, 1*(3), 89-98.

Morath, P. (2000). *Success @ e-business: Profitable Internet business and commerce.* London: McGraw-Hill.

NNI-Nolan Norton Institute. (1999). *Electronic commerce: The future is here.* Research Report. Retrieved September 2, 2000, from *www.kpmg.com.au*

Odedra-Straub, M. (2003). E-commerce and development: Whose development? *Electronic Journal of Information Systems in Developing Countries, 11*(2), 1-5.

OECD. (1999). *Defining and measuring e-commerce: A status report.* Retrieved April 1, 2004, from *http://www.oecd.org/dataoecd/12/55/2092477.pdf*

Oxley, J., & Yeung, B. (2001). E-commerce readiness: Institutional environment and international competitiveness. *Journal of International Business*, 4th Quarter, *32*(4), 705-724.

Palacios, J. J. (2003). The development of e-commerce in Mexico: A business-led boom or a step toward the emergence of a digital economy? *Information Society, 19*(1), 69-80.

Pare, D. (2002). B2B e-Commerce services and developing countries: Disentangling myth from reality. *Association of Internet Researchers International Conference,* October 13-16, The Netherlands.

Poon, S., & Swatman, P. M. C. (1999). An exploratory study of small business Internet commerce issues. *Information and Management, 35*, 9-18.

PricewaterhouseCoopers .(2000). *Introducing emm@.* Retrieved January 1, 2004, from *www.pwcglobal.com*

Raymond, L., Bergeron, F., & Rivard, S. (1998). Determinants of business process reengineering success in small and large enterprises: An empirical study in the Canadian context. *Journal of Small Business Management, 36*(1), 72-86.

Rogers, E. M. (1995). *Diffusion of innovation* (4th ed.). New York: The Free Press.

Selhofer, H. & Mentrap, A. (Eds.) (2004). *A pocket book of e-Business indicators: A portrait of e-Business in 10 sectors of the EU Economy*, Luxemburg: European Commission. Retrieved September 1, 2004, from *http://europa.eu.int*

Stoddard, D. B., & Jarvenpaa, S. L. (1995). Business process redesign: Tactics for managing radical change. *Journal of Management Information Systems, 12*(1), 81-107.

Straub, D. (1997). The effect of culture on IT diffusion: E-mail and fax in Japan and the US. *Information Systems Research, 5*(1), 23-47.

Tabor, S. W. (2000). Electronic commerce adoption and success: A study of organizational factors and influences. In *Proceedings of the America's Conference on Information Systems* (pp. 669-671). [CDROM].

Tarafdar, M., & Vaidya, S. D. (2004). Adoption of electronic commerce by organizations in India: Strategic and environmental imperatives. *Electronic Journal of Information Systems in Developing Countries, 17*(2), 1-25.

Tigre, P. B. (2003). Brazil in the age of electronic commerce. *The Information Society, 19*, 33-43.

Todd, A. (1999). Managing radical change. *Long Range Planning, 32*(2), 237-244.

Van Akkeren, J. K., & Cavaye, A. L. M. (1999). Factors affecting the adoption of e-commerce technologies by small business in Australia-an empirical study. In *Proceedings of the Information Outlook 1999 Conference.* Retrieved April 1, 2004, from *www.acs.org.au/act/events/io1999/akkern.html*

Wigand, R. T. (1997). Electronic commerce: Definition, theory and context. *The Information Society, 13*, 1-16.

Wilson, R. F. (1999). E-commerce readiness assessment tool. *Web Commerce Today.* Retrieved April 1, 2004, from *http://www.wilsonweb.com/wct2/issue23.htm*

Young, D., & Benamati, J. (2000). Differences in public Web sites: The current state of large U.S. firms. *Journal of Electronic-Commerce Research, 1*(3), 1-13.

Zaltman, G., Duncan, R., & Holbek, J. (1973). *Innovations and organizations.* New York: John Wiley.

Zhu, K., & Kraemer, K. L. (2002) e-Commerce metrics for net-enhanced organizations: Assessing the value of e-commerce to firm performance in the manufacturing sector. *Information Systems Research, 13*(3), 275-296.

Chapter XI

An Overview of E-Commerce Security and Critical Issues for Developing Countries

Pierre F. Tiako, Langston University, USA

Irénée N. Tiako, University of Yaoundé, Cameroon

Abstract

Deficient information and communication technologies (ICT) infrastructure, lack of awareness on e-commerce issues, network payment and secure transaction services present enormous challenges to developing countries. For e-commerce to be a viable tool for trade in developing countries (DC), a "secure infrastructure" which makes possible the electronic exchange of financial transactions is a necessary prerequisite. There has been significant research on e-commerce security, although most of these studies have focused on developed countries. Less attention has been paid to underdeveloped countries that face different circumstances due to the above challenges. From technological perspectives, this work discusses important issues of e-commerce security for developing countries in regards to lack of fraud repression and justice in those countries.

Introduction

When the Internet broke with its military origins and its use for interconnecting researchers among university campuses, it offered all kind of services (Tiako, 2003a) such as multimedia from where one can read newspapers, listen to radio, purchase products, etc. Almost all countries in the world adopted the Information Technology and Communication over the Internet. It is the most effective means and economic way to get information, to be trained, to advertise, to buy products and services at lower prices.

United Nations Children's Fund (UNICEF), United Nations Educational, Scientific and Cultural Organization (UNESCO) and several other structures of United Nations (UN) consider ITC as an integral part of the strategy to reduce poverty. United Nations Conference on Trade and Development (UNCTAD) report on electronic trade and development published in 2003 defined the sectors in which developing countries are able to exploit the possibilities of electronic commerce to maximize their profits. Electronic commerce will provide developing countries with the possibility to increase their market space by offering online services to international customers, without traditional intermediaries.

The Internet provides developed countries with opportunities to de-localize services requiring heavy labor in developing countries. The benefits are then used to diversify their economy and to improve the living standard of their citizens.

Currently, e-commerce revenue in developing countries is miniscule in comparison to that found in advanced economies (Hoffman, 2000). Most e-commerce transactions occur between customers and sellers in the progressive economies of the United States, Canada, and Western Europe (Montealegre, 2001). Most of the world's population, however, exists outside the borders of these countries. The state of e-commerce in developing countries is similar to what it was in the United States in the mid-90s (Hawk, 2001). Predictions are variable, but they tend to point to significant growth of Internet access among businesses and consumers in many developing countries within the next five to ten years (McConnell, 2000).

E-commerce is one of the Internet's applications that include Internet services (Tiako, 2003a) for trade and finance. E-commerce is done among companies or between companies and individuals, and involves crucial financial transactions (Gritzalis & Gritzalis, 2001). The economic stakes are so significant during financial transactions that e-commerce systems must sufficiently be protected to prevent possible financial embezzlement. Such fraudulent activity can cause losses of a company's money (Zhu, 2002), equipment or goods to profit other companies or individuals. That is a crucial factor that can heavily discourage developing countries' businesses to join the e-marketplace (Tiako, 2003b).

Because emerging countries do not have enough structures to track, reprimand and use the law to compensate for the loss of funds (Summers, 1994), equipment or goods, an Internet system for preventing or reducing any kind of fraud that could appear during e-commerce transaction is necessary. For this reason, dealing with confidentiality and security issues (Mukti, 2000; Nugent & Raisinghani, 2002) are essential in e-commerce infrastructure for developing countries.

Consequently, an exploratory study conducted in the Brazilian market (Critical Success Factors, 2004) shows that the factors regarding security provided by the online stock broker was ranked as the single most important factor in the process of choosing an online stock broker. In the same investigation, 85% of respondents classified security certification of the electronic broker's Web site as very important. This supports the idea that security issues are fundamental in e-commerce for developing countries (Jennex, Amoroso & Adelakum, 2004). Several works have been done in all facets of e-commerce for developing countries, but very few tackle issues of e-commerce security technology in order to pre-empt the weakness of justice and arbitraries in these countries.

This work presents an overview of e-commerce securities and its critical issues for developing countries from technological perspectives.

Bases and Principles for E-Commerce Transactions

In order to buy goods or services over the Internet, e-commerce transactions are qualified as business-to-business (B2B) when they involve companies, business-to-consumers (B2C) when transactions are between companies and consumers. The transactions are also qualified as private-to-private (P2P) or consumer-to-consumer (C2C) when they take place between private individuals. An example of P2P is online auction sales. P2P or C2C transactions can be decomposed into C2B and B2C. Since C2B is the same as B2C, the issues of e-commerce security are those of B2C.

This work focuses on B2C for which transactions are opened to security vulnerability due to interactions among unknown entities doing business occasionally. This chapter does not deal with security issues among known partners doing business together over the long term (see illustration in Table 1).

Table 1. Entities making business occasionally

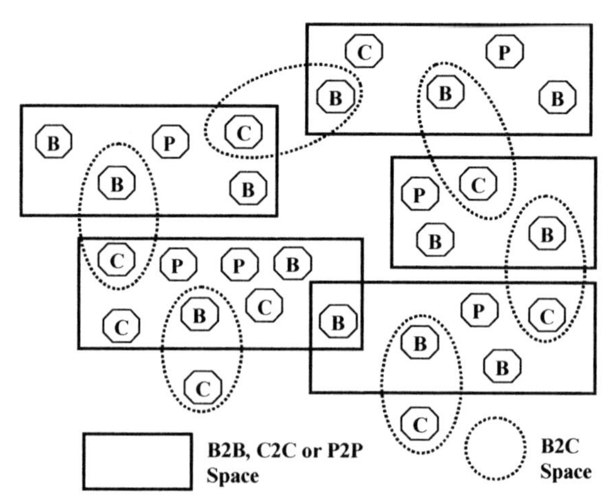

Security, technical and commercial issues relating to e-commerce transactions in DC are generally previously negotiated by contracts or interchange conventions and not implemented in the Internet network infrastructure. The Internet infrastructure in DC is made up of local and wireless data-processing equipments of: (1) active networks, (2) ad hoc networks, (3) programmable networks and (4) content-delivery networks.

Active networks process user data provided by companies or customers. Ad hoc networks calculate the shortest path from one location to another, meaning the path from a company's Web server location to a customer's machine. Programmable networks help to choose priority of flows in the shortest path, implying that services or goods and information are exchanged among customers and companies. Content-delivery networks help to store transient and persistent data exchanged over the Web (Tiako, 2005) in the companies' or customers' computer memory and cache.

Digital patrimony of a customer, intellectual properties of companies and classified information of developing countries will still be accessible if appropriate security measures are not taken to prevent their potential value from being accessed through the Web. The attempts of espionage or hacking against a company, risks of sabotage of information processing systems, recognizing critical functions, and alerts of data-processing wars with great scale are dangers which should be taken into account when establishing e-commerce Infrastructure for development.

Critical Issues of E-Commerce Security in DC

Security is the art of sharing privacies. Privacies are dissimulated in a safe environment, which can be just an individual's memory. When privacy is digital information forwarded over the Internet, such as a cryptographic key (Biryukov & Shamir, 2001; Ko et al., 2000), it must be quantified, which requires another level of privacy. If all digital information conveyed on the Web were confidential, there would remain only open secrets. To deal with confidentiality, secure entities are set up on the Web to start privacy. These entities are connected to all parts of the Web, relayed by cryptographic protocols (Biryukov & Shamir, 2001) containing spatially and temporally secrets. These secrets are tied cryptographically with short messages that contain names and rules for generating certificates of security insurance. These certificates install credibility into the system and thus make it possible for customers and merchants to deal with confidence in the electronic marketplace (Tiako, 2003b).

To improve the level of security (Nugent & Raisinghani, 2002) in e-commerce transactions for developing countries, it is important for each government to officially establish "cyber-notary" agencies in their cities. A cyber-notary agency will attest with respect to all parties involved in a business and in particular for the recipient of a digitally signed message that the holder of the assigned private key, is the correct person. The main duty of a cyber-notary is to check the identity of the customer or merchant in question. That is done by the physical appearance of the customer or a merchant representative at the cyber-notary agency to present legal documents. On this basis, an electronic certificate (a kind of passport for the virtual world) is established, containing the name of the customer/representative and public key (Ko et al., 2000) which identifies the customer/representative with respect to the third parties. To guarantee its authenticity, this certificate is signed electronically by authority of certification before being placed at the disposal of all in a freely accessible register on the Web.

The authority of certification must finally, if necessary, begin to suspend or to revoke a certificate immediately if the holder reports the loss of private key. In such a system, authenticity, integrity and confidentiality are guaranteed, which is crucial to reinforce the confidence of users on the network (Zhu et al., 2001) and to facilitate proof of transactions performed (Gritzalis & Gritzalis, 2001). Today, attacks on the Internet are numerous, and it is impossible to eliminate them all by the usual methods of safety which simply aim at ruining these attacks by reinforcing protections and by eliminating vulnerabilities as soon as they are known.

Additionally, it is useful to use methods of intrusion tolerances to detect intrusions and correct their effects before they cause damage (Gomes & Bridis, 2001). The reinforcement of security (Haber, 2001) uses mechanisms of authentication and the recording of all information which makes it possible to detect attacks and to identify attackers. Under these conditions, more and more customers' and companies' information are provided, transmitted and stored among partners.

Information provided is generally protected by legislation whose application depends on the goodwill of each party. It is advisable to develop technologies that protect privacy, without keeping the culprits unpunished.

Security Policies in E-Commerce for DC

Security policies are sets of rules that specify authorizations, prohibitions and obligations required to agents (customers, merchants and applications) involved in the electronic marketplace (Tiako, 2003b). Security policies' requirements are presented in Table 2.

Table 2. Security policies' requirements

Confidentiality:	No consultation of prohibited information;
Integrity:	No creation, modification or destruction of prohibited information;
Availability:	No prevention to agents to access their legitimate information, services or resources within the system.

The security policies for e-commerce in developing countries should not be static, but dynamic, i.e., configurable and tailored according to: (1) customer profiles, (2) information flows exchanged among agents in the marketplace (Tiako, 2003a), and (3) the context and localization of involved agents.

Security issues within the same the framework have to pull together: (1) information and communication companies with software engineering enterprises and (2) telecommunication companies and wireless communication providers.

Security issues must also integrate protocols to protect software embedded within agents. Security issues for e-commerce infrastructure development are concerned with three dimensions: (1) customers involved in the market place; (2) the company and/or the legal entity and (3) systems, governments and multinational entities.

Developing countries are made up of different communications and telecommunication systems. Security issues have to deal with heterogeneity of existing systems, their multimedia nature for convivial interaction with the Web and existing wireless networks providers. Security policy models have to take into account infrastructures, architectures, implementations and their uses for development.

Security for e-commerce infrastructure in DC has to provide: (1) policies where requirements for agents match those of their developing environment, (2) flexibility to configure the system according to context adapted to a profile of agents and the market situation, with a level of confidentiality, integrity and availability, and (3) design of appropriate cryptographic protocols (Biryukov & Shamir, 2001).

Security will be based on confidentiality, access control, data integrity, identification and non-repudiation. Confidentiality will make intercepted information unusable to all those who are not recipients. Access control will restrict access to data and servers to only authorized agents. Data integrity (Tiako, 2004) will consist in checking that this data was not damaged by fraud. Identification will verify the authenticity of involved partners and the origin of their message. Non-repudiation will allow agents not to deny the contents of information sent or received.

In addition to the above functions, security (Nugent & Raisinghani, 2002) should include authenticity functions to control agents and protect their privacy.

Authenticity in E-Commerce Infrastructure Development

The infrastructure should be able to identify: (1) individuals wishing to connect to the system through a data processing device (smart card, computer, and unspecified electronics component) or (2) electronic devices that are components or machines of the system interacting with them.

In the first case, authentication consists in verifying the identity provided by the individual. Authentication involves the components of: (1) access control, which consists in managing and checking rights that the security policy provides to individuals, and (2) imputability, which consists of recording the success and failure of accesses by users in the system, in order to establish responsibilities or detect anomalics within the system.

Authentication Cases

There exist several means of authentication that can be used in developing countries:

Case 1: Personal Computer Authentication

Providing a password carries out authentication from a personal/home computer. Customers should use appropriate passwords with at least eight characters. Customers are aware that a password is personal and is not to be shared with others. They have to change their password more often.

Case 2: Smart Card Authentication

Authentication using smart card calls upon few mechanisms: (1) identification of the customer's four digits PIN code and (2) authentication of the merchant and its processing equipment in the system. For purchases of small amounts, cryptographic techniques using a public key are used. For larger amount, cryptographic techniques with secret key are used.

Case 3: Online Authentication

For online purchasing, the customer authenticates merchants using a digital certificate made available on the customer's machine or by verifying the signature of the merchant's Web site. For that, the system uses Secure Socket Layer (defined below) protocols marked by a closed lock available in most browsers. Generally, a merchant does not need to physically authenticate the customer, but just to consider credit card information provided that includes card number, validity data and customer address.

Case 4: Tele-Procedure Authentication

This case involves services such as a tax department where individuals or companies have to interact with their online system for payment. The system improves the traditional security level for online payment. It identifies a customer using keys generated on its personal computer at the time of registration to the system. The keys that match the one provided to customers have to be submitted by merchants to the appropriate certification agency. The protocols used offer a mutual authentication to government agency and users.

Cryptography for Electronic Commerce

An essential part of electronic commerce security is cryptography. Cryptography is based on coding. Various methods were invented for coding information exchanged on networks. One can classify these methods in the following three categories (see illustration on Table 3).

Traditional Coding

The principle of traditional coding is to change the characters of the alphabet with others in accordance with a key defined by a table of correspondence between characters. The same key will be used to decode the message. To function, the table of correspondence must remain secret. Substitutions and transpositions are methods used to code and decode information in this approach.

Modern Coding

It represents information numerically. This method of coding rests on symmetrical and asymmetrical algorithms with key. Coding is symmetrical when the same key is used to code and decode information. Coding is asymmetrical when one key is used to code information and another key is used for its decoding.

Future Coding

All the systems studied previously take for granted that numerical communications can be heard without detection because most intrusions preserve the integrity of data

Table 3. Coding and cryptographic techniques

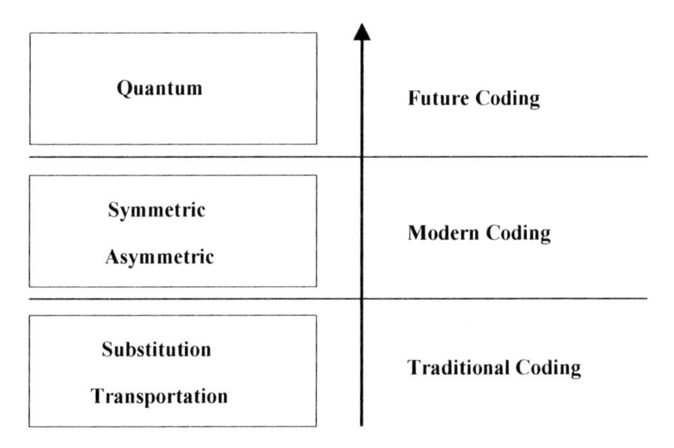

exchanged among entities of the system. The coding of the future is based on quantum systems and Heisenberg's principle of uncertainty (Schneier, 1996). According to this principle, any measure to a quantum system disturbs it. The least intrusion on a quantum transmission channel will alert legitimate users.

Protocols for Electronic Commerce

A protocol is a technology used in the network and telecommunications to make computers communicate. There are two principal communications protocols aiming at the online transactions with credit card: Secure Sockets Layer (SSL) protocol (SSL 2004) and Secure Electronic Transaction (SET) Protocol (SET 2004). Added to that is the NetBill (2004) protocol which is specialized for transactions of small amounts.

SSL is the current standard for secure transactions on the Internet. This protocol helps users to identify and authenticate servers with which they interact. SET is a security protocol of the electronic transactions developed by Visa and MasterCard. It uses the electronic signature of a consumer to initiate a transaction between merchant and the banks. NetBill provides mechanisms to transfer goods and to carry out financial transactions between the merchant and the consumer.

The main benefit of using NetBill for electronic commerce in emerging countries is that the protocol it supports provides financial transactions for small amounts. Such a protocol provides security for both consumers and merchants. It guarantees that the goods are delivered when payment is carried out and that merchants receive payment when goods are delivered.

Related Work and Future Trends

Public Key Infrastructure (PKI) is a proposed technology solution for e-commerce security in developing countries (Ntoko, 2002). It includes a security policy, certification authority, registration authority, certificate distribution system and PKI-enabled applications. Unfortunately the cost of putting this strategy in place (building a PKI, running associated certificate authority and establishing necessary relationships for validation authority and attribute authority) is prohibitive for developing countries. The technologies needed to deploy such an infrastructure present some major challenges to developing countries.

For instance, considering the absence of payment services by local banks in most developing countries, the objective is to connect PKI to ICT infrastructure in industrialized countries in order to provide a solution for e-commerce infrastructures and services to the business sector in developing countries. PKI is therefore a limited solution due to its heavy link to ICT infrastructure in industrialized countries, which is a barrier to sovereignty in developing countries.

The factors affecting the implementation of e-commerce in the global context of Mexico have been identified (Garcia-Murillo, 2004). Using theories of institutional economics and resource-based theory of the firm as a framework of analysis, the contribution focuses on the rules of behavior that have prevailed in Mexican business transactions and are likely to affect the adoption of e-commerce. The study only covers infrastructure, supply and demand factors and does not tackle the issues of e-commerce security.

An investigation of payment and delivery methods provided by B2C sites in a global context including Russia, India and Latin America shows that low credit card penetration and poor delivery systems are widely viewed as serious problems for B2C e-commerce in developing countries (Hawk, 2004). This work just studies the similarities and differences in how sites from the three regions handle payment and delivery and does not deal with issues of e-commerce security.

In future work, we will model interactions of communication among entities for electronic commerce security. Several approaches exist to model such systems: finite automata, neuron networks, object-oriented approach and multi-agents systems (Tiako, 2004b). Future work will study advantages and disadvantages of each of these approaches before specifying the system that more readily adapts to security measures for electronic commerce in developing countries.

For example, the interaction between entities of such a system is of two kinds: (1) reactions of the agents with respect to the other agents and (2) cooperation among agents. Modeling using multi-agent systems seems to be more adapted to deal with security issues than other approaches mentioned above. Those other approaches just focus on managing entities of the system rather than dealing with cooperation or collaboration among these entities. Future work will also study the tools for modeling and verification of communication interactions among entities of the system.

Conclusion

Several studies presented above have been done looking at basic infrastructure requirements for e-commerce in developing countries. They found that major problems restricting the expansion of e-commerce in a global context include security concerns and payment issues. Electronic commerce is an increasingly viable tool for commercial transactions in developing countries. The financial stakes being highly significant, one expects more and more fraudulent activities using information sent between entities of electronic commerce.

For e-commerce to be truly a tool for trade in developing countries, a "secure infrastructure," which makes possible the electronic exchange of financial transactions, is a necessary prerequisite. From technological perspectives, this work discusses important issues of e-commerce security for developing countries in regards of lack of fraud repression and justice in those countries. In this chapter, we presented bases and principles for e-commerce transactions before describing sets of rules that specify authorizations, prohibitions and obligations required in the e-marketplace for developing countries.

References

Biryukov, A., & Shamir, A. (2001). Structural cryptanalysis of SASAS. Advances in cryptology: Eurocrypt 2001. *International Conference on the Theory and Application of Cryptographic Techniques, Innsbruck, Proceedings*, Austria (pp. 395-405).

De Campos, A. M., & Joia, L. A. (2006). Critical success factors for e-brokerage: An exploratory study in the Brazilian market. In S. Kamel (Ed.), *Electronic Business in developing Countries: Opportunities and challenges* (pp. 193-213). Hershey, PA: Idea Group Publishing.

Garcia-Murillo, M. (2004). Institutions and the adoption of electronic commerce in Mexico. *Electronic Commerce Research, 4*, 201-219.

Gomes, L., & Bridis, T. (2001). FBI warns of Russian Hackers stealing U.S. credit card data. *The Wall Street Journal.*

Gritzalis, S., & Gritzalis, D. (2001). A digital seal solution for deploying trust on commercial transactions. *Information Management and Computer Security, 9*(2), 71-79.

Haber, L. (2001). Shoring up security. *Network World*, 53-56.

Hawk, S. (2004). A comparison of B2C *e*-Commerce in developing countries. *Electronic Commerce Research, 4*, 181-199.

Hoffman, T. (2000). UN: Global *e*-Commerce challenges abound. *Computerworld, 34*(28), 12-14.

Jennex, M.E., Amoroso, D., & Adelakum, O. (2004). *e*-Commerce infrastructure success factors for small companies in developing economies. *Electronic Commerce Research, 4*, 263-286.

Ko, K.H., Lee, S.J., Cheon, J.H., Han, J.W., Kang, J.S., & Park, C. (2000). New public-key cryptosystem using braid groups. Advances in cryptology. *Crypto 2000: 20th Annual International Cryptology Conference, Proceedings,* Santa Barbara, CA (pp. 166-183).

McConnell, B. (2000). Risk e-Business: Seizing the opportunity of global *e*-Readiness. Retrieved from *www.mcconnellinternational.com*

Montealegre, R. (2001). Four visions of *e*-Commerce in Latin America in the year 2010. *Thunderbird International Business Review, 43*(6), 717-735.

Mukti, N.A. (2000). Barriers to putting businesses on the Internet in Malaysia. *Electronic Journal of Information Systems in Developing Countries, 2*(6), 1-6. Retrieved from *http://www.ejisdc.org*

NetBill. (2004). Miscellaneous documents. Retrieved from *http://www.ini.cmu.edu/netbill/*

Ntoko, A. (2002). E-business: A technology strategy for developing countries. International Telecommunication Union, *e*-Strategy Unit, United Nations, Geneva, Switzerland. Retrieved from *http://www.itu.int/ITUD/e-strategy/publications-articles/wmrcjune00/ntoko.html*

Nugent, J. H., & Raisinghani, M. S. (2002). The information technology and telecommunications (or *e*-Business) security imperative: Important issues and drivers. *Journal of Electronic Commerce Research, 3*(1).

Schneier B. (1996). *Applied cryptography*. John Wiley.

SET. (1997). SET secure electronic transaction specification, Book 1: Business Description.

SSL. (2004). Miscellaneous OpenSSL Documents. Retrieved from *http://www.openssl.org/*

Summers, B. J. (1994). *The payment system: Design, management, and supervision.* Washington, DC: International Monetary Fund.

Tiako, P.F. (2003a). E-commerce approach for supporting trading alliances. In A. Frew, M. Hitz, & P. O'Connor (Eds.), *Information and Communication Technologies 2003, Proceedings of the International Conference (ENTER 2003),* Helsinki, Finland (pp. 92-491). Springer.

Tiako, P.F. (2003b). Web-services modeling for e-Marketplace. In *2003 Symposium on Applications and the Internet Workshops (SAINT 2003), Proceedings*, Orlando, FL (pp. 111-115). IEEE Computer Society.

Tiako, P.F. (2004). *Dealing with multiple data sources across organizational enterprises.* Working Paper. Submitted to International Journal of Information Technology and Management.

Tiako, I.N. (2004b). Sécuriser le Commerce Electronique par les Systèmes Multi-Agents. Master Philosophy Thesis. National Advanced School of Engineering. ENSP-University of Yaoundé I, Cameroon.

Tiako, P.F. (2005). Data sharing over the Web. In Rivero, Doorn & Ferraggine (Eds.), *Encyclopedia of database technologies and applications (forthcoming)*. Hershey, PA: Idea Group Reference.

Zhu, D. (2002). Security control in inter-bank transfer. *Journal of Electronic Commerce Research, 3*(1).

Zhu, D., Premkumar, G., Zhang, X., & Chu, C.H. (2001). Data mining for network intrusion detection: A comparison of alternative methods. *Decision Sciences Journal, 32*(4), 635-660.

Chapter XII

Venturing the Unexplored: E-Readiness Assessment of Small and Medium Enterprises in Egypt

Nagla Rizk, The American University in Cairo, Egypt

Abstract

This chapter adopts a micro approach towards assessing the e-readiness of small and medium enterprises in the textile industry in Egypt. The analysis focuses on evaluating firms' level of connectivity, awareness and usage of ICTs and identifies barriers to e-readiness. The chapter adds to the literature on e-readiness assessment, which is mostly macro in nature. The micro approach focuses on ICT awareness and usage patterns, and has the advantage of capturing variations at the firm level. Based on connectivity alone, neither small nor medium firms under study were found close to being e-ready. However, extending e-readiness to include the use and awareness of ICTs, one may conclude that e-readiness is in general directly proportional to size. Medium-sized firms, therefore, are potentially e-ready. The chapter argues that raising SME e-readiness would require heavy investment in human capital and raising awareness in addition to upgrading levels and types of connectivity.

Introduction

A number of studies have been conducted toward assessing countries' e-readiness — namely, their preparedness for the digital world. Assessments were based on combinations of indicators such as e-connectivity, human capital, business climate, leadership and others. Quantitative and qualitative indices were devised and used to evaluate and rank countries on the e-readiness scale.

While providing insight into the overall e-readiness of countries on the macro level, few studies have attempted to evaluate e-readiness from a micro perspective, i.e., studies conducted at the sectoral or firm level. In particular, a small number of studies have undertaken an assessment of the adoption of information and communication technologies (ICTs) in small and medium enterprises (SMEs) in the United States, Australia, and in some European, Asian and African countries. However, none was done for any of the countries in the Middle East and North Africa region. This paper takes on this endeavor. It is the first attempt to adopt a micro approach to assess e-readiness of SMEs in an Arab country. The objective of the research is to assess the e-readiness of SMEs in the textile, specifically garment, industry in Egypt, and in particular, evaluate their preparedness for electronic commerce. The research is exploratory in nature, and represents a first step towards more extensive research to assess the preparedness of different sectors in Egypt for embracing and internalizing ICTs.

The paper is divided into four sections. Following the introduction, the second section is a brief review of the literature and methodologies used for e-readiness assessments, highlighting Egypt's macro e-readiness score and ranking as presented in these studies. The section also includes a brief coverage of micro studies assessing ICT adoption and use in SMEs. The third part includes the micro study, starting by a brief description of the textile sector and the firms selected, to be followed by results of the field research and an assessment of the firms' e-readiness. The fourth section presents conclusions and recommendations.

Background: E-Readiness – From Macro to Micro Analysis

The literature on macro e-readiness assessment has taken two approaches. The first group of studies undertakes a quantitative assessment, whereby countries are assigned numerical scores depending on how well they have performed on specific components of the e-readiness measure. A weighted average is calculated based on the relative importance accorded to these components. This approach has been adopted by, among others, the Economist Intelligence Unit (EIU) in devising the EIU E-Readiness Indices (Economist Intelligence Unit, 2001, 2002, 2003 & 2004), the Center for International Development at Harvard measuring the Network Readiness Index (Center for International Development, 2001-2 & 2002-3), the International Data Corporation in calculating the Information Society Index (International Data Corporation and World Times, 2000,

2001, 2002, 2003 & 2004), the United Nations Conference on Trade and Development in measuring the UNCTAD ICT Development Indices (United Nations Conference on Trade and Development, 2003), and the United Nations Development Program in devising the Technology Achievement Index (United Nations Development Program, 2001).

The second group of macro studies concentrates on qualitative assessments, evaluating components such as connectivity, human capital, applications, sophistication of use, and geographical dispersion. Assessments often highlight suggestions for improvements in specific components. Among these are the studies undertaken by McConnell International (2000 & 2001), Wolcott et al. (2001) and the Computer System Policy Projects (CSPP) Readiness Guide (Computer System Policy Projects 2001). A detailed coverage of quantitative and qualitative indices and their respective components and relative weights is offered in Table 1.

In both quantitative and qualitative macro assessments, Egypt's e-readiness has been rated as modest. Egypt was included in all five quantitative studies, and was accorded a very modest macro e-readiness score. Table 2 shows that Egypt ranked amongst the lowest 10-14 countries in a sample of 60-64 countries assessed for e-readiness by the Economist Intelligence Unit between 2002 and 2004. Egypt was also included in the lowest groups according to the Network Readiness Index and the Information Society Index. In 2001, Egypt ranked 57th among 72 countries assessed by the United Nations Development Program (2001) on the basis of the Technology Achievement Index, placing the country in the third category among "dynamic adopters" right before "marginalized," and superseded by "leaders" and "potential leaders" (p. 45). Egypt was also covered by the McConnell qualitative studies undertaken in 2000 and 2001, with a recommendation for improvement needed for e-leadership and information security, and "*substantial*" improvement required in connectivity, human capital, e-business climate (McConnell International, 2000, p. 11; McConnell International, 2001, p. 13).

An overall look at macro assessment tools leads to two major conclusions. First, a common parameter in macro assessments is the inclusion of some measure of physical infrastructure/usage (e-infrastructure) and education (including knowledge of ICTs). These represent the lowest common denominator in assessing the macro e-readiness of countries, and are complemented by policy and economic environment settings. Second, while providing general insights into the countries' e-readiness, macro studies suffer a major drawback: the choice of components and their relative weights may vary from one country to the next. Relative measures and country rankings may ignore internal variations within a country, and as such could be misleading.

Micro studies are therefore recommended as they capture many of the factors that may escape macro analysis, and hence offer a more accurate picture of comparable indicators created for micro e-readiness assessment. While mostly qualitative in nature, a number of studies were conducted to assess ICT adoption and use in SMEs. Studies by OECD (2000c) look into ICT use by SMEs, and identify benefits and barriers. They focus on samples of SMEs within and across member countries (p. 11). OECD (2000c) research has shown that in addition to high costs, the lack of awareness and qualified human resources were the main barriers to exploiting the opportunities of e-commerce for SMEs in OECD countries (p. 3). In agreement with the OECD research, lack of awareness came along with the lack of trained human resources as major constraints for SMEs' adoption of ICTs in

Table 1. E-readiness: Macro indices

	Index	Component
Quantitative Indices	**Economist Intelligence Unit (EIU) E-readiness Index (2003)***	Connectivity & Technology
		Business Environment (20%)
		Consumer and Business Adoption 20%
		Legal and Policy Environment (15%)
		Social and Cultural Infrastructure (15%)
		Supporting e-services (5%)
	Network Readiness Index (Center for International Development, Harvard University, 2001-2002)**	Network Access
		Networking Learning
		Networked Society
		Networked Economy
		Network Policy
	Networked Readiness Index (Center for International Development Harvard University, 2002-2003)**	Network Use (1/2)
		Enabling Factors(1/2)
		Networked Economy (1/4)
		Network Policy (1/4)
		Networked Society (1/4)
		Network Access(1/4)
	Information Society Index (IDC, 2000-2003)***	Computer infrastructure
		Information Infrastructure
		Social Infrastructure
		Information Infrastructure
	Technology Achievement Index (UNDP, 2001)^	Creation of technology (1/4)
		Diffusion of recent innovation (1/4)
		Diffusion of old innovation (1/4)
		Human Skills (1/4)
	ICT Development Indices (UNCTAD, 2003)^^	Connectivity
		Access
		Policy
		Usage/Telecom Traffic
Qualitative Indices	**McConnell International (2000-2001)^^^**	Connectivity
		E-Leadership
		Information Society
		Human Capital
		E-Business Climate
	Readiness Guide (Computer Systems Policy Projects CSPP, 2001)~	Network Infrastructure
		Networked Places (access)
		Networked Applications
		Networked Economy
		Networked world
	A framework for Assessing the Diffusion of the Internet (2001)~~	Pervasiveness
		Geographical Dispersion
		Sectoral Absorption
		Connectivity Infrastructure
		Organizational Infrastructure
		Sophistication of Use

Sources
* Economist Intelligence Unit (2003).
** Center for International Development, Harvard University (2001-2; 2002-3 & 2003-4)
*** International Data Corporation and Worldtimes (2000-2003)
^ United Nations Development Program (2002)
^^ United Nations Conference on Trade and Development (2003)
^^^ McConnell International (2000-2001)
~ Computer Systems Policy Projects (2001)
~~ Wolcott, Peter, Press, L., McHenry, W., Goodman, S., & Foster, W. (2001)

studies conducted for other countries, e.g., the United Kingdom (Meikle & Willis, 2003), and Botswana (Duncombe & Heeks, 2001).

An important point raised by micro studies is that the use of ICTs is more important than mere connectivity. According to the OECD (2000c), ICT adoption is only "part of the story" (p. 3). A similar conclusion was reached by Jutla, Bodorik and Dhaliwal (2002) in a study on government support for the e-readiness of SMEs in Canada, the Netherlands,

Table 2. Macro e-readiness - Egypt and selected countries

	Economist Intelligence Unit (EIU) Index*					
	2002		2003		2004	
Countries	Score (out of 10)	Rank (out of 60 countries)	Score (out of 10)	Rank (out of 60 countries)	Score (out of 10)	Rank (out of 64 countries)
Sweden	8.32	4	8.67	1	8.25	3
Finland	8.18	10	8.38	6	8.08	5
Hong Kong	8.13	14	8.2	10	7.97	9
India	4.02	43	3.95	46	4.45	46
Korea	n.a.	n.a.	n.a.	n.a.	n.a.	n.a.
Tunisia	n.a.	n.a.	n.a.	n.a.	n.a.	n.a.
Nigeria	2.97	55	n.a.	n.a.	n.a.	n.a.
Egypt	3.76	48	3.72	51	4.08	51

	Network Readiness Index (NRI) **					
	2002		2003		2004	
Countries	Score	Rank (out of 75 countries)	Score	Rank (out of 82 countries)	Score	Rank (out of 102 countries)
Sweden	5.76	4	5.58	4	5.2	4
Finland	5.91	3	5.92	1	5.23	3
Hong Kong	5.23	14	4.99	18	4.61	18
India	3.32	54	3.89	37	3.54	45
Korea	4.86	20	5.1	14	4.6	20
Tunisia	n.a.	n.a.	4.16	34	3.67	40
Nigeria	2.1	75	2.62	74	2.92	79
Egypt	3.2	60	3.13	65	3.19	65

	Information Society Index (ISI)***				Technology Achievement Index (TAI)^	
	2002		2003		2001	
Countries	Score	Rank (out of 55 countries)	Score	Rank (out of 55 countries)	Score	Rank (out of 72 countries)
Sweden	7087	1	989	2	0.703	3
Finland	6422	8	934	4		
Hong Kong	6255	11	825	16	0.455	24
India	1331	18	250	51	n.a.	n.a.
Korea	5596	51	850	12	0.666	5
Tunisia	n.a.	n.a.	n.a.	n.a.	0.255	51
Nigeria	n.a.	n.a.	n.a.	n.a.	n.a.	n.a.
Egypt	1478	53	337	47	0.236	57

Sources:
* Economist Intelligence Unit (EIU) (2002, 2003, 2004)
** Center for International Development (2001-2; 2002-3, 2003-4)
*** International Data Corporation and Worldtimes (2000-2003)
^ United Nations Development Program (2001)

Norway and Singapore. The authors rightly argue that in general, "countries appear to be focused on connectivity metrics as opposed to usage metrics" (p. 10).

Another point raised by micro studies is that barriers related to the SMEs themselves are complemented by a set of "external barriers" related to the surrounding environment, e.g., the lack of "an installed base of potential consumers," security and privacy issues (OECD, 2000c, p.12). Pease and Rowe (2003) found that the lack of software standards, advice and assistance, as well as electronic authentication issues were among such "external" barriers to electronic commerce in SMEs in Australia (p. 2). Jutla et al. (2002) argue that SMEs are also affected by "globalization pressures arising from e-commerce operations," which means that they will "have to acquire international trade knowledge" (p. 1).

One can, however, distill a comforting thought from the above work. It does seem that barriers to e-readiness of SMEs in developed countries do not seem much different from those expected for developing countries. In fact the analysis of Jutla et al. (2002), which covered developed countries, concludes that "the vast majority of SMEs in most countries are still at the very early stages of the adoption curve of electronic business practices." They add that although the use of sophisticated e-business software applications by SMEs is not measured in most countries, "informal evidence suggests that uptake of e-applications such as knowledge and content management is low" (p. 10). The challenges are, after all, faced by all. This could provide an incentive for strengthening the e-readiness of SMEs in developing countries.

A final point emerging from this literature is that there is a strong need for more research in this area. In particular, there is an apparent need for more micro studies (both quantitative and qualitative) assessing the e-readiness of SMEs in developing countries. It is from this conviction that I move to the current study.

E-Readiness of Textile SMEs in Egypt

The present research is a pilot study to assess the e-readiness of a group of SMEs in the textile industry, specifically the garment industry in Egypt. As with macro and micro studies, the present work starts off by assessing connectivity infrastructure, which is a necessary but not sufficient prerequisite for e-readiness. While some macro studies complement connectivity assessment by evaluating the macro policies and business environments, the micro approach focuses on ICT awareness and usage patterns, and has the advantage of capturing variations at the sectoral or firm level. This is the approach that is adopted in the present work. As such, this study will assess the firms' level of connectivity, awareness and usage of ICTs in marketing, production and management, and identify the nature of, and barriers to, ICT usage by these firms. Analysis is done through size grouping of firms, in an attempt to test whether "size matters," i.e., if smaller firms are less prepared to embrace ICTs.

Why SMEs in Egypt's Textile Sector?

Egyptian textile SMEs were chosen for three reasons. First, SMEs generally stand to gain large potential benefits from ICTs. Second SMEs represent a large share of Egypt's economy. And third, these particular firms operate in a sector that has an established history of comparative advantage in Egypt.

To begin with, ICTs offer many potential benefits for SMEs. ICTs reduce transactions costs, remove barriers to entry, and as such effectively reduce the optimal size of the firm (Economist, 2000, p. 2). By allowing effective networking, ICTs offer small firms an opportunity to overcome the competitive advantage of larger firms gained due to economies of scale. Jutla et al. (2002) argue that SMEs possess an entrepreneurial spirit and a willingness to experiment and innovate in terms of business models and operations, more so than larger organizations with established hierarchies. SMEs thus stand to witness strong potential benefits from improving their e-readiness by government initiatives (p. 2).

In addition, ICTs also allow the use and management of supply chain networks, which in turn facilitates procurement, inventory control, supply processes management, production costs monitoring and quality control (Economist, 2000, p. 3). They also offer SMEs an excellent tool for marketing and distribution, which facilitates responsiveness to market demand and customization of offerings. ICTs also provide SMEs with an opportunity for innovation and the emergence of new products and services (OECD, 2000c, p. 2). ICTs help small entrepreneurs overcome information poverty. Entrepreneurs hence become more connected, more certain, less risk-averse, and more capable of making well-informed decisions (Pease & Rowe, 2003, p. 3). By empowering the small entrepreneur (Cecchini & Shah, 2002, p. 9), ICTs offer the potential for increasing exports, promoting growth as well as human development.

This opportunity can be very relevant to Egypt, where SMEs represent a significant share in the number of firms, production and employment generation. According to a study prepared in 2003 by the SME Development Unit of the Ministry of Foreign Trade, SMEs represented almost 99% of the number of companies in the private non-farm agricultural sector in Egypt (Abdel Maksoud & Youssef, 2003, p. 8). Unfortunately, the rest of the data available on Egypt's SMEs is rather old, going back to 1996. Official data for this year indicate that SMEs provided 80% of jobs and generated 80% of the value added in the non-agricultural private sector (Ministry of Foreign Trade).

In line with this, the textile sector was chosen because it is one of Egypt's traditional industries with an established history of comparative advantage. The sector accounts for 27% of Egypt's industrial production and employs around 25% of the country's manufacturing labor force. Ranking as the second largest industry after processed food, the textiles industry contributes 11% of Egypt's manufacturing GDP and 3% of total GDP. The sector is also the largest non-oil exporting sector, with exports reaching $1 billion in 2003. These represented almost 15% of Egypt's total commodity exports (American Chamber of Commerce, 2004, p. 20-21) and 50% of total Egyptian exports to the United States (American Chamber of Commerce, 2004, p. 56). In particular, the garment industry represents the bulk of Egypt's textiles industry. Garments account for around 80% of Egypt's textile exports, produce an annual output valued at $3 billion, and employ around one million workers (Egyptian Exporters Association).

Moreover, it is of interest that world online trade in the garment industry has expanded significantly in the past few years. It has been estimated that over 7% of world apparel sales conducted in 2003 took place online, and that online sales of apparel are projected to reach 18% of world sales in 2010 (American Chamber of Commerce, 2004, p. 20). This provides an opportunity for developing countries that possess a comparative advantage in this industry and gives hope to the potential benefits of raising the e-readiness of SMEs in this sector. In the case of India, for example, a positive correlation was found between ICT adoption and export intensity of firms in the garment industry (Lal, 2001, p. 169).

Finally, utilizing ICTs in the textile and garment sectors allows for benefits to be gained by "old economy" sectors from the advances made in the "new economy." It has been argued that the importance of the new economy lies mainly in its impact on increasing productivity in the traditional old economy sectors (Economist, 2000, p. 2). This is an opportunity for Egypt to catch up with the information revolution, even if only on grounds of ICT use.

The Study

Field research was conducted on a sample of 36 firms specialized in the textile, specifically garment, industry and located in the greater Cairo region. All firms chosen are exporting firms. They were selected from a data base of over 120 exporting companies which represent around 55%-60% of the total number of garment exporting companies, and are responsible for approximately 95% of Egypt's exports of garments. All firms chosen are located in greater Cairo, since this is where 80% of all factories in the garment industry are located (American Chamber of Commerce, 2004, p. 22). The definition of "small" and "medium" here is taken to refer to 30-199 workers and 200-999 workers respectively. Large firms are included for comparison (more than 1,000 workers).

A total of 14 small companies, 17 medium, and 5 large companies were surveyed. These represent 43% and 13% of the respective populations: 40 firms in the first two categories, and 39 in the third. Ideally, equally sized samples of firms would have been more representative of the size distribution of the populations, but this was not feasible due to practical limitations. In-depth reviews with higher management and Information Technology specialists were combined with a detailed questionnaire which they were asked to complete. The questionnaire requested information on variables such as e-infrastructure, human capital, actual and perceived use of ICTs and barriers to implementing ICTs.

Given that the sample size is small, the results below are taken as providing preliminary indicators rather than grounds for broad generalizations. Acknowledging that, some interesting insights can be drawn nonetheless. The following is an explanation of the survey results on firms classified by size.

Small Firms

First, while all small firms have telephone connections, they have modest levels of personal computers (PCs) intensity and Internet connectivity. None of the small companies has more than ten employees connected to the Internet, and all of them have less than five PCs in management. They rely mostly on dial-up for Internet connectivity (Figure 1). Small firms also have the lowest percentage of software ownership (Figure 2).

Figure 1. Connectivity infrastructure

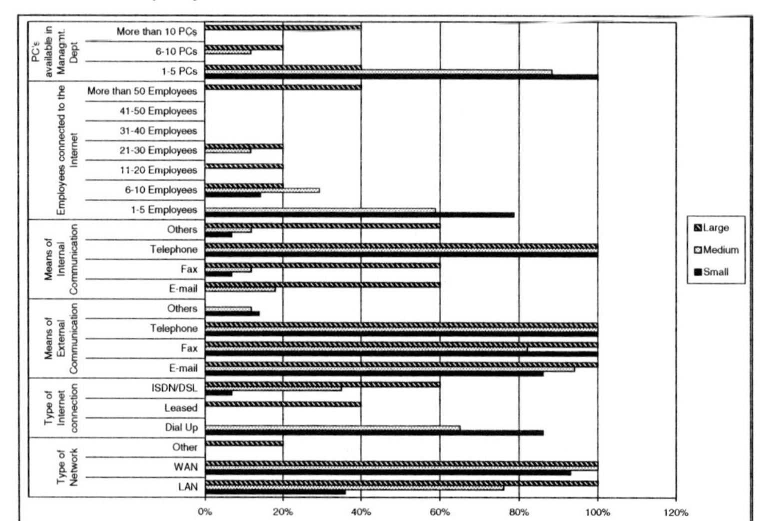

Figure 2. Use of software in management, production and marketing

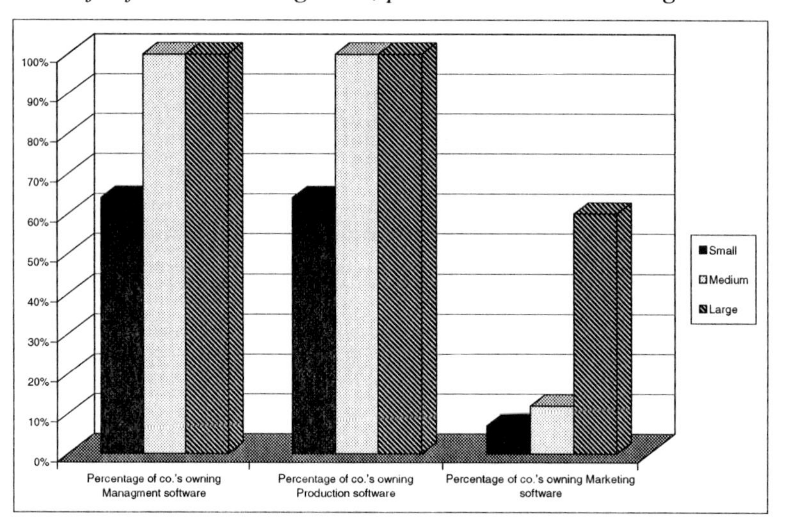

While more than one-third of the small firms agreed that e-commerce will improve sales (Figure 3), their use of ICTs for sales transactions is limited to sales and e-procurement and at very low levels (14% and 7% respectively). None of the small firms uses ICTs for quality control, order tracking or maintenance (Figure 4). Although the use of ICTs by small firms for marketing, research and offers surpasses medium and large firms, in absolute terms this use is modest and offers much room for expansion (Figure 4).

Figure 3. Perceived importance of ICT in production, management and e-commerce

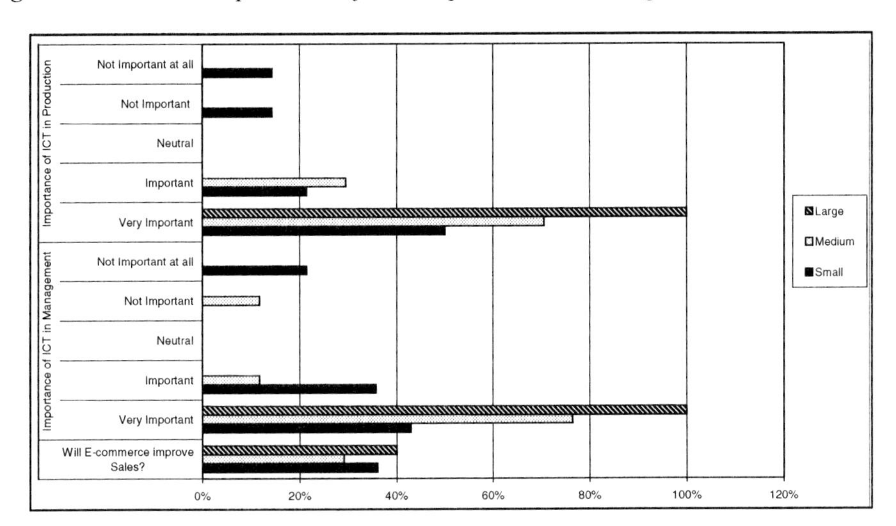

Figure 4. Use of ICT in marketing, production and management

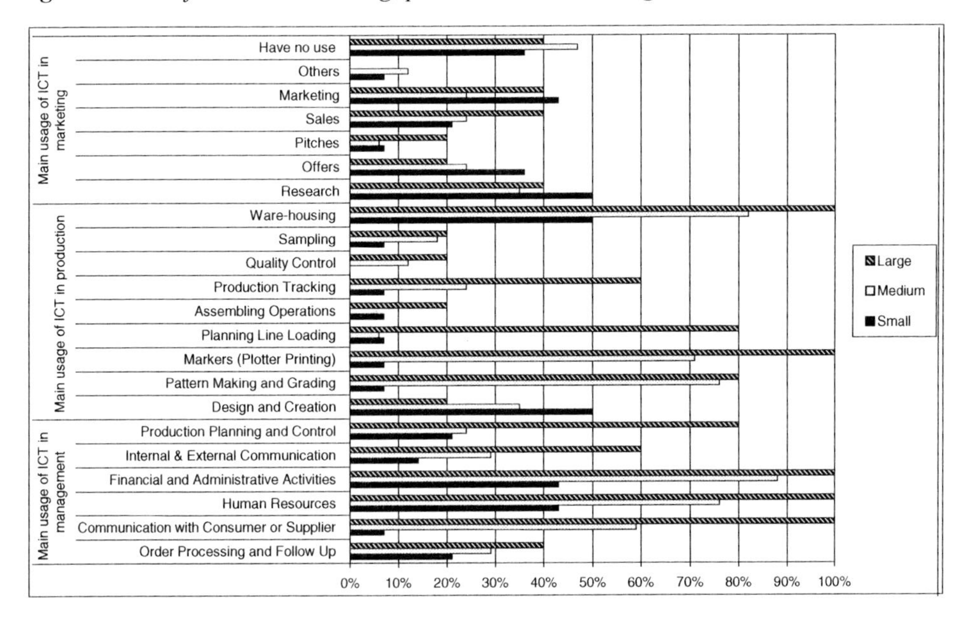

In line with this, limited awareness came as the top barrier to implementing ICTs in marketing (Figure 5). Limited awareness (the "lack of need" expressed by almost 30%), is actually followed by "unqualified personnel," then "limited budgets." This comes contrary to original expectations that the cost constraint would be the highest barrier to implementing ICTs for small firms.

Next, small enterprises expressed a relatively high degree of perception of the role of ICTs in management. About 70% perceived the role of ICTs in management as at least

Figure 5. Barriers to developing ICT in marketing, production and management

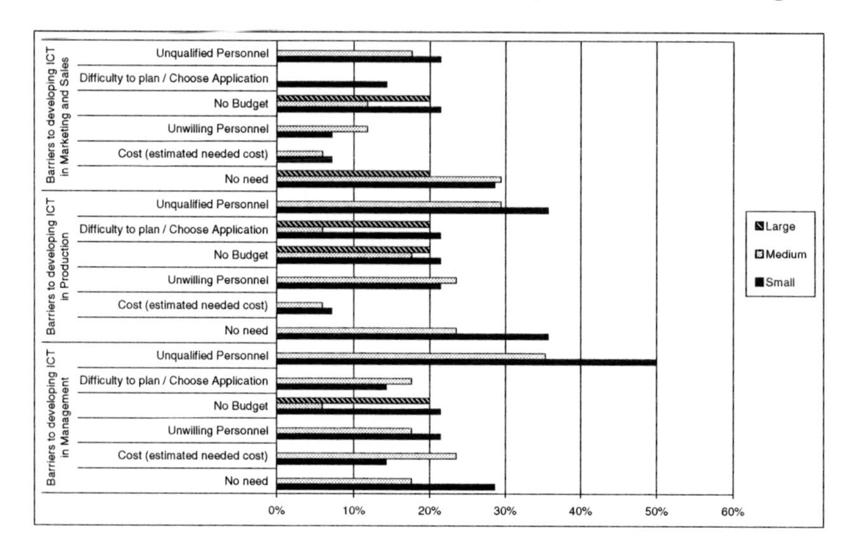

important (Figure 3). Nevertheless, in practice, hardly more than 40% of the firms used ICTs for any one aspect of management (Figure 4). ICTs were mainly used for human resources and financial and administration activities.

When asked about barriers to implementing ICTs in management for small firms, the top barrier was "unqualified personnel" (50% of firms), followed by lack of awareness ("no need," 29% of firms). Budgetary constraints followed (Figure 5). Again we witness the need for training, upgrading human capital, and increasing the level of awareness of the role of ICTs for small firms.

When asked about the importance of ICT in production, again 70% of the small firms mentioned it was at least important (Figure 4). Still, with the exception of using ICTs for design and creation and for warehousing (50% of firms for each category), the use of ICTs by small firms for other aspects of production (e.g., sampling, tracking, planning line loading, etc.) is very low (Figure 4). It is interesting that the percentage of small firms using ICTs for design was higher than that for medium and even large enterprises. Again, "unqualified personnel" and "no need" come as the top barriers to implementing ICTs in production for small firms (36% each) (Figure 5). This is a confirmation of the need for upgrading human capital and increasing awareness of the role of ICTs for small firms.

Based on the above, it is feasible to conclude that based on connectivity, e-infrastructure, and ICT use in marketing, production and management, the level of e-readiness of the small firms under study is very modest. For these small firms to be e-ready, there is a strong need for expanding the use of ICTs in marketing, production, and management which would be fuelled by upgrading their weak e-infrastructure along with raising awareness and upgrading human capital. These seem to be the pressing needs for e-readiness for the small enterprises surveyed.

Medium Firms

Medium firms also have modest connectivity. While all medium firms use telephones, they all have less than 30 employees connected to the Internet and less than ten PCs in management (almost 90% have less than five PCs) (Figure 1). Compared to firm size, these numbers deem the e-infrastructure for medium firms relatively more modest than small firms. Medium firms are, however, relatively high on owning management and production software, in fact as high as large firms. This comes in sharp contrast to the relatively low percentage of their ownership of marketing software (Figure 2).

Medium-sized firms may be a step ahead of their small counterparts in that they have ICTs included in all sales transactions (quality control, order tracking, maintenance, sales, and e-procurement), albeit with a small percentage (never exceeding 20% of the firms) (Figure 4). In fact medium-sized firms are the only ones using ICTs in quality control. This might be a promising scenario as these firms also have a relatively high share of the "export only" segment. It is, however, disappointing that when asked if e-commerce was expected to improve sales, the least level of awareness came from medium firms (Figure 3).

In line with this, the level of use of ICTs by medium firms for marketing is low. More than 45% of the firms have no use for ICT in marketing (Figure 4). Only 10% of medium firms owned marketing software (Figure 2). It is perhaps no surprise that lack of awareness came as the number one barrier to using ICTs for marketing for medium firms (Figure 5). As in the case of small firms, "unqualified personnel" came as the number two barrier to implementing ICTs in marketing. Again there is a strong need for raising awareness and for upgrading human capital for raising the e-readiness of medium-sized firms.

The level of perception of the importance of using ICTs in management is a step higher for medium-sized firms. Almost 90% rated it as at least important (Figure 3). In practice, medium firms use ICTs mostly for financial and administrative purposes, as well human resources (Figure 4). Like small firms, there is a need for medium-sized firms to extend the use of ICTs to other components of management (e.g., order processing and follow up, internal and external communication). Like small firms, medium firms ranked "unqualified personnel" as the number one barrier to using ICTs in management (about 35% of firms) (Figure 5).

In production, all medium firms viewed ICTs as at least important (Figure 3). Actual use, however, did not reflect that. While more than 80% of medium firms use ICTs for warehousing, more than 70% for markers (plotter printing), and 60% for pattern making, a small portion of firms utilize ICTs for other aspects of production (e.g., design and creation, sampling, line loading, production tracking, quality control, etc.) (Figure 4). There is room for more use of ICTs in production for medium firms to reflect the relative high degree of awareness. Like in management, the number one barrier to implementing ICT in production for medium sized firms is "unqualified personnel" (Figure 5).

Based on the above, medium-sized firms are generally one step ahead of small firms in that they have a higher degree of awareness and implementation of ICTs in management and production. Awareness and use of ICTs in marketing is still modest, so is the level of e-infrastructure and connectivity. Barriers to ICT implementation remain the same: the lack of qualified personnel and the lack of awareness. Again the key to raising these firms' e-readiness is upgrading the human capital and raising awareness along with improving their e-infrastructure.

Large Firms

Large firms connect to the Internet using either leased lines or Integrated Services Digital Network (ISDN) and Digital Subscriber Line (DSL). Both ISDN and DSL are superior to dial-up in quality, speed, and scope of Internet use. The ratios are 40% leased lines and 60% ISDN/DSL (Figure 1). Despite that, and given that they employ more than 1,000 employees each, 60% of large firms have less than 30 employees connected to the Internet and less than ten PCs in management. Relatively speaking, these numbers point to a relatively "more modest" e-infrastructure for large firms compared to small and medium enterprises. This is an interesting scenario, and leaves us wondering if connectivity was actually inversely proportional to firm size.

Most of the large firms, however, owned management and production software. Their ownership of marketing software is not as widespread. Less than 60% of firms owned marketing software (Figure 2). This has implications on the overall level of awareness of the role of ICTs in marketing as opposed to management and production.

Indeed, while only 40% of large firms thought e-commerce could improve sales, all 100% percent of large firms believed ICT to be very important in management and in production (Figure 3). This mirrors the scenarios witnessed for small and medium enterprises, where a relatively stronger weight was placed on the importance of ICTs in management and production as opposed to sales transactions and e-commerce.

As far as actual use is concerned, Figure 4 shows that 40% of large firms have no use of ICTs for marketing purposes. In management, on the other hand, all large firms use ICT for financial, administrative and human resources. All large firms also use ICTs for communication with consumers or suppliers, which is a clear difference from the other two groups of firms. Almost 80% of large firms use ICT for production planning and control, and 60% use ICT for internal and external communication.

Large firms also use ICTs relatively more extensively in production. All large firms use ICTs in warehousing and markets (plotter printing). A large proportion (80%) use ICTs in pattern making and planning line loading, 60% use ICTs in production tracking, and 20% use ICTs in sampling, quality control, design and assembling operations (Figure 4).

Unlike small and medium enterprises, large firms brought up budget concerns as the number one barrier to the use of ICT in management and in production (Figure 5). In fact, "no budget" was the only barrier given to the use of ICT in management, and a significant barrier along with technical difficulties hindering the use of ICT in production for large firms. This echoes the relatively stronger awareness of the role of ICTs in management and production, and perhaps a realization of the required investments.

Large firms' relatively limited awareness of the role of ICTs in marketing shows up again in the only incidence of expressing "no need" as a barrier to implementing ICTs in marketing (Figure 5). Limited awareness was, however, brought up along with the budget constraint barrier (20% each).

In line with this, one may conclude that positive signs for a higher e-readiness for large firms are the stronger channel of Internet connectivity and the degree of awareness and use of ICTs in management and production. A negative sign, however, is the relatively weak e-infrastructure and the low level of awareness of the role of ICTs in marketing and the potential benefits from e-commerce. Human capital was never brought up as a barrier. It seems that the priority for raising the e-readiness of larger firms is strengthening the e-infrastructure and increasing awareness of the role ICTs in marketing.

Synthesis of Survey Results

A synthesis of the above results brings to the fore a number of interesting conclusions. First, while all firms surveyed are low on connectivity, it seems that size does not actually matter as far as the *level* of connectivity is concerned. In fact, large firms are doing relatively worse on the numbers of PCs and Internet users. However, it does seem that the *type* of connectivity, Internet network and *use* of e-mail for communication is generally proportionate to size. Large firms have either ISDN/DSL or leased lines, while small firms rely relatively more on dial-up. The use of the Internet and e-mail for communication increases as firm size increases (Figure 1). It will be important that large firms utilize the high quality networks that they have access to, partly by increasing the necessary infrastructure, namely PCs and the number of connected people. For small and medium firms, however, there is a need to increase the present e-infrastructure, and/or expand usage of ICTs.

Next, in all firms and with varying degrees, ICTs are mostly used for conventional purposes, namely management and production, specifically financial, administrative and

Figure 6. Target market and transactions involving ICT

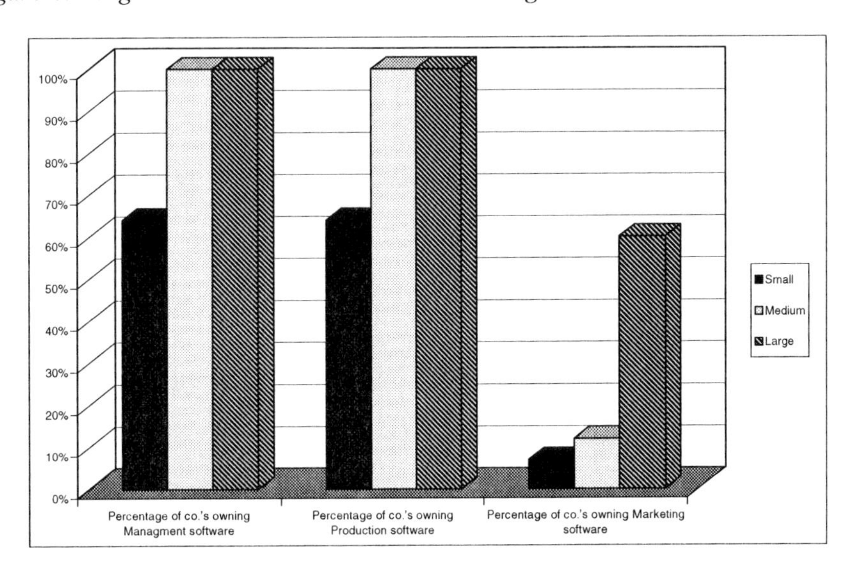

human resource management. This, in part, reflects the relatively higher awareness of the role of ICTs for traditional use (management, then production) as opposed to awareness of the potential benefits of innovative ICT usage for marketing and e-commerce. Indeed, with the exception of some components of marketing for small firms, e.g., research, the use of ICTs for marketing is generally less than ICT use in management and production.

Overall, firms' usage of ICTs is generally proportional to size. Involving ICTs in sales transactions (Figure 6), ownership of marketing, management and production software (Figure 2), and the use of ICT for management (Figure 3) are all directly proportional to size. With the exception of design and creation, one may make a similar conclusion about the use of ICTs in production. The case for marketing is not as clear cut (Figure 3).

Contrary to expectations, budgetary concerns are not the leading barrier to ICT implementation for small and medium enterprises. The lack of qualified personnel was the number one barrier for implementing ICTs in both management and production for both small and medium firms. Limited awareness came next. Lack of awareness and qualified human resources emerging as major barriers to SMEs is reminiscent of the findings of micro studies done for other countries, developed included (OECD, 2000c; Meikle & Willis, 2003; Duncombe & Heeks, 2001). For large firms, budgetary concerns emerge, sometimes as the only barrier. This may imply the conclusion that human capital and awareness need to be satisfied as prerequisites before affording the luxury of worrying about costs of ICT investment.

Conclusion

The present study is a preliminary effort to assess the e-readiness of small and medium enterprises in the textile garment sector in Egypt. Large enterprises were surveyed for comparison, albeit the limitation of a small sample size. The research is exploratory in nature, and is an attempt towards more in-depth work in an area of research that has not been ventured for Egypt or any other Arab country. While scratching the surface, this pilot study still provides some useful insights into the e-readiness of firms as assessed from a micro perspective.

Based on connectivity alone, neither small nor medium firms under study were found close to being e-ready, and large firms present a modest potential (Table 3). However, extending e-readiness to include the use, awareness and barriers to ICTs, one may conclude that e-readiness is in general proportional to size, and hence larger firms are the most e-ready.

Nevertheless, one could make an argument for medium firms. First, large firms could end up being more "locked in" to specific technologies, and the switching costs might be a concern. Moreover, and despite their modest connectivity, medium firms are generally one step ahead of small firms in that they have a high degree of awareness and some base of implementation of ICTs in management and production. In fact medium firms may possess a reasonable level of dynamism and awareness, which could provide a promising potential for engaging in e-commerce, and hence they could be the most e-ready, relatively speaking.

*Table 3. A preliminary e-readiness assessment - summary**

		Small	Medium	Large
Connectivity/E-infrastructrue	Level	L	L	L
	Type	L	L	H-M
	Software Ownership	L	M-H	H
ICT Use in Transactions	Awareness	L	L	L
	Use	L	L-M	M
ICT Use in Marketing	Awareness	L	L	L
	Use	L-M	L	M
ICT Use in Production	Awareness	M	H	H
	Use	L	M	M-H
ICT Use in Management	Awareness	M	H	H
	Use	L	M	H
Barriers to ICT use in Marketing		Awareness/Personnel	Personnel/Awareness	Awareness
Barriers to ICT use in Production		Personnel/Awareness	Personnel	Budget
Barriers to ICT use in Management		Personnel/Awareness	Personnel	Budget/Technical Difficulties
E-Readiness**	Connectivity/ E-infrastructure	L	L-M	M
	ICT Awareness	L-M	M	M
	ICT Use	L	L-M	M
	Overall***	L	M	M

* L: Low; M: Medium; H: High

** A preliminary ranking is done based on the average for each category, e.g. connectivity is taken as the average of the three sub-components: level, type and software ownership.

*** Overall E-Readiness is an average of the above three cells: connectivity, awareness and use.

Based on that, increasing the e-readiness of medium-sized firms would require heavy investment in human capital, to be complemented by raising awareness and upgrading levels and types of connectivity. For small firms, priority should be directed to increasing awareness of the role of ICT, together with improving e-infrastructure and human capital. It would be beneficial for small and medium firms to work in clusters in order to benefit from economies of scale. There is a wide scope for public/private partnerships to raise the level of e-readiness for small and medium enterprises in the economy. Such projects should be placed as priorities on the development plan and donor support agenda.

A final point emerging from this study is that the SMEs surveyed are low on their e-readiness not only because of the low level of their e-infrastructure, but because of the more serious barriers related to awareness and human capital. This finding is reminiscent of the point raised in other micro studies (e.g., OECD, 2000c; Jutla et al., 2002). By the same logic, SMEs development in general and their e-readiness in particular will be affected by traditional old economy challenges such as financing issues, legal infrastructure, policy setting and the business environment. This is also reminiscent of "external" barriers as raised in other studies (OECD, 2000c; Pease & Rowe, 2003; Jutla et al., 2002) and mentioned earlier in this chapter. The role of the government will be a crucial factor in alleviating such barriers. One might then extend the micro e-readiness concept to include such old economy challenges that will affect SME e-readiness. Given that the Information Society entails maintaining a smooth interaction between the "new" and "old" economy, as well as developing the "e" and the "non-e" components, a comprehensive micro index for e-readiness is a challenge that is worth pursuing. This will be the subject of future research.

References

Abdel Maksoud, S., & Youssef, M.A.A. (2003). Information and communication technology for SMEs in Egypt. SME Development Unit, Ministry of Foreign Trade. Retrieved September 1, 2004, from *http://www.sme.gov.eg/*

American Chamber of Commerce in Egypt (2004, August). *The textile and clothing industry in Egypt.* Business Studies and Analysis Center.

Bridges.org. (2001). Comparison of e-readiness assessment models. Retrieved May 1, 2004, from *http://www.bridges.org/*

Buonanno, G., Gramignoli, S., Ravarini, A., & Tagliavini, M. (2002). ICT diffusion and strategic role within Italian SMEs. In F. Tan (Ed.), *Global perspective of information technology management* (pp.163-178). Hershey, PA: IRM Press.

Cecchini, S., & Shah, T. (April, 2002). Information and communication technology as a tool for empowerment. World Bank Empowerment Sourcebook: Tools and Practices. Retrieved January 15, 2004, from *http://www.cefe.net*

Center for International Development, Harvard University. (2001-2; 2002-3; 2003-2004). Global information technology report: Readiness for the networked world. Retrieved January 15, 2004, from *http://www.weforum.org*

Cloete, E. (2003). SMEs in South Africa: Acceptance and adoption of e-commerce. In S. Lubbe & J. Van Heerden (Eds.), *The economic and social impacts of e-commerce* (pp. 121-134). Hershey, PA: Idea Group Publishing.

Computer Systems Policy Projects. (2001). The CSPP readiness guide for living in the networked world. Retrieved May 1, 2004, from *http://www.cspp.org*

Duncombe, R., & Heeks, R. (2001). Information and communication technologies and small enterprise in Africa: Lessons from Botswana. Results of a research project undertaken by the Institute of Development Policy and Management. Retrieved September 1, 2004, from *http://www.sed.manchester.ac.uk*

E-readiness assessment: Who is doing what and where? (2002). Retrieved September 1, 2004, from *http://www.bridges.org*

Economist. (2000, March 30). A thinker's guide. Retrieved December 21, 2003, from *http://www.economist.com*

Economist Intelligence Unit. (2001). The 2001 e-readiness rankings. Retrieved May 15, 2004, from *http://www.intelligentcommunity.org*

Economist Intelligence Unit. (2002). The 2002 e-readiness rankings Retrieved May 15, 2004, from *http://verdi.unisg.ch/org*

Economist Intelligence Unit. (2003). The 2003 e-readiness rankings. Retrieved May 15, 2004, from *http://graphics.eiu.com*

Economist Intelligence Unit. (2004, April 6). Scandinavia consolidates lead in fifth annual economist intelligence unit e-Readiness rankings. Retrieved May 15, 2004, from *http://www.store.eiu.com*

Egyptian Exporters Association. (n.d.). Textiles and made ups. Retrieved September 1, 2004, from *http://www.expolink.org*

Friedrich Ebert Stiftung. (1999). *The Internet for SME's in Egypt: Potential and challenges.* Cairo: Friedrich Ebert Stiftung.

International Data Corporation and World Times (2000-2003). *Information society index.* World Paper. Retrieved December 14, 2003, from *http://www.worldpaper.com*

IntraPairot, A., & Srivihok, A. (2003). The e-commerce of SME's in Thailand. In T. Thanasankit (Ed.), *E-commerce and cultural values* (pp. 199-219). Hershey, PA: Idea Group Publishing.

Jutla, D., Bodorik, P., & Dhaliwal, J. (2002). Government support for the e-Readiness of small and medium sized enterprises. *Proceedings of the 35th Hawaii International Conference on System Sciences.* Retrieved September 1, 2004, from *http://csdl.computer.org*

Kong, W.C. (2003). The implementation of electronic commerce in SME's in Singapore. In T. Thanasankit (Ed.), *E-commerce and cultural values* (pp.51-74). Hershey, PA: Idea Group Publishing.

Lal, K. (2001). The determinants of the adoption of information technology: A case study of the Indian garments industry. In M. Pohjola (Ed.), *Information, technology, productivity, economic growth.* Oxford: Oxford University Press.

MacGregor, R.C., & Bunker, D.J. (2000). EDI and small/medium enterprises. In S. Rahman & M. Raisinghani (Eds.), *Electronic commerce: Opportunities and challenges* (pp.142-151). Hershey, PA: Idea Group Publishing.

McConnell International. (2001). Ready, Net, Go!. Retrieved December 7, 2003, from *http://www.mcconnellinternational.com*

McConnell International. (2000). Risk e-business: Seizing the opportunity of global e-readiness. Retrieved December 7, 2003, from *http://www.mcconnell international.com*

Meikle, F., & Willis, O. (2003). E-business development issues in UK SME's. In G. Gingrich (Ed.), *Managing IT in government, business and communities* (pp. 164-174). Hershey, PA: IRM Press.

Ministry of Foreign Trade, Egypt. (n.d.). SME Statistical Information. Retrieved January 17, 2004, from *http://www.sme.gov.eg*

OECD. (2000a). Global forum: Knowledge economy and digital economy. Bologna Charter 2000. Retrieved January 21, 2004, from *http://www.oecd.org*

OECD. (2000b). OECD small and medium enterprise outlook: Enterprise industry and services. Retrieved January 21, 2004, from *http://www.insme.info*

OECD. (2000c, June 14-15). Realizing the potential of electronic commerce for SMEs in the global economy. Conference for Ministers responsible for SMEs and Industry Ministers, Bologna, Italy. Retrieved January 21, 2004, from *http://www.conferenzabologna.ipi.it*

Pease, W., & Rowe, M. (2003). Issues faced by small and medium enterprises (SMEs) and their take-up of e-commerce in Australian regional communities. Retrieved May 15, 2004, from *http://www.usq.edu.au*

United Nations Conference on Trade and Development. (2003). Information and Communication Technology (ICT) Indices. Retrieved May 15, 2004, from *http://www.unctad.org*

United Nations Development Program. (2001). Today's technological transformations creating the network age. Human Development Report 2002. Retrieved January 21, 2004, from *http://hdr.undp.org*

Wolcott, P., Press, L., McHenry, W., Goodman, S., & Foster, W. (2001). A framework for assessing the global diffusion of the Internet. *Journal of the Association for Information Systems.* Retrieved January 21, 2004, from *http://jais.aisnet.org*

World Bank. (2001). E-readiness as a tool for ICT development. Retrieved January 21, 2004, from *http://www.infodev.org*

Chapter XIII

Offshore Outsourcing: An E-Commerce Reality (Opportunity for Developing Countries)

Purva Kansal, Panjab University, India

Amit Kumar Kaushik, Panjab University, India

Abstract

In an attempt to influence their pace of development, developing countries around the world try and influence the rate of investment (especially foreign private investments) in their economy. These countries attempt to influence investor decisions by matching and changing their portfolio with that of foreign investors' needs. However, to make the country portfolio impressive, a country requires massive investment in infrastructure and other portfolio variables which brings countries at an impasse. This chapter discusses the viability of increasing income as a way out. This leads to another important issue as to how to increase revenue of a country with its limited portfolio of strengths. Recent developments in information technology and the Internet have led to a simple solution to this - offshore outsourcing. Outsourcing as a strategy has been around for many years. Traditionally, companies used to outsource their activities to independent suppliers who were best, but the choice was made from the suppliers located in the vicinity of the outsourcing company for easier coordination and control of the activities of the partner. However, due to developments in e-commerce, distance has become a relative term. Exchange of information in a fraction of a minute, irrespective of physical distance, has made it possible for companies to widen their horizons and look for independent suppliers in different nations — offshore outsourcing.

This allows them to compliment their company portfolio with a variety of created and locational assets thereby allowing them a competitive advantage. As for countries, offshore outsourcing is an opportunity for increasing income. This chapter, while discussing this strategy, also highlights India as a country which has managed to exploit this opportunity successfully.

Introduction

Keynes (1978) gave an income equation. It said:

$$Y = C + I$$

where Y is income, C consumption and I investment.

It states that level of income, output and employment in an economy depends upon effective demand which, in turn, depends upon the expenditures on consumption goods and investment goods. Out of consumption and investment, Keynes states that consumption is stable in the short run. Fluctuations in income therefore are due to changes in investment. Thus, according to Keynes, investment plays a strategic role in determining the level of output, income and employment at a given point of time. Therefore, every country is trying to increase the level of investment in its economy so as to increase the level of development in its economy.

Traditionally, loans from international banks and aids from developed economies were major sources of investment for the developing economies. However, due to losses suffered by banks in recent recessions and recessions in developed economies, these traditional sources of finance have declined. This led to increased dependence of these countries on alternate sources of finance like Private Foreign Capital. Therefore, in an attempt to fuel their development programs, these developing economies are zealously involved in framing policies and strategies to attract Foreign Direct Investment (FDI) (a more reliable source of Private Foreign Capital (Mac Dougall, 1960; Hymer, 1976; Buckley & Casson, 1976; Caves, 1971).

"What attracts/motivates companies to invest abroad?" is a question which plagues the policy makers of developing countries. Researchers have tried to predict FDI tendencies and investor behavior. Works of researchers like Altzinger and Bellak (1999), Goldhar and Ishigami (1999), Penna (2000), Partharthy (1999), UNCTAD (1999), Bajpai (2001), and Goyal (1990) give valuable insight into FDI tendencies. While some like Kathuria (1999), Korwar (1997), Meldrum (2000), and Graham (1997) predict sectoral behavior of FDI, and others like Dunning (1993), Vernon (1996), Levi (1990), Di Mauro (2000), Porter (1985), Sorensen (1998), Caves (1971), Uberoi (1993), WTO (1998), and Rugman, Lecraw, and Booth (1986) give insight into investor behavior.

The work of these researchers has helped countries understand their investor behavior. Armed with this new found understanding these countries were able to experiment with their policy structures and attract FDI. These countries, in a bid to attract more investment, started catering to the whims and fancies of these foreign investors. However, this only led to increased bargaining power of companies and a decrease in the effectiveness of policy initiatives.

Moreover, due to their increased bargaining power, companies evaluate not only policies (because they can be influenced), but the portfolio of a country as a whole. The comparison is not only in terms of components of the portfolio but this portfolio is compared with that of other countries. For example, a pharmaceutical company looking to invest in Asia would not only evaluate an industry-specific environment like manufacturing facilities, infrastructure, availability of factors of production, permit system, etc., in India but compare it with that of China, Malaysia, Philippines, etc. Thus, *a country has to target a dual strategy to attract FDI, i.e., it has to increase the degree as well as the number of variables in its portfolio so as to overshadow the portfolio of competing nations.*

However, to bring about his change a country requires massive injections of investment. One of the ways of acquiring these investment injections is to support industries which are increasing revenues earned by a country. This would result in increased income as well as lead to a spill-over effect. Simply speaking when a tide comes in it raises the whole boat along with it. Similarly, above average rate of return in an emerging industry would lead to increased revenues in support industries as well.

However, the important issue here is how to increase revenue of an industry in a developing country with its limited resource portfolio. The opportunity to solve this question lies in recent changes in the environment.

Environmental Trend: Opportunities For Developing Nations

These days, the quest of countries to attract more FDI is matched by the desire of companies to increase their competitiveness. **Competitiveness** is governed by a company's ability not only to increase value and decrease cost (Kansal & Keshni, 2004) but also to do more than the competitor. *Companies these days measure competitive success more in terms of inventory turnover and speed of serving the market.*

The secret to competitiveness is to nullify the advantages of the leader by quickly creating reputation, economies of scale, learning curve and preferred access to suppliers or channels. This task becomes complex when we consider the additional pressures of globalization, slowdown of global economies and increased opportunity cost of already limited factors of production. To remain profitable and retain competitive advantage companies have learned to focus their energies, both in terms of time and resources, on their core competence. However, this does not mean that a company can neglect the rest of their activities. Companies have realized that the key to success is in achieving superiority in each and every process performed by it for value addition. Only then would it be able to win over customers and defeat the competition.

Here the twist is to find core competencies, internalize them and gain expertise in them and leave the rest to outside specialized suppliers. This not only helps in minimizing the cost but also improves the chances of the process being handled by experts. Therefore achieving an "and" strategy, i.e., decrease cost and increase value and do so more then the competitor. Porter's value chain provides better insight into this strategy.

Value Chain: Exploiting "and" Strategy

Value chain is a high-level model explaining how businesses receive raw material as an input, add value to the raw material received through process and sell finished goods to the customers. Porter (n.d.) says that a firm performs hundreds and sometimes thousands of processes during value addition. However, analyzing all of them becomes difficult. Therefore, for simplification purposes, he categorized these processes into primary and secondary activities. The primary activities are involved with a product's physical creation, its sale and distribution to buyers, and its service after sales. Main activities under the primary category are:

1. **Inbound Logistics:** involve relationships with suppliers and include all the activities required to receive, store, and disseminate inputs.
2. **Operations:** are all the activities required to transform inputs into outputs (products and services).
3. **Outbound Logistics:** include all the activities required to collect, store, and distribute the output.
4. **Marketing and Sales:** activities inform buyers about products and services, induce buyers to purchase them, and facilitate their purchase.
5. **Service** - includes all the activities required to keep the product or service working effectively for the buyer after it is sold and delivered.

The other category of processes, i.e., secondary activities, is performed to provide support for primary activities to take place. These are:

1. **Procurement:** is the acquisition of inputs, or resources, for the firm.
2. **Human Resource management:** consists of all activities involved in recruiting, hiring, training, developing, compensating and (if necessary) dismissing or laying off personnel.
3. **Technological Development:** pertains to the equipment, hardware, software, procedures and technical knowledge brought to bear in the firm's transformation of inputs into outputs.
4. **Infrastructure:** serves the company's needs and ties its various parts together. It consists of functions or departments such as accounting, legal, finance, planning, public affairs, government relations, quality assurance and general management.

These primary and secondary activities are directed towards fulfilling two goals, that is:

- Providing a unique combination that meets customer needs at a price that cannot be matched by the competitors.
- Having a sequence of participants work together as a team, each adding a component of value to the overall process.

Therefore, we are concerned not only with physical flow but also with information and financial flow. The typical cost of a manufacturer's product is about 10% labor, 50% direct material and 40% overhead (Ayers, 2001). The overhead figure includes a lot of information handling and on a proportional scale, resources consumed by this information handling function are several times more than directly related to production.

What porter's value chain teaches us is that each of these activities has to be performed efficiently and effectively to achieve the objective of an "and" strategy. Only then will the company be able to increase its margins.

Though, in today's highly competitive environment every company wants to adopt ab "and" strategy, yet it is not practically possible for an individual firm to possess all the resources and capabilities required to achieve competitive superiority in each of the value chain processes. Even if a company possesses resources and capabilities, high opportunity cost associated with them usually demotivates a company from investing across the value chain. Therefore, an incumbent firm concentrates on those value chain processes which are valuable, rare, non-imitable and organizationally exploited (Barney, 2002). It is these processes which deliver the most value and earn the maximum margins and lead to competitive advantage. Therefore companies, at least conceptually, self-assess and find out what processes of the value chain they perform well and find others who excel or have core competence in the rest of the processes. The idea is called the strategy of outsourcing.

Offshore Outsourcing: Opportunity Due to E-Commerce

Business process outsourcing (BPO) is the contracting of one company by another to execute a business process end to end (Baukley, 2001). It is defined as "strategic use of outside resources to perform activities traditionally handled by internal staff and resources" (Meyers, 2002).

Thus, outsourcing is about integrating two value chains as one so as to gain competitive leverage.

Outsourcing is not a new concept. It has been in use since the turn of the 20th century, when Ford decided that instead of owning rubber plantations to produce its own tires it should outsource them (Prasuna, 2003). Outsourcing of manufacturing has been in practice for many years now.

Figure 1.

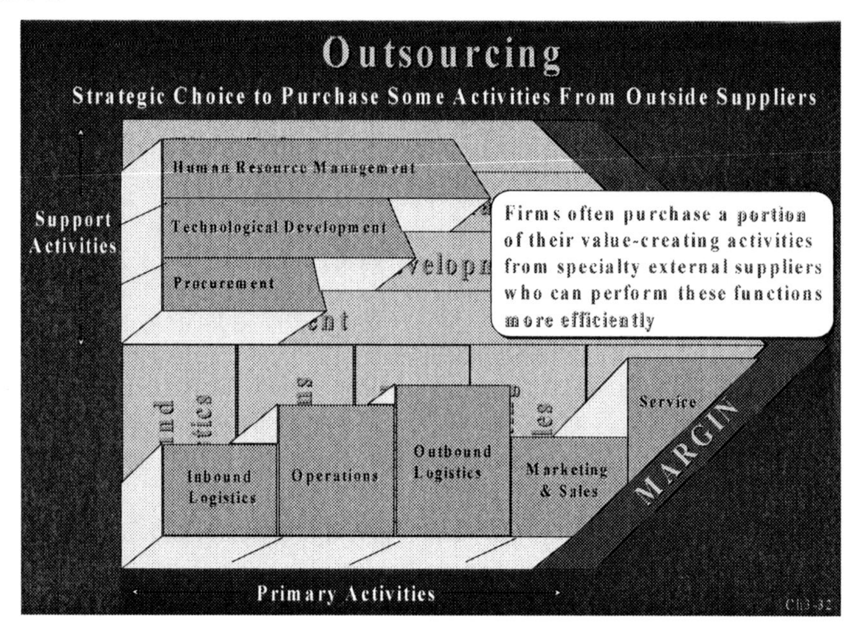

Traditionally, outsourcing meant finding a external supplier but one which was physically close to the outsourcing company so as to facilitate control and coordination of that value chain with that of outsourcing companies (Figure 1). Therefore, the three elements of the value chain (i.e., physical flow, information and finance) were controlled by controlling physical distance between partners.

However, due to increased competition companies have been forced to look for partners who help them improve both the physical and financial aspects of the value chain by complimenting their strengths. Companies look for partners who add a unique set of created and locational assets to the outsourcing companies value chain. This search has led to companies exploiting the option of off-shore outsourcing.

Off-shore outsourcing is the practice of hiring an external organization to perform some or all business functions in a country other than the one where the product will be sold or consumed.

The idea is to benefit from each country's set of created and locational assets called OLI factors (Dunning, 1993) (i.e., Ownership, Locational and Internalization factors). These are unique to a particular area. The strategy works on the fundamental that the domestic companies, over the years, by using the OLI factors develop capacity to perform different business process or to perform a business process differently. A company that integrates its value chain with that of an external supplier can gain synergistic competitive leverages across the value chain. The beauty of the strategy is that a company is able to do so without having to use its internal resources. The strategy further helps the outsourcing company to increase its flexibility by allowing it to adjust to fluctuating demand and needs while paying for only what it needs. Though the mutual benefit is the most important motivation for an outsourcing alliance, some other popular reasons for off-shore outsourcing of activities are:

- To improve its focus it terms of activities it specializes in, thereby reducing operational cost and improving efficiency
- To reduce or to control opportunity cost by redirecting assets from inefficient activities to efficient or core competencies.
- Fundamentally, outsourcing requires the outsourcing company to accomodate its partner's value chain in its own and vice versa. This in turn requires some drastic and open-minded structural changes from both of the parties of the strategy. These structural and attitudinal changes help companies to increase their flexibility and adaptability to changing environment dynamics. Moreover, this also helps companies to share risks with the partners.
- The idea behind outsourcing is to find what a company is good at and outsource the rest to the best. This allows companies to access world class capabilities and resources which might not be available internally. This helps a company cater exponential value to a customer at a fraction of increase in cost. In today's competitive world, this means the difference between acquiring and being acquired.

Thus, off-shore outsourcing is a tough decision for a company as it involves:

- Massive changes in the existing organization structure.
- Massive changes in the existing and acceptable process flows.
- Multiple trade-offs.

Therefore, once implemented outsourcing means a set of changes which are difficult to reverse. Moreover, when adopted in long term it means sharing of information with the provider, which can lead to interdependency among partners and in some cases it might result in dilution of any competitive advantage that a firm has.

Therefore, as and when a company invests time and effort in looking and choosing a partner, it would like to get a competitive advantage with a multiplier effect (i.e., would like to make sure that the benefit of the efficiency in the outsourced activity is reaped across the value chain over a period of time. Practically, this requires trust between the partners and development of feeling of cooperation and mutual benefit (Figure 2). The partners have to be convinced that they are involved in a positivist sum game and that they are in an interdependent relationship where they both need each other to survive. To achieve this state of mutual trust there needs to be a free and more importantly timely flow of information.

Figure 2. Long term implications of outsourcing

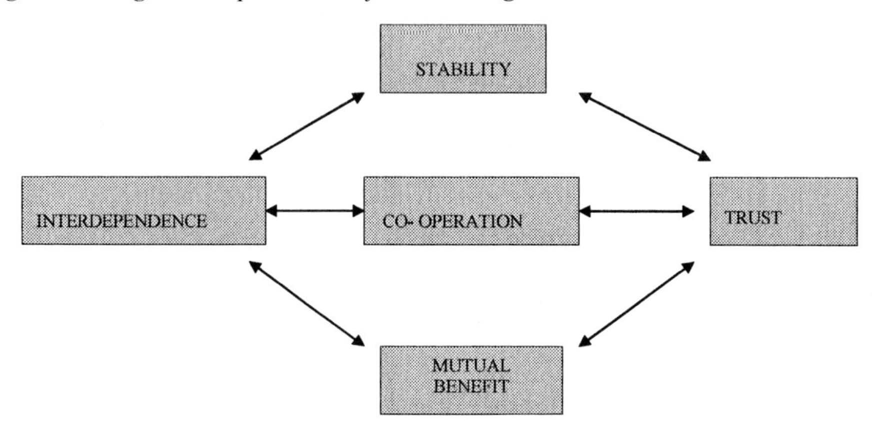

Therefore, a company needs to consider the information viewpoint in addition to physical and financial flow for their value chain. The information viewpoint seeks to improve linkage not only within the company's value chain but also with the partner's value chain by implementing computer applications. This targets the philosophy that irrespective of physical distance, the right information should be available to the right person at the right time.

Until recently, the idea discussed was well perceived and thus most of the activities were either performed in-house or were outsourced to a local supplier who could be monitored closely. Thereby, inefficiencies in linkages and information flow were overshadowed by a good control system. However, linkages and information flow across value chains have come to limelight due to increased physical distance between the company and its outsourced process provider. The increased distance has influenced the level and type of monitoring and control possible. This has also brought to the limelight the high costs of "unintegrated value chain". Moreover, every new country these companies outsource to adds a new and unique set of external environment variables to their portfolio of opportunities and threats. These variables have led to increased insecurities of these companies.

To decrease their insecurities and to reap benefits of outsourcing strategy, these companies felt the need to integrate the partner's value chain with their own. This gives the competitor, customer or another company an illusion of being one value chain (Figure 3).

To achieve this degree of integration, companies need to target three components (i.e., information, organization linkage, and coordination). Information refers to free flow of knowledge across the partners. This enables partners to develop a learning organization attitude and learn from each others mistakes. Thereby, this achieves an increased degree of uniformity in processes across a value chain. This in turn helps in integration. However, due to a lack of infrastructural facilities in developing countries, this strategy seemed far fetched to many. However, the recent developments in e-commerce have made flow of information possible in a fraction of a minute, irrespective of physical distance. This thereby makes the correct information available to the right person at the right time.

These e-commerce changes have helped companies to achieve this information linkage in its value chain despite the infrastructural limitations in the developing countries.

Organization linkages have been defined as a source of competitive advantage by Porter (1985). He argues that competitive advantage is resultant of linkages in the value chain (i.e., what is the relationship between the way one value activity is performed and the cost of performance of another). For example in a fast food chain, the timing of promotional campaigns can influence capacity utilization. Therefore, an activity can be a basis for competitive advantage by itself, but over time single activities are easily copied by competitors. It is important not only to excel in each individual activity but also to manage the linkages between activities. The linkage between activities is more difficult to copy and is therefore more likely to be a basis for sustainable advantage than a single activity.

Once first two elements are there, coordination elements of integration leads to best leverage. Coordination means:

- Having clear demarcation of roles and increasing the action ratio to the information provided across the value chain by improving understanding of the information.
- Reducing ambiguity in communication and increasing accessibility across the chain irrespective of physical distance between the partners.
- Reducing action time by increasing availability of information.

The resulting integration means that the rate of delay between the external environment event, which influences the market service rate and rate of inventory turnover, and its communication to decision makers is almost instantaneous. Which in turn means increased flexibility across the value chain.

A survey conducted by Tanainecz in 2000 best outlines the benefits of an integrated value chain. He says that according to the participants of the survey major benefits of an integrated value chain are increased sales, cost savings, inventory reductions, accelerated delivery times, etc. The respondents indicated that due to information sharing 41% companies experienced increased sales, 62% experienced cost savings and 60% experienced improved quality (Table 1).

E-commerce is a tool which is used these days to achieve the required speed and efficiency in information sharing. It helps a company increase operational efficiency, by complimenting its linkage effectiveness not only within a company, but also with outside independent suppliers and customer. This leads to a sustainable competitive advantage which is difficult to copy. Moreover, e-commerce also helps achieve the target of efficient bi-directional communication.

These days it is possible to outsource any activity of the value chain (i.e., IT, Administration, Distribution and Logistics, Finance, Manufacturing, Sales and Marketing, etc.) to virtually any country. *E-commerce has made distance a relative term.*

This is acting as a huge opportunity for developing countries. Due to the irrelevance of distance, the country's created, locational assets have become the decisive factors. One country which has exploited this opportunity successfully despite its limited portfolio of infrastructural variables is India. For example, infrastructure in India remains below the

Table 1. Value chain survey: Respondents major benefits areas from sharing information

	Percentage of Companies in Excellent or Very Good Chains	Percentage of all Companies
Increased Sales	41%	26%
Cost Savings	62%	40%
Increased market Sales	32%	20%
Inventory reductions	51%	35%
Improved quality	60%	39%
Accelerated delivery times	54%	40%
Improved Logistics management	43%	27%
Improved customer service	66%	44%

Source: G. Taninecz, Forging the Chain, *Industry Week, May 15 2000, p44. Recovered from Robbins, Stephen, P., Decenzo, David, A.,* Fundamentals of Management: Essential Concepts and Applications *(4th Ed), New Delhi: Prentice Hall of India.*

Developed World Standards. National security issues like Kashmir and terrorism scare International Investors and Security of Customer data and labor laws in India remains every company's concern. Still the revenues earned by India from this industry are increasing.

Case India

Today, if you are in London or New York and you phone your bank or credit card company, chances are you would get a number that is more than likely puts you through to India. Companies like Citibank, Dell, Sun Microsystems, LG, Ford, Oracle, GE Capital, British Airways, etc., all have their call centre operations in India.

Why? Motivation can be put across best by the example of the World Bank. By outsourcing back-office operations to India from Washington, D.C., the World Bank was able to achieve three-fold results. Firstly, it was able to slash its costs by 15% due to India's lower wage structure. Second it was able to chop down a back log of accounts receivables and expense forms from 100s of items to just a handful. Lastly, Indian accountants helped the bank get smarter by re-engineering the business (Indian Management, 2002).

Figure 3. Integration of multiple value chains as one

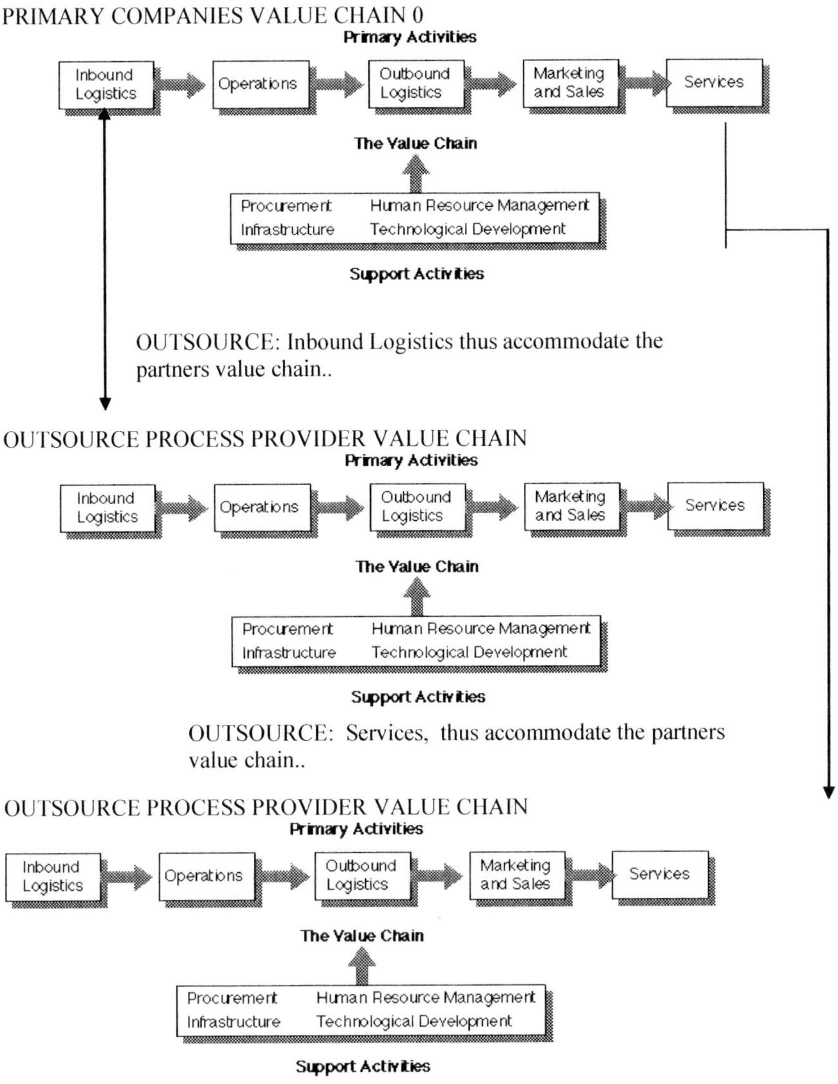

PRIMARY COMPANIES VALUE CHAIN 0

Primary Activities

Inbound Logistics → Operations → Outbound Logistics → Marketing and Sales → Services

The Value Chain

Procurement Human Resource Management
Infrastructure Technological Development

Support Activities

OUTSOURCE: Inbound Logistics thus accommodate the partners value chain..

OUTSOURCE PROCESS PROVIDER VALUE CHAIN

Primary Activities

Inbound Logistics → Operations → Outbound Logistics → Marketing and Sales → Services

The Value Chain

Procurement Human Resource Management
Infrastructure Technological Development

Support Activities

OUTSOURCE: Services, thus accommodate the partners value chain..

OUTSOURCE PROCESS PROVIDER VALUE CHAIN

Primary Activities

Inbound Logistics → Operations → Outbound Logistics → Marketing and Sales → Services

The Value Chain

Procurement Human Resource Management
Infrastructure Technological Development

Support Activities

Therefore, companies enter India motivated by its OLI factors portfolio, that is:

- Highly skilled, English-speaking workforce.
- Cheaper workforce than their Western counterparts. According to Nasscom, the wage difference is as high as 70%-80% when compared to their Western counterparts. If a bank shifts work of a 1,000 people from U.S. to India it can save about $18 million a year due to lower costs in India. According to Mckinsey, giant U.S. pharmaceutical firms can reduce the cost of developing a new drug, currently

Table 2. Network readiness index rank: 45 (2003-2004) (102 countries)

Readiness Component Index	50
Usage Component Index	44
Environment Component Index	44
Main telephone lines in operation in 2002 (growth % 1999-2002 - 56%)	41,420,000
Cellular Mobile Phone subscribers, 2002 (growth % 1999-2002 - 573%)	12,687,640
Internet Users (estimated) 2002 (growth % 1999-2002 - 492%)	16,580,000

Source: World Economic Forum, The Global Information Technology Report. *2003-2004.*

estimated at between $600 million and $900 million by as much as $200 million if development work is outsourced to India (Knowledge Base, 2003).

- Lower attrition rates than in the West. The attrition rate in India is 12%-35%, while in the U.S. it is 70%-120% in call centres (Prasuna D.G., 2003).
- Dedicated workforce aiming at making a long-term career in the field.
- Round-the-clock advantage for Western companies due to the huge time difference.
- Lower response time with efficient and effective service.
- The Internet has driven marginal cost of transferring data to zero and telecom costs are falling.
- Mature process expertise and Six Sigma culture of quality control.

However, what one needs to ponder over is, why have these OLI factors been noticed only now? They have been in the portfolio of India for years.

This can be attributed to what the World Economic Forum calls network readiness. The *World Economic Forum defines Network Readiness as degree of preparation to participate in and benefit from information and communication technology developments*. The Index calculates the capability of an economy. The capability includes 48 variables, like presence of skills to use Information and Communication Technology (ICT) within individuals, access and affordability of ICT for corporations, government use of ICT for its own service and processes, Internet users, cellular mobile phones, etc. The Network Readiness Index ranks India at 45th out of total of 102 countries for 2003-2004. This rank has improved over the years. India was ranked 37th out of 82 countries in 2002-2203 and 54th out of 75 countries in 2001-2002 (World Economic Forum, 2003-2004).

It is because of this development that OLI factors came into the limelight and the passive industry became one of the most active ones. As a result of India's network readiness, the companies were able to overcome their insecurities, integrate their value chain activities and gain advantage of OLI factors. This change has opened up multiple opportunities for India to attract the off-shore outsourcing industry.

Outsourcing: Opportunities for Host Economy: India

The advantages reaped by the Indian economy as a result of the off-shore outsourcing industry have been both direct as well as in spill-over form. Some of the direct advantages are:

1. **Increased employment:** On the hiring front, the industry absorbed about 74,000 people in 2003 (BPOindia.org, 2004) and 245,100 people were employed at the end of March 31, 2004 against 171,100 last year. The industry witnessed a hiring growth rate of about 40%-42%. The future of this industry is even brighter. As per the Nasscom report, BPO is expected to provide employment to 1.1 million people by 2008 (Prasuna, D.G., 2003). The size of the Indian BPO market is likely to be around $9-$12 billion by 2006 and will employ around 400,000 people, ICRA said in its Indian BPO industry report. The quantum of absorption is matched by the salaries. In the outsourcing industry, salaries vary from 8,000 Rs for a customer care representative to Rs 12 Lakhs for training heads. The rise of the BPO sector adds jobs in the ancillary businesses as well, including transportation, catering, housekeeping, etc.
2. **Increased investment opportunities for private domestic companies:** McKinsey & Co. predicts the global market for IT-enabled services to be over $140 billion by 2008. Indian market size estimates of BPO are:

Table 3. These $142 Billion can be broken up and shown as below

Customer Interaction Services	33.0
Finance & Accounting Services	15.0
Translation, Transcription & Localization	2.0
Engineering & Design	1.2
HR Services	5.0
Data Search, Integration & Management	44.0
Remote Education	18.0
Networking Consulting & Management	15.0
Web site Services	5.0
Market Research	3.0
Total	141.2
In that the opportunity for India will be $ 17 Billion.	

Source: NASSCOM McKinsey Study - India IT Strategies

a. Nasscom has estimated that the Indian ITES industry will gross over $5.7 billion by 2005 (based on a conservative year-on-year growth of 65%).

b. Nasscom-McKinsey in 1999 estimated that by 2008 it will be $17 billion, revised to $21-$24 billion by 2008. India can capture 25% of the global BPO off-shore market and 12% of the market for other services such as animation, content development and design services.

c. Gartner estimates $1 billion (2002), $1.2 billion (2003), and $13.8 billion by 2007. Gartner does not incorporate animation, medical or other (legal) transcription services, GIS, market research, data search, research and development, network consultancy and other non-business processes in its estimates on the ITES market size and potential (Indian Management, 2002).

Motivated by these Industry growth figures, private domestic companies and foreign companies are investing in the industry. This in turn is leading to increased investment in the economy. For example, Infosys Technology launched Progeon in April 2002. Progeon received a $5 million investment from Infosys and a $20 million infusion from Citigroup Investments (Op_cit Indian Management, n.d.) These massive injections of investments, in turn, are leading to multiplier effect on income and employment in India, thereby, leading to a more sustainable growth rate and development.

3. GDP and revenue shows that Indian revenues from BPO are estimated to have grown 107% to $583 million (Knowledge Base, 2003) According to the Nasscom report, the contribution of the BPO segment to countries' GDP is expected to be around 3% by 2008. If BPO grows at the expected rate from 2 billion to 10 billion over the next four years, then the contribution could be as high as 15%-20% (Prasuna D.G., 2003). Table 2 gives details of Indian BPO revenues up to 2007, as per Gartner Dataquest.

Table 4.

Revenue \ Year	2002	2003	2004	2005	2006	2007	CAGR
Offshore BPO Revenue	1,322	1,825	3,017	6,439	12,563	24,230	78.91
Indian BPO Revenue	912	1,205	1,961	3,928	7,412	13,811	69.35
Total BPO Market	110,167	121,687	131,171	143,090	157,033	173,070	9.45
CAGR in % 2002-07		Figures in $ million					

Source: Gartner Dataquest (May, 2003)

The outsourcing industry also creates some indirect opportunities for the host economy, called spillover effects. *Spillover effects include increased opportunities for exports, improvement in standard of living, increased management skills, technology revolution, increased awareness among Indian consumers, improved job opportunities and work environment.* Moreover, the increased disposable income of the public has led to increased consumption of other products, which has increased the market size for other industries. Thereby, making India's market size as one of the primary variables attracting FDI.

4. Opportunities for Exports shows that off-shore outsourcing is also opening up opportunities for India in the export sector. India lays claim to producing more than $6 billion in software exports in 2002-2003. Moreover, the total Indian technology export market is estimated to be $12.5 billion, with predictions of growth to $50 billion by 2009 (Ehr, 2004).
5. Off-shore outsourcing to India has exposed many corporates to India's portfolio of created, locational assets. Many of these companies, either to avoid long-term bottlenecks of off-shore outsourcing or to take advantage of a booming industry, eventually end up setting up off-shore operations in India. This has led to increased foreign investment in the Indian economy. For example, Analog Devices will invest $50 million over the next five years to expand its India operations. While Cadence Design Systems also plans to invest $50 million over the next three years in the region, to scale up product development and high-end customer support. As many as 230 multinational companies have arrived in Bangalore since 2001 for R&D work according to NASSCOM, which projects an additional $1.5 billion in investment over the next three years (NASSCOM, n.d.).

According to Nasscom, nearly half of the Fortune 500 companies outsource various functions to Indian firms.

Summary

The increased pressures of globalization, severe competition and the general slowdown of global economies have forced companies to rethink their strategic position. To remain profitable and retain competitive advantage companies have learned the need to focus their energies both in terms of time and resources on core competencies, leaving the non-core processes to specialized providers. Today it does not seem out of place for a company in the US to outsource its process to India. Technology has made the process virtually hassle free. The distance of thousands of miles does not matter as even at that distance it is possible to communicate, coordinate and control operations.

Web-based tools have made it possible for companies to evaluate and examine their value chains and those of their service providers. This allows them to seek opportunities to outsource non-core activities and gain cost as well as competitive advantage. This new found interest and need of outsourcing could prove a golden opportunity for developing countries. This could lead to an increase in income and in turn investment opportunities for these countries.

References

Altzinger, W., & Bellak, C. (1999). Direct versus indirect foreign direct investment impact on domestic exports and employment. Retrieved March 1, 2004, from *www.wv.edu/gee!wp09.pdf*

Ayers, J. B. (2001). *Handbook of supply chain management.* WA: St. Lucie Press/APICS series on Resource Management.

Bajpai, N., & Sachs, J. (2001). The decade of development: Goal setting & policy changes in India. Retrieved May 12, 2004, from *www.cid.harvard.edu/cidwp/046.pdf*

Barney, J. B. (2002). *Gaining and sustaining competitive advantage* (2nd edition). New Delhi, India: Pearson Publishers.

Baukney, H. (2001). *Be prepared for business process outsourcing.* Retrieved March 14, 2004, from *www.cio.com/archive/050101/bpo.html*

BPOIndia.org. (2004). Attrition in India BPO industry. Retrieved October 4, 2004, from *http://www.bpoindia.org/research/attrition.shtml*

Buckley, P., & Casson, M. (1976). *The future of MNE's.* London: MacMillan.

Caves, R.E. (1971). International corporations: The industrial economics of FDI. *Economica*, *38*, 1-27.

Chopra, P.N. (2000). *Advanced economic: Theory (micro and macro analysis)* (4th ed.). Ludhiana, India: Kalyani Publishers.

Di Mauro, F. (2000).*The impact of economic integration on foreign direct investment and exports : A gravity approach.* Retrieved November 20, 2000, from Web site for the Centre for European Policy Studies (CEPS) *www.ceps.be/pubs/2000/wd/156pt2.htm*

Dunning, J. (1993). *Multinational enterprises and the global economy.* UK: Addison-Wesley.

Ehr, M. (2004). *Indian outsourcers make big gains.* Retrieved October 23, 2004, from Web site for Network World Outsourcing Letter *http://www.nwfusion.com/newsletters/asp/2004/1018out1.html*

Goldar, B., & Ishigami, E. (1999,May 29-June 4). Foreign direct investment in Asia. *Economic Political Weekly*, *24*(22).

Goyal, S.K. (1990). *Multi national corporations & Indian policy framework.* Retrieved May 12, 2004, from *www.isidev.nic/vsisidev/mnc/pdf*

Graham, E. M. (1997). *Working together: Foreign direct investment & trade.* Retrieved June 12, 2002, from *www.cipe.org/ert/e25/gratiae25.html*

Hymer, S. (1976). *The international operations of national firms: A study of foreign direct investment.* PhD dissertation, Massachusetts Institute of Technology. (Originally published 1960.) Cambridge, MA: MIT Press.

Indian Management. (2002, September). Weighing the pros and cons of outsourcing. *Indian Management*, 42-44.

Kansal, P., & Keshni, A. (2004). Supply chain management and customer relationship management – Development and integration – Focus.In H.S. Kehal (Ed.), *Corporate strategies in a digital world* (pp. 359-380).Hershey, PA: Idea Group Publishing.

Kathuria,V. (1999). Spillover effects of technology transfer to India : An econometric study. Retrieved May 10, 2004, from *www.isidev.nic.in/vsisidev/wroking.htm*

Knowledge Base. (2003). Recovered Octover 4, 2004, from *http://www.bpoindia.org/knowledgeBase/#indian-market-size*

Korwar, A. (1997). *Creating markets across the globe.* New Delhi, India: Tata McGraw Hill.

Levi, M. D. (1990). *The growth and special problems of multinational corporations. International finance. The markets and financial management of multinational business* (2nd ed. - International Edition). Cambridge, MA: McGraw Hill.

Mac Dougall. (1960). The benefits and cost of private investment from abroad: A theoretical approach. *Economic Record, 36*, 13-35.

Meldrum, H., & Duncan (2000). *Country risk & foreign direct investment.* Retrieved May 12, 2001, from *www.findarticles.com/m1094/135/59964458/p1/articles.jhtml*

Meyers, J. T. (2002). *Why company's outsource services: Opportunities and challenges.* Retrieved March 10, 2004, from *http://oeiwcsntsl.omron.com/pdfcatal.nsf/0/6B03B04F72D268E1862S6B7D007AC07C/$File/outsourcingarticle.pdf?openelement*

NASSCOM. (2003). *India Tech boom.* Retrieved October 23, 2004, from *www,nasscom.org/artdisplay.asp?Art_id=234*

Op_cit Dunning

Op_cit Indian Management by Arrangement with Knowledge@Wharton

Op_cit Knowledge Base

Op-cit Porter

Op-cit Porter, Michael E., Pg 33-52.

Op-cit Prasuna D.G.

Parthasarthy, R. (1999). Best states to invest in. *Business Today*. Retrieved May 2, 2004, from *www.indiatoday.com/btoday/19991222/cover2.html*

Penna, A. (2000). *Delhi tops in states business ranking.* Retrieved March 13, 2003, from *www.indiainfo.com/2000/09/01/01delhi1.html*

Porter, M. E. (1985). *Competitive advantage – Creating and sustaining superior performance.* Canada: The Free Press.

Prasuna, D.G. (2003, January). BPO: Advantage India. *Chartered Financial Analyst*, 19-26.

Rugman, A. M., Lecraw, D. J., & Booth, L. D. (1986). *International business- Firms and environment* (International Edition). Cambridge, MA: McGraw Hill.

Sorensen, O. J. (1998). *Ways of choosing international market entry.* Retrieved June 13, 2002, from *www.aalborguniversity.edu/issno908-16666.pdf*

Uberoi, R. (1993,December 9). Creating conducive environment for FDI. *The Hindu.* Retrieved June 1999 from *www.prakesh.org/economy/ecohin1993120901.htm*

UNCTAD. (1999). The least developed countries '99 report overview. Retrieved June 21, 2003, from *www.unctad.org/en/pub/LDC 1999.pdf*

Vernon, R. (1996, May). International investment and international trade in product life cycle. *Quarterly Journal of Economics*, 190-207.

World Economic Forum. (2003-2004). *The global information technology repot: Towards an equitable information society.* New York: Oxford University Press.

WTO. (1998). FDI flows and global integration. Retrieved July 22, 2003, from *www.wto.org/english/thewto_e/minist_e/english/aboute/22fact_e.htm*

Chapter XIV

Adoption and Implementation Obstacles of E-Banking Services: An Empirical Investigation of the Omani Banking Industry

Abdulwahed Mohammed Khalfan, Public Authority for Applied Education & Training, Kuwait

Abdullah Akbar, Public Institution for Social Security, Kuwait

Abstract

All forms of business organizations are being drawn into e-commerce and e-business. Electronic commerce is a powerful new way of conducting business and one that has presented many opportunities for enterprises and consumers. Yet, despite its many advantages there is evidence that it has not been adopted in full and has not reached its full potential. The global nature of e-Business provides developing countries with a unique opportunity in market places that were beyond their reach. It has the potential to reduce physical trade obstacles, increase market access and trade efficiency that can provide a competitive stimulus for local entrepreneurs and organizations in developing countries. Interests in e-commerce and electronic markets have been largely been focused on North America and Europe, and yet there are many interesting developments happening in some of the developing countries which have received little attention. This study aims to address these issues. This research chapter explored the potential impeding factors that could inhibit the wide adoption and use of electronic commerce applications in the Omani banking industry. Data were collected using a survey questionnaire complemented with using semi-structured interviews, and review of internal banking documents. The participants in the study were IS executives and managers. The results provide a pragmatic picture on the adoption and use of e-commerce applications in this country's core sector. One of the major inhibiting factors in this study is the lack of top management support. Top management perspectives and attitudes toward e-commerce adoption and use play an important role in developing internal IS/IT competencies and success. Information privacy and security issues were identified as serious inhibiting factors on the successful adoption and use of electronic banking applications in the financial sector. Other factors such as power relationships (e.g., conflict between managers that arise during the process of IS/IT adoption), and lack of investment in e-commerce applications were found less significant as inhibiting factors. Although this is a context specific research, the findings should be relevant to other businesses in Arab countries in particular and developing countries in general when contemplating their e-commerce strategies.

Introduction

E-business is a new way of doing business that involves connectivity, transparency, sharing, and integration. It requires the integration and alignment of business processes, technology, and people with a continuously evolving e-business strategy (Helmi, 2002). E-business involves fundamentally rethinking the business model to transform an organization into a digitally networked enterprise. Electronic commerce is a new way of conducting business that brings innovation in every part of the industrial value chain (Kalakota & Whinston, 1996).

The World Wide Web (WWW), the first and current networked global implementation of a hypermedia computer-mediated environment (CME), is increasingly being recognized as an important emerging commercial medium and marketing environment (Hoffman & Novak, 1997). The convergence of technological, economic, and organizational forces is resulting in the widespread use of e-commerce. Technologically, the power of desktop systems, networking infrastructures (Internet, intranet, and extranet), data warehouses, and the browser technologies are enabling organizations to use World Wide Web for true corporate-wide information processing (Lack, 2000).

The global nature of e-business provides developing countries with a unique opportunity in market places that were beyond their reach. It has the potential to reduce physical trade obstacles, increase market access and trade efficiency that can provide a competitive stimulus for local entrepreneurs and organizations in developing countries. Banks in the Arab Gulf region in general have been quite slow to launch e-banking services.

While they are convinced that online services reduce overhead significantly, a mixture of customer insecurities, technology investment costs and a lack of market-readiness have all conspired to make e-banking "unattractive." There is a strong argument that the electronic commerce is going to become the dominant form of interchange between business and consumers. Moreover, computing, telecommunications and consumer electronics are merging into a complex and vast constellation of information infrastructures that provides a totally new platform for delivering digital services based on Internet protocols and IP-based technologies. E-commerce is already causing fundamental changes to the traditional economic market place, affecting every aspect of how business is and will be conducted. The four principal factors that are bringing about the era of e-commerce are: (1) reducing transaction cost; (2) providing better services to customers; (3) meeting consumer demand; and (4) creating efficient transactions (Kalakota & Whinston, 1996,1999). E-commerce has the potential to open up global trading, with few limitations to growth. The literature and research surrounding e-commerce activity has been dominated by an examination of its relevance to large enterprises like those of the banking sector.

With the advent of the Internet and the World Wide Web in 1990, economic entities such as business firms and consumers are converging on the new frontier for economic exchange: e-commerce. In reality, e-commerce has many synonyms, such as online commerce, Internet commerce or trading and digital commerce. For example, IBM uses the term "e-business" to encompass all the implications of electronic, Internet-related technologies for business functions, from HRM to marketing to corporate strategy. It is obvious that there are often overlaps between e-business and Internet marketing activities, since one of the most important objectives of Internet marketing is doing business online. E-commerce has become a key issue for business over the past few years, with increased awareness about the use of computer and communications technologies to simplify business procedures and increase efficiency (ECIR, 1996).

A resource-based view of organizations maintains that organizational performance is linked to resources. In this view, the resources are firm-specific, rare and difficult to substitute or imitate (Barney, 1991). In general, the resources of an organization are composed of its technologies, capital, knowledge, employees, information systems, structure, rules, and procedures that are controlled by the organization. Researchers in the IT field have shown that IT is an important resource for improvements in the performance of enterprises (Chatfield & Bjorn-Anderson, 1997). IT viewed in the context of a resource-based perspective shows that the IT resources of the organizations that are participating in global e-commerce have an impact on their performance (Huda et al., 2000).

A broad perspective of e-commerce is taken so that it includes not only business-to-business and business-to-consumer e-commerce activities, but also intra-organizational e-commerce activities such as workflow and enterprise resources planning (ERP) systems and their integration with external electronic commerce systems such as online shopping malls. Choi and Lin (1995) state that in general, organizations have had the disadvantage of only being able to access local markets, whereas large companies have grown to cover worldwide markets. Advocates of the virtual market, Choi and Lin (1995) believe that e-commerce is the "great equalizer" in terms of its potential to expose small companies to global markets.

The chapter is structured as follows. The next section sets out some definitions of e-commerce, current and future trends on e-commerce, background on Internet banking trends and benefits, Internet banking service innovations, and developing countries' difficulties associated with e-commerce applications. Research objectives and background on Oman are discussed next. The research methodology is then described and the results of the analysis of the data are presented. The final section presents the conclusions and summarizes the research.

E-Commerce Definition

There appears to be no simple and straightforward generic definition of the term. Some examples of what IT authors define as e-commerce are given below.

According to Bloch et al. (1996, p. 93), e-commerce is, "The buying and selling of information products and services via computer networks. It is the support for any kind of business transactions over a digital infrastructure." Another definition of e-commerce is provided by Angell and Heslop (1995, p. 1): it is "the process of conducting business on the Internet using a combination of tools and establishing a server presence for users." According to Anon (1996) e-commerce is "using electronic information technologies to improve business relationships between trading partners." It can be used in a variety of business environments: Business-to-Business, Business-to-Consumer, Business-to-Government, and Government-to-Constituent. Adoption on the other hand is defined as the process through which organizations or individuals decide to engage in e-commerce activities.

Current and Future Trends in E-Commerce

The dollar value of e-commerce transactions continues to increase at a phenomenal rate. Forrester Research predicts that by the year 2004, the value of goods and services traded between organizations over the Internet will represent 4% of the global economy (approximately $1.3 trillion). Firms seeking to do business overseas must use different strategies such as customizing cultural elements, languages, and currencies to make visiting their Web site a meaningful experience in each market. The statistics of e-commerce are overwhelming. For example, 90% of top US banks were expected to have full-service Internet banking sites by 1999. About 2.3 million Americans bank online today and that number is growing quickly (Dixon, 1999). In addition, it has been predicted that by 2000, 1,000 banks worldwide would have Internet sites and within five years, one in every five US households would participate in Internet banking (Dixon, 1999). On the other side, however, Andersen Business Consulting surveyed nearly 150 banking, lending, asset management, brokerage and insurance firms between April and May of 2001 to identify leading e-business practices and opportunities in the financial services industry. According to Andersen's findings, the banking industry still has much to do to meet e-business customer expectations and needs. Those banks that do not improve their online capabilities will quickly fall behind their more innovative competitors (Neckopulos, 2001). In addition, according to the Gartner Research Group, the vast majority of Internet bank users (21.6 million) merely checked accounts online. Only 6.6 million used services such as bill payments online (Hechinger, 2001). The Boston Consulting Group (BGC) estimated that the value of goods sold over the Internet in the Asian markets to increase to US $1.3 billion by 2003. The study had also reported that 66 million Asians are hooked up on the Internet either motivated by cheaper access charges or more interesting things to purchase online.

Background on Internet Banking Trends and Benefits

The Internet is a relatively new channel for delivering banking services. However its early form, "online banking services," which required a PC, modem and software provided by the financial services vendors, were first introduced in the early-1980s with the "Homelink" service offered by the Bank of Scotland and the Nottingham Building Society in the UK (Tait & Davis, 1989). However, generally it failed to get widespread acceptance and was discontinued.

In its very basic form, electronic banking can mean the provision of information about a bank and its services via a home page on the World Wide Web. A more sophisticated Internet-based service provides customers with access to their accounts, the ability to move their money between different accounts, make payments or apply for a loan or mortgage, and so on. The term "Internet banking" will be used in this chapter to describe the latter type of provision of services by a bank to its customers. The customer may be an individual or another business. With the rapid growth of other types of electronic services, mainly on the Internet, since the mid-1990s, banks have renewed their interest in electronic modes of delivery. A large number of organizations are currently announcing the launch or development of transactional Internet banking (Liao & Cheung, 2002). Similarly, Filotto and Omarini (2000) estimated that 60% of retail banking transactions would be performed online in 10 years' time. A report from Cap Gemini mentioned that large banks expect that 25% of transactions might be online by 2005 (Whyte, 2000).

As in the electronic trading environment, the real role of Internet banking is to provide electronic services, low cost and an information-rich environment. Electronic trading represents the front-end services (i.e., the retail branch). Accordingly, banks are replicating the branch experience online, even to the extent of creating a "3D virtual branch" (Retail banking, 1997) for their customers to navigate through. These services innovate specialization in the degree of services, mobilize efficiency, increase productivity, reduce complexity and lower costs. Information systems are taking over the branch's responsibilities to provide "non-stop 24-hour online banking" in order to provide maximum flexibility. The beauty of this system is that the customer can obtain services at any time from anywhere around the world. For example, the Bank of Scotland set up a statement "open a branch in your own living room" on its opening page, in order to give the customer more confidence.

Many banks are eager to utilize this channel (the Internet) because of the numerous potential benefits that are argued to be associated with it. In terms of Internet banking trends, the dollar value of e-commerce transactions continues to increase at a higher rate.

Forrester Research predicts that by year 2004, the value of goods and services traded between organizations over the Internet will represent 4% of the total global economy (i.e., approximately $1.3 trillion). Global Internet banking services are expected to reach 7.3 trillion by year 2004 (Lack, 2000). Andersen Business Consulting surveyed nearly 150 banking institutions between April and May of 2001 to identify leading e-business practices and opportunities in the financial services industry. The findings showed that the banking industry still has much to do to meet e-business customer expectations and needs. According to the Gartner Research Group, the vast majority of Internet bank users (21.6 million) merely checked accounts online. Only 6.6 million used services such as bill payments online (Hechinger, 2001).

The use of IT within banking may benefit in reducing costs by replacing paper-based methods and also by providing services to the customers through remote banking (Pigni et al., 2002):

- The Internet lets companies bypass others in the value chain.
- Firms can use the Internet to deliver new products and services for new and existing customers.

Internet Banking Service Innovations

Banks are gradually becoming more aware of the importance of electronic banking (e-banking) in this era. The Internet has emerged as a key competitive arena for the future of financial services. Many organizations are using e-business as a vital part of their competitive strategy with the influx and media interest in electronic information. This topic was chosen to extend the current investigation into the implications of Internet banking in today's world. It fills a gap in the literature by examining the potential barriers that lie behind e-commerce's huge growth and expansion and provides a framework for further research to be conducted in this area. In short the study aims to present a systematic discussion of the issues underlying the slow IT adoption in the Omani banking sector.

The banking sector, traditionally one of the most conservative sectors, has been obliged not only to embrace the development of IS/IT technology, but in many ways, to the lead the way. To date, little research has been carried out on the subject of how the banks make best use of IS (Alsindi & Mayhew, 2001). Adoption is the process through which organizations or individuals decide to engage in e-commerce activities in their daily business.

As in the electronic trading environment, the real role of Internet banking is to provide electronic services, low cost and an information-rich environment. Electronic trading represents the front-end services (i.e., retail branch). Accordingly, banks are replicating the branch experience online, even to the extent of creating a "3D virtual branch" (Retail banking, 1997) for their customers to navigate through. These services innovated specialization in the degree of services, mobilize efficiency, productivity and reduce complexity and lower cost. Information systems take over the branches responsibilities to provide "non-stop 24 hours online baking" in order to enable the customers' maximum flexibility. The beauty of this system is that customer can obtain the services at any time from anywhere around the world. For example, Bank of Scotland set up a statement "open a branch in your own living room" at the opening page, in order to give the customer more confidence.

Internet banking can not replace the branch activities 100%. There are still some services, such as withdrawal of cash and deposits, which cannot be implemented because "the cash stops at the till" (Polovina et al., 1999). This means that, the innovated services on the Internet are tailored specifically to the network environment, and electronic capability (Network-centric banking, 1998). This technology created high demand within the banking boundaries. According to a survey by Meridian Research of Needham, 60% of USA banks offers WWW pages and 90% are predicted to by the year 2000 (Anonymous, 1998).

The provision of new, innovative services includes the following functions: transfer money between the same branches of the bank; set up bill payments online to cover a wide range of firms; standing orders; direct debits; pay bills; check balances and statements; view transactions with a search and sort facility; print out full bank account details for long time period; order cheque books and deposit books; copy information into other financial management packages such as Microsoft Money; and loan applications or interest-rate calculator.

The above services reduce the amount of human resources, cost, time and efforts and increase the processing power.

Developing Countries' Difficulties Associated with E-Commerce Applications

Developing nations are far behind the more advanced Internet economies, not just in the number of Internet Service Providers (ISPs), hosts connected to the Internet, number of individual users online, Internet diffusion ratios, and number of organizations leasing line connections. This imbalance also extends to content in terms of the number of Web sites in the developing countries (DC), the amount of local language content, and the use of online content by key sectors. According to the International Telecommunications Union report (Challenges to the Network: Internet for Development, 1999), there are more Internet hosts in Finland than in all of Latin America and the Caribbean. There are also more hosts in New York than in all of Africa.

The World Bank's annual development report, Knowledge for Development (1998), stresses the importance of leveraging new media technologies like the Internet in developing countries in areas like learning, and the training and retention of skilled workers in financial institutions. But the main obstacle that faces developing countries is that they lag behind in communication infrastructure, technical know-how, and information processes about the economy and the environment. The lack of adequate banking infrastructure is seen as one of the problems faced by developing countries in building e-business solutions.

The global nature of e-commerce potentially provides developing countries with a unique opportunity to compete in market places that were beyond their reach. E-commerce has the potential to reduce physical trade barriers, and increase market access and trade efficiency and could provide a competitive stimulus for local products and entrepreneurs.

Research Objectives

There is growing awareness of the need to understand MIS issues from a global perspective (Palvia et al., 1992). This chapter uses the results of a research project which should be of interest from both a research and policy-making standpoint. It will provide Omani banking organizations with a framework to evaluate their use of IT on e-commerce and alert them to potential areas of business opportunities and weaknesses. The results of the study should also be useful for multi-nationals operating or contemplating operations in the Middle East and transnational organizations responsible for regulating IT.

This research chapter, therefore, has two main objectives:

1. To identify the factors that could inhibit the wide adoption and use of electronic commerce applications in the Omani banking industry, and
2. To rank the importance of such inhibiting factors on the decisions to adopt and use e-commerce applications in Omani banks.

In line with these two objectives, the following research question was developed:

> What are the barriers/inhibitors in terms of <u>industrial</u> and <u>managerial</u> dimensions that affect the wider use of e-commerce applications in the Omani banking industry?

Background on Oman

The Sultanate of Oman is a developing country, which is located on the south east of the Arabian Peninsula. The Sultanate of Oman is the second largest country among the Gulf states after Saudi Arabia. It is spread over 309,500 square kilometers and has a 1,700-kilometer coastline. Since the accession of Sultan Qabbos in 1970 to the throne, there has been remarkable progress in different fields such as economy, politics, education, and the civil services.

The modernization and restructuring of the administrative systems of the civil service as a result of the continuous progress have been vital in term of the changing circumstances and future requirements. The Sultanate of Oman enjoys a stable political, economic, and social system. Although Oman has achieved remarkable progress, both socially and economically with implementation of five-year Development plan, the country is watching closely the new challenges facing the country with the fluctuation of oil prices.

The country is sparsely populated with 7.40 inhabitants per kms. Of the 2.29 million population, about 600,000 are expatriates. A very strong feature of the country is its young population. About 50% of the population is below the age of 15 years.

Oman is classified by the World Bank as an upper middle-income developing country. The production of oil and the export of crude oil is the principle activity of the Omani economy contributing to approximately 32% of GNP.

Research Methodology

The lack of systematic research in this area justifies the exploratory nature of the study, as this research study may be characterized as exploratory (Yin, 1994). The goal of the field study was to identify the factors causing barriers to the execution of business-to-consumer e-commerce applications and expansion. The theoretical framework and relevant literature review guided us to formulate the field study.

This section outlines the research methodology used in this study. It covers the research approach, questionnaire design, sample, and semi-structured interviews.

Research Approach

In this study, a methodological triangulation approach is adopted. According to Leidner and Jarvenpaa (1993, p. 28): "Case study research is appropriate in situations where the research question involves a 'how', 'why', or exploratory 'what' question, the investigator has no control over actual behavioral events, and the focus is on contemporary as opposed to historical phenomenon." Following the recommendations of Yin (1994), the data in this study were collected thorough survey questionnaires and semi-structured interviews supported by looking at organizational documents. These different data collection methods were used to provide more detailed data about the phenomenon under investigation. The unit of analysis defines the boundaries of the case study research. In this study, the units of analysis were the IT functions within the selected Omani banking organizations.

Questionnaire Design

The questionnaire is designed to obtain a view of the potential barriers of e-commerce expansion in Omani banks. The questionnaire consists of two categories of close-ended questions on issues to help identify inhibiting factors in both dimensions of the study (i.e., industrial and managerial). The inhibiting issue factors are developed based on the literature review (Liao & Cheung, 2001, 2002; Liao et al., 1999). Lists of six and eight barriers are used for the managerial and cultural/environmental perspectives respectively. Participants are asked to rank the perceived importance of each of the listed barriers to adopt and use e-banking applications in Oman using a 5-point Likert scale (ranging from strongly agree to strongly disagree). The research instrument was validated using procedures recommended by Straub (1989), i.e., instrument review by an expert in the field, a pilot test, internal reliability, and statistical conclusion validation. The questionnaire was written and administered in English. The use of English was deemed appropriate because English is the business language in Oman.

The mean score and standard deviation for each perceived inhibiting factors were calculated. Then, the factors were ranked according to their mean score to identify the order of perceived importance of the potential impeding factors that could inhibit the wide adoption and use of e-commerce applications in the Omani banking industry. The standard deviation value is used as a measure of the spread of opinions. The reliability of the hampering factors used in the two multi-scale questions were determined by using the Cronbach alpha procedure: the managerial hampering factors (6 items) Cronbach alpha 0.744, and the industrial hampering factors (4 items) Cronbach alpha 0.739.

Sample

The commercial banking system in the Sultanate of Oman consists of the Central Bank of Oman and a total of 16 commercial banks of which only seven banks have their headquarters locally. The other nine banks are branches of foreign banks. The Central Bank of Oman and the seven locally headquartered banks are the target population for the study. They were all contacted and all agreed to participate in the study. The banks are National Bank of Oman, Oman Arab Bank, BankMuscat, Bank Dhofar AlOmani AlFaransi, Oman International Bank, Commercial Bank of Oman and Industrial Bank of Oman.

Thirty-nine questionnaires were distributed to the IT departmental heads and a randomly selected group of IT staff with administrative duties in the participating banks. Twenty-eight questionnaires were returned with a response rate 71.8 % of the total sample. The data was analyzed using the SPSS package. Since the study is exploratory and trying to identify broad trends, simple mean scores and frequencies were derived to draw conclusions. The reported results in the next section are based on responses from the 28 questionnaires' respondents and the 8 interviewees.

Semi-Structured Interview

Putins and Petelin (1996) state that interviews are an extremely important form of communication in society. They are a means by which information is exchanged between individuals and successful communication is achieved. Mintzberg (1979, p. 587) states that "semi-structured interviews provide a controlled framework which facilitates analysis but also allows for the collection of 'soft' anecdotal data." Marshall and Rossman (1989) provide a framework for matching research purpose and strategy with research methods and data-capture techniques. The researchers suggest that when the research study has a descriptive and exploratory focus, as in the case of this study, appropriate research strategies are field studies comprising of in-depth interviews.

Interviews were conducted throughout the period of data collection. A number of questions were developed to provide "some" structure for the interview and to ensure coverage of all issues related to e-commerce growth. Further, the questions themselves were grouped to address the various dimensions of the phenomenon under study.

In total 8 people, selected in light of their qualifications and involvement in their organizations, were interviewed including departmental heads and senior IT managers. The interviews were semi-structured. Interviews were recorded to free the interviewer from note taking and to increase the accuracy of data collection.

After the data were collected, analysis began with the transcription of the interviews. All data obtained from interviews and documents were consolidated and linked together to create a picture of the entire e-commerce adoption process. A content analysis (Remenyi, 1992) was used to discover important patterns from the data. All data obtained from the interviews and the organizational documents (such as organizational flowcharts and information on IT plans) were combined to create a picture of IT processes in the study sample.

Research Analysis and Discussion

Discussing the Industrial Hampering Factors

The following factors, as shown in Table 1, have been identified as existing within the banking industry and were consequently inhibiting the incorporation of e-commerce by that business.

Banks in the Arab Gulf region in general have been quite slow to launch e-banking services. While they are convinced that online services reduce overhead costs significantly, a mixture of customer insecurities, technology investment costs and a lack of market-readiness have all conspired to make e-banking "unattractive" (The Gulf Business, 2001). IT managers of Oman banks have welcomed the innovative technology, but at the same expressed cautions and warrants that the new technology might not be very much "suitable" for the Omani environment. One manger was quoted as saying, "Worldwide, e-commerce is growing at an exponential rate, and fast replacing several traditional delivery channels. In Oman, as well, with the growing population of personal computers, people are looking actively at this channel. Secondly, several businesses in Oman especially the banking sector are looking at the Gulf Cooperation Council (GCC) and other markets to expand their business. In this context, e-business is extremely relevant to Oman in the days to come."

The first factor was identified as "industry is not yet ready for e-Commerce." The factor had a mean value of 4.24 and captured 88% of the respondent's agreement. It seems that the banking industry in Oman as a whole is not yet ready for adopting e-business. An IT senior employee made his comment on this particular point by saying: "the banking industry in Oman is definitely looking at the Internet as a new, innovative, and effective delivery channel for extending a plethora of banking services over the Web. However, at the same moment the road ahead is not, as I see it, very clear for industry as a whole to make a top decision to go ahead with such strategy of implementing e-banking services. It looks to me it takes some time to clear up the ambiguities."

The second most cited factor was "security problems of virtual banking systems," which was selected by 80% of the respondents. The mean value was 4.04. This finding was consistent with Mukti (2000) who found security is the most important barrier to e-commerce expansion in Malaysia. Security is perhaps the most feared problem on the Internet. Banks and customers take a very high risk by dealing electronically. Increasingly, the pundits of e-banking are raising the stakes on online security (Whyte, 2001). Customers are still reluctant to transfer their money through telephone lines, especially if the e-bank is a dot-com with no physical branches. To add to the picture, as one IT executive noted to the author that "a secure computer system is not simply a set of passwords and encryption procedures, it also comprises the day-to-day business processes and functions." Another IT senior staff stated that "the core issues of security are data confidentiality and privacy, data availability, and authority and accountability. These are the key issues that the top management in our bank is not confident if our current computer systems are able to maintain such high standards of security." Ian McKenzie, business development director for Internet security at ESOFT Global says there is no single approach that solves all information security challenges. "The technologies and protocols that are the foundation of security approaches can be abstract, complex and difficult to grasp. New protocols, technologies and strategic alliance are announced with alarming regularity that have the potential to change the complexion of the information security marketplace," he says. Further, e-business raises these issues beyond the level of mere support to being critical for success. "Any organisation that uses the Internet for e-mail or a Web site needs to treat security more strategically and see it as a source of competitive advantage," he adds. Therefore, complex technical issues, such as security might be better supported by a specialist partner (Computer Weekly, September 14, 2000).

The third factor was "no-competitor has done e-Commerce." It is widely known that competition makes better services and creates more options for customers. There was no bank which took the initiative to make e-business available as a virtual system. As Table 1 shows, this factor had the mean value of 3.56 and attracted 68% of the respondents' views.

The fourth factor was "high transaction fee for using e-Commerce applications." This factor is not a major inhibitor but would be considered in the final round factors hampering the e-commerce decision.

Other related issues as government support were raised during the interviews, as one IT executive noted to the author that government support should be emphasized and increased through a number of "new" measures. Alshawaf (2001) suggested that government policies that enhance the ability of organizations to compete in the marketplace have a strong positive influence on technology development strategy at the corporate level. Another issue raised through the discussions was that more investments are needed in establishing more communications and networking to make a stronger base of national IT infrastructure in the country.

Table 1. The industrial factors

Factor	Rank	Mean	Std Deviation	Scale
Industry is not yet ready for e-commerce	1	4.24	.88	1-5
Security problems of virtual banking systems	2	4.04	1.34	1-5
No competitor has done e-commerce	3	3.56	1.08	1-5
High transaction fee for using e-commerce applications	4	3.20	1.00	1-5

The Managerial Hampering Factors

By and large, IT managers in Omani banks value the emerging technology of e-commerce but at the same time expressed some cautions, doubtful that the new technology might not be very "suitable" for the Omani business environment. One manager was quoted as saying, "Worldwide, e-commerce is growing at an exponential rate, and fast replacing several traditional delivery channels. In Oman, as well, with the growing population of personal computers, people are looking actively at this channel. Secondly, several businesses in Oman, especially in the banking sector, are looking at the Gulf Corporation Council (GCC) and other markets to grow their business. In this context, e-business will be relevant to Oman in the future."

After calculating the mean score for each hampering factor from the managerial perspective, the 6 factors were ranked in descending order. Table 2 identifies each factor's rank, mean score, and standard deviation. The top most ranked hampering factor in the list is "lack of top management support." The factor has a mean value of 4.08 and captured 72% of the respondents' agreement on the higher ends of the Likert scale. This is somehow consistent with findings of other IT researchers on the region. For example, Abdulgadr and Alangari (1994) attributed the failure (or little utilization) of IT systems in developing countries to many factors including lack of top management support of IT, especially in the Arab Gulf region.

The second ranked hampering factor on this list is "dependence on external vendors." This factor has a mean value of 4.00 and captured 70% of the respondents' agreement on the higher ends of the Likert scale. There are various motivations and rationales behind outsourcing. Cost reduction, overcoming the shortage of skilled human resources, flexibility, meeting the requirements of leading-edge technology, and so on can be considered as some advantages of outsourcing. At the same time, it is thought that lack of skills is the most serious problem, which is pushing organizations in Oman to opt for IT outsourcing. One IT manager commented that, "since all banks are fully automated with fairly high levels of investment in technology, they have to set up infrastructure required to apply e-commerce applications."

There are several technology solution providers in the GCC who can be asked to set up an e-business/e-banking service. Another alternative for banks is to tie up with a third-party service provider (vendor) that would enable banks to provide the e-business service to its customers on a pure variable cost basis without requiring any substantial up-front investment. Such service providers are also available within the region. Summarizing, one IT manager made it clear that, "The current significant shortage of computer manpower locally needs to be considered as an impediment to the adoption of e-commerce. The IT skills needed are not available locally and can always be hired (outsourced) from outside until local manpower equipped with the necessary skills sets takes over, and that's likely to take a very long time!" Another IT managers commented on this point: "The lack of reliable IT vendors and IT manpower in Oman does seem to a be a strong factor in the slow adoption of e-commerce in Omani financial institutions. However, we should bear in mind that this is to a large extent a worldwide problem."

The third perceived hampering factor in the list is "Internet banking projects are NOT aligned with organizational objectives." As Table 2 shows, this factor had a mean value of 3.88 and attracted 64% of the respondents' consent on the higher ends of the Likert scale. All IT projects must be linked or aligned with the organizational and business objectives and goals of any organization. Internet Projects management is a vital part of EC implementation strategy. Such projects must be carefully planned and executed. Internet projects need to be business-driven by a cross-functional project team, and a rapid decision-making process is also necessary to help ensure that the project does not fall behind (Filotto & Omarini, 2000).

The fourth perceived hampering factor in this list is "limited technological knowledge." This is considered important mainly because IT technology changes so rapidly that to keep up with all the changes would require a solid follow-up to the latest technology developments as well as attending conferences and training programs. In addition, bringing consultants to the organization would also increase its technological knowledge. One IT manager addressed this issue by saying, "Unlike other industries, most banks in Oman are highly technology-driven and automated, but at the same time keeping up with all the rapid changes is problematic." He added that, "Banks also realize the power of technology in driving down transaction-processing costs substantially." Another IT manager pointed out that "the banks in Oman have diversified their IT infrastructure. Some are operating on old legacy systems while a few others are already on the Internet-ready platform AS/400 with state-of-art networks and communication infrastructure."

The last two hampering managerial factors, namely, "lack of knowledge on the importance of EC projects" and "managers preference of 'traditional' ways of doing business," were not perceived as important with no support by respondents on the higher ends of the Likert scale.

Other important issues raised based on the outcome of the interviews were:

- Power relationships

Conflict between managers can often emerge during the process of EC adoption in an organization because of different perspectives on roles and responsibilities in the process, as well as differences of opinion on priorities, and so on.

- Severe shortage of IT skills

Lack of IT manpower does seem to be a very strong factor in slow rate of Web adoption. Lack of reliable IT vendors has also been a problem in the Omani environment. As one manager put it, "We try to 'import' IT skills, sometimes across the border from a neighboring country (United Arab Emirates)." However, this strategy also has limitations. There is a real need to address "soft skills" issues in Oman, as Whyte (2001) has recognized that the important skills needed in e-commerce software systems implementation are team building and communication skills.

- Lack of financial resources needed to adopt e-commerce applications

Lack of financial resources was stated by interviewees as a minor constraint to e-commerce adoption. From the perspective of the interviewees, the cost involved in developing the computer systems and networking infrastructure needed for Internet banking was not justified when the likely return on investment (ROI) was considered.

- Education and expertise

The education and experience of both banks and employees/managers tends to be narrow. Managers with narrowly focused education and expertise might therefore be less likely to appreciate the value of Internet banking.

- IS/IT external expertise and services available in the market

Table 2. The managerial hampering factors

Factor	Rank	Mean	Std Deviation
Lack of top management support	**1**	4.08	1.01
Dependence on external vendors	**2**	4.00	1.08
Internet banking projects are NOT aligned with organizational objectives	**3**	3.88	1.36
Limited Technological knowledge	**4**	3.60	1.08
Lack of knowledge on the importance of e-commerce projects	**5**	3.36	1.19
Managers prefer the "traditional" ways of doing business	**6**	3.01	1.35

Some organizations reported that it was not easy to find IS/IT expertise and good IT consultation services. This problem was said to be more common during the introduction of e-commerce and Java applications.

One manager summarizes the situation as follows: "Is it the right time for Omani banks to launch e-commerce services through the Web? Considering both internal factors - business vision, marketing plan, product suitability and management style - and external factors - competition, consumer behavior and the industrial environment - can help a bank to identify whether it is ready to launch e-commerce services and its related applications."

Conclusion

This study can be considered as an attempt to identify inhibiting factors that have been responsible for the slow uptake of e-commerce applications in the banking industry of Oman, as there has been no literature published on any aspects of e-commerce in the context of Oman. These factors need to be identified, and then addressed, so that the banking sector in Oman can exploit the potential advantages of e-commerce to remain competitive.

The objective of this study, therefore, was to examine a number of factors, both internal (i.e., managerial) and external (i.e., industrial), that are responsible for the slow utilization of e-commerce applications. The results provide a pragmatic picture about the adoption of e-commerce applications in the core financial sector domain of Oman.

From the industrial aspect, it was found that "industry is not yet ready for e-commerce," "security problems of virtual banking systems," "no competitor has done e-commerce" were respectively the main barriers. On the other hand, it was found that the main obstacles for the organizational factors were: "security issues," "e-commerce applications is not part of the strategic planning of the organization," and "low level of existing software." Additionally, from the interviews, it was found that severe shortage of IT skills, and the absence of a formal IT strategy aligned with business and organizational objectives were probably another set of critical reasons for the slow adoption of e-commerce in Oman. Others have pointed out to the lack of government support in providing more investments in establishing new national IT technological infrastructure is a sound reason for slow uptake of e-commerce in Oman.

Regarding the managerial aspect, it is found that "lack of top management support," "dependence on external vendors," "Internet banking projects are NOT aligned with organizational objectives" are (in that order) the main perceived barriers. On the other hand, it is found that the main perceived obstacles among the group of cultural/environmental barriers are: "privacy and security issues," "lack of computer ownership among people," and "lack of regulation and legislations on e-commerce (legal issues)."

Additionally, from the interviews, it was found that e-commerce applications were not part of the strategic planning of the organizations under study, and that severe shortage of IT skills, and power-relationship conflicts were also critical reasons for the slow adoption of e-commerce in Oman.

This study provides corporations in the financial sector with a starting framework to assess their use of IT for e-commerce applications. Such an assessment should help these corporations to identify and address areas of strengths and weaknesses in their IS systems. Any significant progress in e-commerce development cannot be achieved without employees well trained in cutting-edge e-commerce technology. Academic institutions should respond to this need to develop specialized IT degrees in e-commerce as well as integrating e-commerce subjects into the general business curricula.

This millennium promises to bring to the world more innovations, more market opportunities, but also more challenges. Developing countries, including Oman, need to be prepared for a more competitive global marketplace that is continuously innovating through technology-enhanced products and services. Today, if utilized properly, the Internet provides a powerful platform for business and socioeconomic development. Thus, the challenge facing developing countries is multifaceted with reference to affording required IT resources, and managing, deploying, and leveraging these resources to achieve optimal economic advancement.

Oman has to formulate a national plan outlining its vision so as to prepare itself for a more competitive market environment. E-commerce expansion promises to offer enormous opportunities for its marketplace, with implications for the labor force, as new employment potentials will be focusing on the information-based economy.

Lastly, despite the exploratory nature of this study and its limited context, the results also provide promising areas for future research. The study provides baseline information on the impeding factors relating to e-commerce adoption and use for a country/region that has been under-researched. Future studies could use cross-sectional samples from organizations in the Gulf region and the Arab world to strengthen the generalization of these findings.

References

Abdul-Gader, A. H., & Alangari, K. (1994). Information technology assimilation in the government Ssector: An empirical study. Final report of funded research, King AbdulAziz City of Science and Technology. Riyadh, Saudi Arabia. *Research project #AR-11-025.*

Alsindi, T., & Mayhew, P. (2001). The key to successful systems development in developing countries: A case study. *BIT World International Conference,* Cairo, Egypt.

Angell, D., & Brent, H. (1995). *The Internet business companion: Growing your business in the electronic age.* Addison-Wesley.

Anon, C. (1996). You too can surf the Internet. Retrieved from *http://www.entrepreneurs.net/hw-financial/research/surf.htm*

Anonymous. (1998, March). Where will Internet banking be in two years? *ABA Banking Journal, 90, 3 561(1).*

Alshawaf, A. (2001). Critical issues of information systems mnagement in Kuwait. *Journal of Global Information Technology Management, 44*(1), 1-26.

Avison, D.E. (1993). Research in information systems development and the discipline of IS. *Proceedings of the 14th of Australian Conference on Information Systems,* Brisbane.

Barney, J. (1991). Firm resources and sustained competitive advantage, *Journal of Management, 17*, 771-792.

Bloch, M., Pigneur, Y., & Segev, A. (1996). Leveraging electronic commerce for competitive advantage: A business value framework. In P.M.C. Swatman (Ed.), *Ninth International Conference on EDI - IOS: Electronic Commerce for Trade Efficiency and Effectiveness,* Bled, Slovenia (pp. 91-112).

Bryman, A. (1995). *Quantity and quality in social research.* London: Routledge.

Chatfield, A.T., & Bjorn-Anderson, N. (1997). The impact of IOS-enabled business process change on business outcomes: Transformation of the value chain of Japan Airlines. *Journal of Management Information Systems, 14*(1), 13-40.

Choi, E., & Lin, E. (1995). *Electronic commerce, WWW, & the Internet.* Retrieved from *http://kinchee.mit*

Dixon, M. (1999). .Com madness: 9 must-know tips for putting your bank online. *America's Community Banker, 8*(6), 12-15.

Electronic Commerce Information Resource (ECIR). (1996). What is electronic commerce? Retrieved from *http://worldserver.pipex.com/year-x/yxwhatis.htm*

Filotto, U., & Omarini, A. (2000). *E-finance and e-commerce. Banks and new competitors.* Roma: Bancaria Editrice.

Gable, G. (1994). Integrating case study and survey research methods: An example in information systems. *European Journal of Information Systems, 3*(2), 112-126.

Hechinger, J. (2001). Check it out. *The Wall Street Journal, 02-01-2001*, R8.

Helmi, A. (2002). Managing e-Business strategies among Asian dot coms. *The 2nd International Conference on Systems Thinking in Management,* Manchester, UK.

Hoffman, D., & Novak, T (1997). A new marketing paradigm for electronic commerce. *The Information Society, 13*, 43-54.

Huda, N., Nahar, N., & Tepandi, J. (2000). Globalization through electronic commerce: A developing country context. *Proceedings of the Business Information Technology Management: Leveraging International Opportunities (BIT2000World Conference),* June 1-3, Mexico City, Mexico.

Kalakota, R., & Robinson, M. (1999). *e-Business: Roadmap for success.* Reading, MA: Addison-Wesley.

Kalakota, R., & Whinston, A. (1996). *Electronic commerce. A manager's guide.* Addison-Wesley.

Kalakota, R., & Whinston, A.B. (1997). *Electronic commerce: A manager's guide.* Addison-Wesley.

Kalakota R., & Whinston A.B. (2000). *Frontiers of electronic commerce* (4th ed.). Addison Wesley.

Kosiur, D. R. (1997). *Understanding electronic commerce*. Washington, DC: Microsoft.

Lack, J. (2000). Banking goes virtual. *Strategic Finance, 81*(10), 36-39.

Leidner, D. E., & Jarvenpaa, S. L. (1993). The information age confronts education: Case studies on electronic classrooms. *Information Systems Research, 4*(1), 24-54.

Liao, Z. (1999). Internet retail banking systems: A Singapore perspective. In W.D. Haseman (Ed.), *Proceedings of the Fifth Americas Conference on Information Systems,* August 13-15, Milwaukee, Wisconsin (pp. 591-593).

Liao, Z., & Cheung, M. T. (2001). Internet-based e-shopping and consumer attitudes: An empirical study. *Information and Management, 38*(5), 299-306.

Liao, Z., & Cheung, M. T. (2002). Internet-based e-banking and consumer attitudes: An empirical study. *Information & Management, 39,* 283-295.

Lynch, D.C., & Lundquist, L. (1995). *Digital money: The new era of Internet commerce.* John Wiley.

Marshall, C., & Rossman, G. (1989). *Designing qualitative research.* Sage Publications.

Mata, F.J., Fuerst, W.L., & Barney, J.B. (1995). Information technology and sustained competitive advantage: A resource-based analysis. *MIS Quarterly, 19*(4), 487-505.

Minoli, D., & Minoli, E. (1997). *Web commerce technology handbook.* New York: McGraw-Hill.

Mintzberg, H. (1979). An emerging strategy of 'direct' research. *Administrative Science Quarterly, 24,* 582-589.

Mowery, D., & Rosenberg, N. (1979). The influence of market demand upon innovation: A critical review of some empirical studies. *Research Policy, 8,* 102-153.

Mukti, N. (2000). Barriers to putting businesses on the Internet in Malaysia. *The Electronic Journal of Information Systems in Developing Countries, 2*(6), 1-6.

Neckopulos, J. M. (2001). Challenge: Integrating the channels. *US Banker, 111*(10), 68-70.

Network-centric banking. (1998, March-April). *Banking Strategies, 74*(2), 56(4).

Oppenhein, A.N. (1992). *Questionnaire design, interviewing and attitude measurement.* London: Pinter Publications.

Palvia, S., Palvia, P., & Zigli, R. (1992). Global information technology environment: Key MIS issues in advanced and less-developed nations. In S. Palvia & P. Palvia (Eds.), *Global issues of information technology management* (pp. 2-34). Hershey, PA: Idea Group Publishing.

Pigni, F., Ravarini, A., Tagiavini, A., & Vitari, C. (2002). Identifying eBanking strategies: The on-line financial services of the Italian retail market. *The 7th Association Information Management Conference,* Tunisia, May 30 - June 1.

Polovina, S., French, T., & Vile, A. (1999). Semiotics for e-commerce: Shared meanings and generative futures. In R. Hackney (Ed.), *Proceedings of Business Information Technology Conference (BITC).*

Putnis, P., & Petelin, R. (1996). *Professional communication: Principles and applications*. Prentice Hall.

Remenyi, D. (1992). Researching information systems: Data analysis methodology using content and correspondence analysis. *Journal of Information Technology, 7*, 76-86.

Retail banking: Caught in a eb? (1997). *Mckinsey Quarterly, 2*,42-55.

Straub, D. (1989, June). Validating instruments in MIS Research. *MIS Quarterly*.

Tait, F., & Davis, R. H. (1989). The development and future of home banking. *International Journal of Bank Marketing, 7*(2), 3-9.

Turban, E., Lee, J., King, D., & Shung, H. M. (2000). *Electronic commerce: A managerial perspective*. London: Prentice Hall.

Whyte, W. S. (2001). *Enabling e-Business*. Chichester, UK: John Wiley.

Yin, R. K. (1994). *Case study research design and methods* (2nd ed.). Newbury Park: Sage.

Chapter XV

Open Sourcing E-Learning for Developing Countries

Ronald M. Lee, Florida International University, USA

Manfred Zielinski, Erasmus University, Netherlands

Ramanathan Somasundaram,
National Institute for Smart Goverment (NISG), Denmark

Abstract

A critical success factor for developing countries is advancing their intellectual capacity. Nearly all aspects of economic development depend on the ability of individuals to extend their understanding of their business, suppliers, customers, laws and society from a local perspective to a broader, more international one that is able to assimilate change brought by new technologies and respond with innovations that bring competitive advantage. The Internet offers countries information access to the entire world — if only one can understand it. That same Internet also offers fantastic educational resources — effectively a global knowledge base. However, much of Internet content is oriented towards the needs and interests of industrialized countries. Here we propose an approach for developing countries to pool efforts to create a digital commons of e-learning resources that are appropriate and relevant to their specific needs.

E-Learning for the Developing World

Education as Antidote to Poverty

> *"Education is the only antidote to poverty."* – A.R. Bawa, Dep Minister Education, Ghana (Bawa, 2003)

While cheap, universally available education is desirable everywhere, it is most desperately needed in the developing world, especially in rural areas where access is most limited. For most people in developing countries, education is not only a goal in itself, but is also instrumental to fulfilling other needs. In addition to the direct costs of education, there are also opportunity cost — time lost for other kinds of employment. In many poor countries, this applies even to children, for they are also working assets of the family (e.g., to work the fields). The benefits of sending children to school must be seen to clearly outweigh the costs. These benefits are seen most clearly as offering new economic opportunities, which is itself an educational challenge.

E-Learning as Enabler for E-Business

There are a variety of related definitions of e-business — most include the notion of business processes that exploit Internet technologies. A somewhat broader notion suggested by the Gartner Group (www.fourthwavegroup.com) is that e-business is "a phenomenon involving a significant change in the cultural, economic, societal and market interactions that people and businesses experience in dealing with one another."

For the purposes here, let us assume that e-business involves the enhancement of economic opportunities via Internet technologies. In this sense, one typically imagines the ability to start new enterprises, opening access to international markets, linking to venture capital, etc. In short, enabling access to (the positive features of) globalization.

But many people in developing countries, this comes as a culture shock. They have lived in relative isolation, familiar only with their local culture and practices. Thus, the transition to e-business requires adaptation and learning about the outside world and its business practices. The focus in this chapter is on e-learning support to facilitate this adaptation to e-business.

A Vision of E-Learning for Development

Like most new technologies, e-learning has had its initial impact mainly in the industrialized countries. The developing world so far lacks the infrastructure and availability of computer resources to make serious investments in e-learning developments. When governments in poor countries face shortages of food and medicine, investment in education may be seen as a luxury.

Around the world, education is normally considered as a responsibility of governments. Every government has a Ministry of Education. The implicit assumption is that educational developments are done "top-down," from national priorities and programs, extending eventually to the local level. In many countries, public education is also complemented by private-sector educational institutions, such as private universities and private secondary schools. Another variation that is appearing in many developing countries, are educational initiatives that are partnership efforts between the government and certain private companies. The companies are typically large, technology-based multi-nationals, such as Cisco and Microsoft. These public-private partnerships have had some success, especially in areas of technical training. While the commercial linkages may be criticized as motivated by globalization interests of the companies (providing lower cost trained labor supply), they also have the clear advantage of linking the educational time investment of individuals to a broadened range of economic opportunities (Kersemaekers, 2002).

There are also examples of educational models that develop "bottom-up," without major investments or direction from governments or corporations. Significant among these is the Committee for Democratization of Information Technology (CDI), a non-profit, non-governmental organization started 1995, to "promote social inclusion, by using information technology as a citizens rights and developing tool" (www.cdi.org.br). CDI works in partnership with low-income community centers and special needs individuals, such as physically handicapped, psychiatric patients, homeless children, and prisoners. The design of CDI has some remarkable merits. Most impressive is the degree of motivation and involvement it achieves among its students. Another stunning characteristic is its "viral growth" where schools tend to replicate into hundreds more. An additional noteworthy characteristic is that the growth of CDI schools is done on a minimal budget, with little or no outside investment. However, it will be noted that the education made available from CDI is basic computer literacy, and a sense of familiarity with the potential applicability of the technology, coupled with aspects of citizenship and social responsibility. CDI does not compete with conventional schooling in offering the broad range of subjects normally associated with primary and secondary education.

Providing the Telecom Structure

One of the immediate concerns in discussions about e-learning in developing countries are the technical requirements of providing computer technology and telecommunications access. As is widely recognized, these are necessary but not sufficient conditions (Warschauer, 2003).

Thanks to widespread de facto standardization of operating systems (Windows, Linux), there is already relatively easy portability of software among hardware platforms. As the earlier CDI example illustrates, it is possible to assemble workable computer labs through heterogeneous collections of computer hardware acquired by donations and creative reengineering. A more serious challenge, especially for remote areas, is Internet access. Like previous generations of telephone technologies, the bulk of telecom infrastructure is built from cable networks. These have relatively easy availability in the cities, but the fan-out from these hubs to remote areas can leave whole regions without access.

However, new possibilities are emerging via satellite (Global VSAT Forum, 2002). Stimulated by the demand for entertainment, satellite TV has been expanding quickly to remote areas. With relatively modest extensions, this same satellite infrastructure can now offer Internet access. For instance, with an initial equipment investment of about $1,000 and monthly connection charges at about $100, one can now get a 500 kb connection in most of South America and soon Africa (https://register.earthlink.net). For instance, satellite access to the Internet is now available in the Amazon jungle (SpaceDaily, 2002). While these prices may out of reach for most individuals in developing countries, they might be sufficiently economical to support a remote cyber-café.

Demand and Supply for E-Learning

The link between educational investment and economic opportunity has always been tenuous, both at the strategic level and at the individual level. Strategically, the more successful examples of economic development, e.g., Singapore and South Korea, have also made heavy investments in education, especially at university levels. However, just where to make those investments is less clear, since building educational infrastructure coupled with the educational process itself often moves more slowly than market demands.

At the individual level, the issue is where to specialize. If one is beginning a four-year university degree, the employment market could shift substantially in the intervening time. For individuals in developing countries, the problem may be the more basic one not of which area to specialize in, but whether to pursue education at all. The choice is more difficult because the relative personal costs are much higher. For a poor young man in a rural area, the cost of secondary education may mean that the family loses its strongest worker for half the time. For a woman (e.g., in Muslim countries), there may be additional stigmata. These relative personal costs are even higher for tertiary education, which may require leaving the home altogether to attend university in a faraway city.

These kinds of constraints are strong arguments for e-learning applications. E-learning can offer anywhere anytime access to education on a self-paced program, that can be customized to individual backgrounds and needs. A farm worker can take courses at night. A father whose education stopped with grade school can take remedial courses without embarrassment. A woman can take special interest courses in the private context of a computer screen. Moreover, the same e-learning facilities can help individuals search for employment opportunities, and the specific knowledge packages that might be relevant for those opportunities. This is a kind of "just-in-time" education.

It is easy to visualize the wide range of benefits that e-learning could offer for people in developing countries. There are however, three main resource constraints:

- **Financial:** these people are poor, and can only pay very small amounts for education beyond what is provided free by the government.
- **Telecom:** in most developing countries, telecom infrastructure is lacking outside the big cities.

- **Content:** the content for online courses needs to be developed to fit the local context. While some subjects (e.g., algebra), may be globally generic, nearly any topic involving social aspects (e.g., management) has cultural dependencies that need to be contextualized. Presentation styles may also need local adaptation. Obviously, course content needs to be offered in the local language.

Of these, perhaps the most challenging is the development of localized courseware content. This needs to be developed, or at least adapted, locally — both to make it locally relevant, but also to make it economically viable, with costs of production that are compatible with the revenues that the local market can produce.

Open Sourcing E-Learning Content

Open-source software systems[1] have now emerged as a real, economically significant phenomenon for producing high-quality, broadly usable software. While the emergent character of open-source projects is still the focus of much study, enough has been learned that some are trying to generalize this to other areas. For instance, a close sibling to open source is open content (Cedergren, 2003), of which there are also a number of highly successful projects. In this chapter, we consider another area of potential application for open-source (and open-content) project models, namely e-learning.

Open-source systems are often admired as a kind of self-organizing project management strategy. Related, but somewhat distinct from this are the aspects of creative synergy in (the more successful) open-source projects. Individuals from different parts of the world, who have no prior acquaintance, can modify and extend each other's creative work to produce an evolutionary product beyond what any individual could achieve. Indeed, several of these projects compete successfully with large-scale corporate projects (e.g., GNU/Linux vs. MS Windows).

To a certain extent, e-learning might be considered simply as another application domain for open-source projects. For instance, various kinds of computer-aided instruction (CAI) software might be produced as open source. On a larger scale, the more complex kinds of software known as learning management systems (LMS) might also be produced as open source. Indeed, there is already a significant open-source LMS project called Claroline (www.claroline.net).

But, unlike most open-source software, e-learning contains a heavy dose of content. In this regard, a major challenge for open e-learning is the development and global distribution of content that is appropriately modularized and adaptable to re-use in a variety of educational contexts. The MIT Open Courseware initiative is mainly a sharing of digital content (http://ocw.mit.edu/).

Additionally, e-learning might also embrace a digital medium intermediate to open source and open content that might tentatively be termed "open lessonware." Open lessonware is animated and interactive digital content. It is typically designed using a graphical user interface (GUI) that is easy to learn by non-technical people — thus distinguishing it from open source. For instance, the teacher/author interface might resemble that for presentation software (e.g., PowerPoint). One can easily pick up and modify someone else's presentation file for a related purpose. One can easily imagine other kinds of authoring software with added functionality devoted to producing interactive lessons for students.

In this chapter, we are primarily interested in this open lessonware kind of e-learning application. Our goal is not to dwell on the technical requirements for such systems, but rather on the emergent aspects of the potential community of teachers that might provide a globally shared library of such lessons. Such a resource could be of immense value, especially for developing countries.

Scenario

Consider the following scenario:

> *A grandmother in Guangzhou, China, creates some flashcards to teach color words to her granddaughter. The older brother converts this to a computer game (including audio in Cantonese), which is put on the Internet. A mother in Singapore finds the game, and modifies it with Mandarin pronunciation, adding some additional vocabulary for flowers. A schoolteacher in Taipei finds the Singapore version of the game, and changes the Chinese characters to the traditional style still used in Taiwan, also adding vocabulary for butterflies. Another teacher in Korea adapts the game structure for Korean, extending the vocabulary to include pictures of birds as well as their calls. A teacher in England finds the Korean version, and adapts it to teach identification of English birds by their calls.*
>
> *And so on...*

This little scenario conveys how interactive e-learning software might evolve.

In actuality, there is relatively little such learning software on the Internet with this evolutionary character. While one may find various freely available programs, e.g., to learn Spanish, seldom is there any creative synergy among the projects. Thus, one may find a variety of lessons for beginning Spanish grammar, but these tend to be repetitions of the same material, with little opportunity for students to extend beyond the basic level. It might be noted that this is also a major failing in the commercial marketplace for educational software. Each new educational product tends to begin from ground zero, with little creative advancement from one vendor's product to another's.

As noted earlier, the functionality we are pursuing shares some features of both open source and open content. For purposes of discussion, let us hypothesize that a new kind of authoring tool is available called LESSON-MAKER, which is for use by teachers to produce interactive lessons. We also imagine a companion program called LESSON-

PLAYER that is used by students for performing these interactive lessons. We assume that both programs are freely available at no cost. (We might also assume that they were developed as open-source projects, but that is not the main issue in this scenario.)

The purpose of LESSON-MAKER is to enable widespread collaborative development of e-learning materials — what we have called open lessonware. The lesson-specific content and interactions are stored in a "script" file, e.g., in an XML format with specialized tags for the various kinds of content and interaction involved in the lesson. These lessonware script files are input to the LESSON-PLAYER software that conducts an interactive lesson with the student. These lessons may be simple matching exercises, like the color words example above, or they might be more complex, such as lessons in trigonometry.

The LESSON-MAKER authoring software should be easy to learn and use by teachers of all grades (primary, secondary, university) throughout the world. It should be easy for a teacher in one part of the world to pick up a lesson from any other country, and adapt it for local purposes. The creation of new lessons will make use of a graphical user interface with various kinds of drawing tools, depending on the desired student interaction. These tools will themselves be modular, allowing the addition of more advanced tools later on, as needs evolve.

What Makes Open-Source Projects Take Off?

Much has been written about open source projects. (See for instance the collected papers on-line at *www.opensource.org/*). Our purpose in this section is to glean a summary of the success factors for open-source projects, to see how they might be applied in the area of open e-learning. But before we can identify success factors, we need to be clear what we mean by success. Obviously, unlike for commercial enterprises, this does not mean profitability, either short or long term. One might characterize success for an open-source project in terms of the degree that it satisfies the needs of its users. However, this is also problematic in that since there are no sales, it is difficult to measure the demand. Furthermore, in many accounts of open-source projects (e.g., Raymond, 1999), the satisfaction of users is regarded more as an extra bonus. What really counts is keeping the developer community active and productive. While this might be measured in terms of continuing updates to the code, it has an implicit reference to some original requirements of the software, for which functionality needs to be maintained throughout changes in the subject software's computing environment. Thus success for an open-source project means sufficient developer involvement to ensure that the software grows to the required functionality, becomes thoroughly debugged and reliable, and continues to be supported in the future.

What then are the success factors for an open-source project? Why do some projects thrive and grow, while others languish? As we have just noted, the key population that needs to be satisfied is the software developers, rather than the end users of the software.

Thus, the success of an open-source project depends heavily on the motivations for developers to participate in the project. Feller and Fitzgerald (2002) describe these motivations at two levels, individual and organization/community. They distinguish three types of motivations: technological, economic and socio-political.

Among the technological motivations for individual developers is what Raymond (2001) called "scratching one's own itch," that is, developers working a problem where they are also users of the result. Another motivation is to elicit help in the debugging and review of the code. On the other hand, it has been observed that some 40% of open-source code is actually paid for by organizations, devoting their programming staff to make contributions to the project. Why? Typically, this is where the software involves infrastructure needed by the organization, but is not its core business. Since other organizations have a similar need, it pays them to collaborate in the development. In this way, they may also benefit from synergies of talent with these other organizations.

There are also various economic motivations for participating in an open-source project, though they tend to apply more to organizations than individuals. For individuals, they may enhance their CV or their technical reputation. For organizations, they are often able to obtain required software at a lower cost, which is especially attractive for organizations in developing countries. It is sometimes cited as a disadvantage that open-source software often requires greater technical expertise on the part of the organization — but this can also be a benefit. For developing countries, this can contribute to the overall human resource capacity development of the country.

Feller and Fitzgerald (2002) also cite various socio-political motivations. For individual developers, these include the "joy of hacking" experienced by many programmers, including a certain ego gratification of their technical virtuosity, and a certain competitive/collaborative satisfaction of sharing this among the technical developer group. Altruism at providing a valuable product ("gift") for the world is also an oft-cited factor. At an organization/community level, the motivations are less for companies, and more for social communities or countries, seeking to share social benefits by providing a public good.

Raymond (1999) summarizes the following factors as favoring an open-source (vs. commercial) project:

- reliability/stability/scalability are critical
- correctness of design and implementation cannot readily be verified by means other than independent peer review
- the software is critical to the user's control of his/her business
- the software establishes or enables a common computing and communications infrastructure
- key methods (or functional equivalents of them) are part of common engineering knowledge

He continues: "In summary, where network effects (positive network externalities) dominate, open source is likely to be the right thing."

Open-Source Governance

The management of open-source projects depends to some extent on the way they were initially created. There are three roughly distinct paths:

- **Individual Effort Pathway:** one passionate developer leads the way, such as Richard Stallman (1985, 1993, 2003) who began the GNU Project (www.gnu.org/fsf), or Linus Torvalds for the Linux kernal.
- **Joint Need Pathway:** several individuals or companies cooperate to solve a problem (e.g., Apache).
- **Orphan Pathway:** after the dot.com bust, many software projects were abandoned — some were continued as open-source projects.

Raymond (2000) comments:

> *"What does 'ownership' mean when property is infinitely reduplicable, highly malleable, and the surrounding culture has neither coercive power relationships nor material scarcity economics? ... Actually, in the case of the open-source culture this is an easy question to answer. The owner of a software project is the person who has the exclusive right, recognized by the community at large, to* distribute modified versions."

These pathways to initial creation tend to influence the kind of governance structure that evolves:

1. **Single leader:** typically the founder of the project
2. **Multi-headed body:** eight wise men for Apache
3. **Member consensus:** based on voting systems

Raymond (2000) continues, " The trivial case is that in which the project has a single owner/maintainer. The simplest non-trivial case is when a project has multiple co-maintainers working under a single 'benevolent dictator' who owns the project. A co-developer who accepts maintenance responsibility for a given sub-system generally gets to control both the implementation of that sub-system and its interfaces with the rest of the project, subject only to correction by the project leader (acting as architect). Some very large projects discard the 'benevolent dictator' model entirely. One way to do this is turn the co-developers into a voting committee (as with Apache). Such complicated arrangements are widely considered unstable and difficult."

Open-Source Repositories

SourceForge.net is the world's largest open-source software development Web site, with the largest repository of open source code and applications available on the Internet. SourceForge.net provides free services to open-source developers.

- Operated by VA Software since 1999
- Offering free hosting and project administration to Open Source developers
- 66,446 projects
- 674,911 registered users
- Offers "Enterprise Edition" of same development environment for corporate use
- Offers consulting about large-scale software development

Other open-source libraries include:

- Freshmeat (http://freshmeat.net), the Web's largest index of Unix and cross-platform software
- BerlioOS Developer (http://developer.berlios.de), a German site, fostering open-source development
- OSDir.com, sponsored by O'Reilly Publishing, promoting stable, open-source applications (http://OSDir.com)
- Metalab, Open Source Metadata Framework, (www.ibiblio.org)

Other Forms of Open Digital Media

What is typically called digital content, or e-content includes essentially all things digital that are not executable code. This includes text, images, sound and video.

Open E-Books

There are several open-content projects that focus on providing a freely accessible library of books. These are normally for classic texts where the copyright has expired.

The oldest of these is *Project Gutenberg* (http://promo.net/pg), originally launched in 1971 by Michael Hart. Project Gutenberg provides a library of books in electronic form (html) and is currently approaching 10,000 online books. However, since the texts provided generally pre-date digital media, there is a substantial conversion problem. The rough conversion of physical book pages to digital media is done by scanning in the pages of the book, and running them through OCR software. However, this still leaves many errors that must be corrected by manual editing.

The need for manual editing has lead to the creation of another remarkable project known as *Distributed Proofreaders,* founded in 2000 by Charles Franks (www.pgdp.net). While it was originally conceived to assist Project Gutenberg, it now has a broader objective to support the digitization of Public Domain books for a variety of projects. It currently involves the participation of thousands of volunteers, to do the editing and corrections after OCR scanning.

Whereas these projects refer to the conversion of books that have already been written, the Wikimedia Free Textbook Project (http://textbook.wikipedia.org) is an open collaboration project to write new books. So far there are 169 book modules ("pages") in this project.

Open Directory Project (Odp)

Probably the largest open-content collaboration is the Open Directory Project (http://www.dmoz.org/), which is operated by Netscape/America Online, part of the media giant Time Warner. About 50,000 volunteers work on a huge hyperlink catalog — analogous to a library's card catalog — for the entire Internet. The governance structure for these volunteer editors is quite hierarchic (Cedergren, 2003).

Open Digital Encylopedia

Another very large open-content collaboration, also numbering in tens of thousands of volunteers, is the open digital encyclopedia known as Wikipedia (http://www.wikipedia.org). Wikipedia's goal is to build a collectively created online encyclopedia, offered as open content. Wikipedia's organization could be described as anarchistic compared to some of the other open-content projects.

Open Course Materials

Complementary to these developments, are the actions of some universities and educational institutions to make their course content materials available publicly via the Internet. The most noteworthy example here is an initiative of Massachusetts Institute of Technology (MIT), to release all course materials for all university classes on the Internet by 2005 (http://opensource.mit.edu).

Open E-Music

The widespread exchange of digital music via peer-to-peer networks (e.g., Napster) has received great notoriety. While the technical logistics of this employ substantial creativity, it has little to do with synergistic creation of new music.

On the other hand, there are some other developments that do have aspects of open collaboration. One of these is so-called lyric databases — online collections of song lyrics. Many of us have a fragment of a dimly remembered song echoing in our heads. But we cannot quite get the rest of it from our memories. As it turns out, you are not alone. There are thousands of others that have had the itch to remember the rest of the song, and have combined efforts to tally these in online collections of song lyrics. As Lessig (2001) points out, "these thousands produced a far better, more complete and richer database of culture than commercial sites had produced."

One could also imagine a parallel evolution of sheet music provided online. At present, this is mainly done as scanned images that are exchanged. One needs a further mark-up language (like HTML) to make this easier. However, an even more robust standard already exists in the form called MIDI (musical instrument digital interface) (www.midi.org). MIDI files not only represent the notes like a musical score, but also other acoustic aspects of tone. A MIDI file can therefore be used to perform the music digitally. While some distance collaborative composition experiments have been done, this remains a largely undeveloped opportunity for open collaboration.

Open E-Movies

A digital repository of more than 10,000 different movies has been collected by Rick Prelinger (www.archive.org/movies/prelinger.php). Many of them are characterized as "ephemeral" or temporary — which has made it legally possible for Prelinger to digitize and offer them as public domain content of the Internet. The variety of movies in the collection includes corporate, educational, and official information movies as well as advertisements. The digitized versions are in the public domain, so they can by re-used by anyone without violating copyright restrictions (Cedergren, 2003).

Open E-Games

"Anyone who makes a distinction between games and learning doesn't know the first thing about either," said Marchall McLuhan (Falstein, 2002).

Computer games are especially interesting, not only as a class of software that is moving towards open source, but also as a model that connects closely with the interests of e-learning.

Open Source Games

Hargreaves (1999) remarks: "Nintendo, Sega, and Sony are already virtually giving away their hardware in order to sell games. ... if I was a game development company I would be looking really hard to find ways of reducing my programmer salary expenses. One obvious approach would be to open source your engine, and let other people improve it for you. ...the most sensible strategy would be a two-pronged approach. You sell a shrink-wrapped product to one set of users, while giving away your sources to another group of people, who are your peers working on different game development projects. ...Open source ideas might not apply very well to the end product of commercial game development, but they work superbly for reusable infrastructure projects like a game engine, where your artists are the users."

Just such an open-source repository seems to be emerging — by analogy to SourceForge, it is called WorldForge. The WorldForge project aims to create an open-source library and tools and resources for people who want to develop online games — more specifically, Massively Multiplayer Online Role Playing Games (MMORPG's). The vision is to "foster an independent community in which many free games can develop and evolve with unique role-playing oriented worlds and rules, running on a wide selection of server and client implementations with a standard networking protocol tying everything together" (www.worldforge.org/media). The WorldForge project has so far not produced any complete, playable games. There are various games under development — but the project is still working on developing the supporting tools. "WorldForge's mission is to produce the necessary tools and technologies that will allow the creation of graphically rich games of cooperation and socialization. We strive to blur the distinction between player and maker, and wish to establish a positive community environment for current and future free game developers."

E-Games for E-Learning

Somewhat along the lines of WorldForge, Silverman et al. (2003) is developing an e-learning game toolbox, for rapid assembly of a certain class of e-games, with an e-learning agenda to them. He remarks, "We envision a future where many games exist that help people to cope with their health issues, child rearing difficulties, and interpersonal traumas. Further, these games will be so compelling and easy to revise, that many players will feel compelled to contribute their own story to the immersive world – a contribution that is both self-therapeutic and that helps others who see some of their own dilemma in that story. This will be an industry that is consumer grown, since they will be the creators of new games for other consumers." This is in marked contrast to the objectives of most commercial e-games, which offer addictive, immersive entertainment but have little education focus, and provide few if any directly re-usable components.

A further dimension to this line of e-game development is the notion of interactive drama. Interactive drama is like game playing, but with the plot line pre-determined by the author, whereas with game playing the story creation is more the responsibility of the player.

Opportunities and Challenges for Open Lessonware

Having now reviewed different forms of open collaborative development in areas of open source and open content, we now consider some aspects more specific to the notion of open learnware proposed earlier.

Interactivity of Content

Even though there is much activity in the area of e-learning, most of it is essentially passive in nature. The student may download texts to read, or perhaps hear a lecture via streaming video. Like a large classroom, these seldom allow the chance for active participation on the part of the student.

By contrast, a fundamental characteristic of the proposed learnware notion is that it is interacts and engages the student. In an article entitled, "What ever happened to programmed learning?" Kimbrough (1997) remarks:

> *"We already know about multimedia and it's called television. It's not engaging and enriching. We know 'Sesame Street' at best is not harmful. It encourages passivity. The natural response to a great show is to sit back and watch it. You're wowed by fireworks, but you're not engaged. Compare the involvement to a play or a book, where you're really involved with the characters, care about them and want to know what happens. [However] there are tremendous successes. They are in a broad class of software called simulation software. What the software does is mimic a real-world system with enough accuracy that's appropriate to your task. A real example is a flight simulator. ...*
>
> *There already are math-tutor programs for children, which diagnose errors and give feedback. The military uses war games. Businesses have management games that simulate parts of the economy. They teach people how their performance can be improved with elementary modeling. There's tremendous work with experimental games in lab situations — games on markets, negotiations, any sort of decision-making that involves coordinating with other decision-makers who may not have the same objectives that you have."*

These sorts of interactive gaming notions are essential to the design of learnware.

Localization and Translation Of Content

We fully intend for learnware to be used globally. But clearly, as regards e-learning, one size does not fit all. There are important differences not only in language, but also in the choice of examples and other references that make the lesson meaningful. For instance, a cow in India has a much different significance than it does in Argentina.

Beyond that, student learning styles may differ. To some extent these may be individual differences, but in other cases they may have a cultural aspect.

While we may think of language in terms of the major languages such as English, Spanish, Japanese – there can be many variations by geographical regional and by social class. To some extent the use of visual imagery can help reduce the burden on written language.

Challenge: Dynamic Content

Beyond the localization and translation of content to regional and social contexts, there is a dynamic aspect to the content itself. While fields like mathematics may be relatively stable in their content, areas like technology or pharmacology need to be continuously updated.

This notion of dynamic content is about capturing current information generated from various sources, converting it into appropriate content, and delivering it to the learner in the required form at the right time. It is also about capturing the relevant collaborative "experience" of various learners just-in-time and disseminating it in the appropriate form — to make learning efficient and effective.

Towards a Digital Commons for E-Learning

What is a Digital Commons?

In medieval times, the commons was a public area in a village, that all may use for grazing their cattle. The so-called "tragedy of the commons" is that it is in each individual's interest to graze his/her cattle as much as possible, leading to overuse which destroys it for everyone. Hardin (1968) writes:

> *"Therein is the tragedy. Each man is locked into a system that compels him to increase his herd without limit — in a world that is limited.* Ruin is the destination toward which all men rush, each pursuing his own best interest in a society that believes in the freedom of the commons. Freedom in a commons brings ruin to all."

The metaphor to a commons is often cited as paradigm of collective goods, where market mechanisms alone do not manage the resource optimally, necessitating government intervention.

Lawrence Lessig (2001) distinguishes a different type of commons that he calls a digital commons. This notion refers to public resources on the Internet for free sharing of information and software, including re-use for new creations. The distinguishing aspect of a digital commons is that the resource is not diminished by usage, hence cannot be overused. Indeed, because usage encourages further creativity and further contributions to the commons, the more the resources of the digital commons are used, the *greater* is their total value. Lessig (2001) elaborates this collective synergy effect:

The new builds on the old, and hence depends, to a degree, on access to the old. Academics writing textbooks about poetry need to be able to criticize and hence, to some degree, use the poetry they write about. Playwrights often base their plays upon novels by others. Novelists use familiar plots to tell their story. Historians use facts about the history they retell. Filmmakers retell stories from our culture. Musicians write within a genre that itself determines how much of the past content needs to be within that genre. (There is no such thing as jazz that does not take from the past.) All of this creativity depends in part on access to, and use of, the already created." It is this kind of communal synergy that we seek to create in the educational domain, especially for developing countries.

Critical Mass Aspects

We find the e-learning community as a fitting example of a collective good. A collective good is characterized by: (i) jointness of supply, and (ii) impossibility of exclusion (Hardin, 1982). The e-learning community is in joint supply for one's consumption of e-learning information does not adversely affect the size of the information e-learning commons. This is unlike disjoint goods such as a pizza where one's consumption of it reduces the availability for the others. We regard e-learning community as open in nature. The openness here denotes that information in the e-learning commons accumulated via voluntary contributions will be made available to all via the Internet.

An e-learning community is not just a collective good, but it is also interactive in nature. This interaction is broadcast and stored unlike in the case of telephones where the interaction is targeted, direct and instant. A key challenge in the development of an e-learning community is that of enticing content from voluntary contributors. The entice-ment is challenging especially in the early stages of community formation when the community's worthiness is low. A volunteer contributing e-learning content at an early stage is less likely to benefit from the small database. It however gets easier to entice voluntary contributions once the database grows in size. Oliver et al. (1985) define this key challenge as the challenge of attaining "critical mass."

In their seminal paper, Oliver et al. (1985) analyze the dynamics of critical mass attainment via accelerating and decelerating the production function. Production function is a curve that plots one's contribution towards the likelihood of a collective good. In the decelerating case, the curve is convex shaped where initial contributions steeply enhance the likelihood of a collective good. In the accelerating case, the curve is concave shaped where early contributions contribute little towards the likelihood of a collective good. Markus (1987) uses the critical mass rationale for analyzing interactive media in which she concludes that the production function for interactive media would mostly be accelerating in nature. We show that the production function for e-learning communities would as well be accelerating in nature through a simple conceptual analysis that follows.

A contribution towards an e-learning community takes the form of an interactive teaching lesson. The value of a community increases with the number of such lessons in the database, the extent of topics covered and the quality of contributions. In order to be regarded as of high value by a wider audience, the community requires a critical mass of interactive teaching lessons. While it is a challenge to predict the critical mass, we can surely claim that adding one, two or five teaching lessons in the early stages does not increase the community's perceived value steeply. However, when there are a few thousand of teaching lessons, a large population would find the community as valuable. This would mean increased participation at least in terms of downloading lessons from the community. While an individual's contribution would add little to the production function, the number of individuals contributing towards the collective good would enhance significantly at this stage, resulting in the production function taking off in an accelerated manner.

The challenge for one developing an e-learning community is to attain the critical mass of content. This can happen only when adequate numbers decide to contribute toward the community. Granovetter (1978), while analyzing the threshold model of collective behavior, bases his explanation on the assumption that one decides to act when the perceived cost is less than the perceived benefits of acting. In other words, one acts when the net benefits equals or exceeds one's threshold to act. The model further assumes the threshold as distributed, implying that potential participants perceive costs and benefits differently. Moreover in a collective good, one's threshold to act is affected by the amount of contribution made towards the collective good. The e-learning community developer by providing incentives can minimize the threshold of a critical mass of potential participants to the extent that they would decide to contribute content. Two types of incentives that can be provided for enticing participation are discussed below.

The first type of incentive can be provided through developing mechanisms for ranking one's contribution. The number of downloads and the content quality as perceived by those who downloaded the content can for instance be ranking measures. However, simply creating these mechanisms alone is inadequate. Efforts should be taken to make relevant administrative bodies and procedures to recognize a contributor's community ranking. The organization of open-source communities inherently includes mechanisms for recognizing one's contribution. A contributor is considered as talented when he or she is a part of the steering committee of an open-source project. For the committee member, this recognition is saleable.

While the first type of incentive is indirect in nature, the second recommends providing a direct incentive via micro-payments. When this logic is implemented, one who has contributed a lesson would get paid a small sum of money for each download. Government and non-governmental organizations can contribute by paying the transaction costs toward education activities. The direct and indirect types of incentives proposed here are complementary in nature.

While critical mass theory as proposed by Oliver et al. (1985) has been applied extensively (Oliver & Marwell, 2001), it as such has not been used as a research methodology tool. An exception to the case is the recent work of Ramanathan (2003), who recommends that we differentiate those who joined a community earlier from those who joined latter in terms of context specific resources and interests. This logic when applied in the e-learning context highlights the need for differentiating those who would potentially contribute first from those who contribute latter. This differentiation when empirically established can assist a developer in enhancing his marketing effectiveness, thereby reducing the likelihood of a community's failure. In the next paragraph, we discuss two contributor types defined in terms of their contribution likelihood.

A basic necessity for contributing towards an e-learning community is the availability of an Internet connection in the academic sector of a nation. This availability can be measured in terms of ratio of computers-per-person at school and university levels. Scandinavian countries such as many developed countries generally have a high level of computer penetration. This enhances the likelihood of obtaining participation from this sector. A normative recommendation then is that academics in the developed countries should primarily be targeted for inducing contribution during the early stages. Developing countries should be contacted when an adequate volume of e-learning lessons is available in the e-learning commons.

The emergence of e-learning communities is affected by not just the collective action aspects, but as well by the institutional factors. The process of creating course contents varies vastly from one institution to another. While the process is highly centralized in the case of Madras University (India), the process is highly decentralized in Scandinavian countries in general. When course content creation is a centralized activity, only a small percentage of the available pool would have access to the e-learning database. In the decentralization case, there will be a large number of users and likely contributors. The developer's objective should then be to encourage decentralization in developing countries.

In summary, the requirements for creating a digital commons for e-learning involve two key aspects:

- Attaining an initial critical mass of contributions from which further contributions can evolve.
- Having sufficient incentives that users (teachers) will not only exploit the commons' top-draw materials as input for their own lessons, but will also have incentives to deposit new resources into the digital commons.

Evolutionary Requirements of Lessonware

An important difference between open-source projects and open lessonware is what is called "forking." In open-sourse projects, this happens when the developers disagree over some design aspect, and a subset decides to go off in their own direction. This is considered bad for open-source projects. In the case of lessonware, however, forking is a common and desirable occurrence — motivated in part by needs of localization, translation, and different learning styles. Whereas in open-source projects, all contributions lead to the latest version of the software, in open lessonware developments, each sub-version continues to have its own localized purpose. Thus, if we imagine a growth metaphor for open-source projects, it might resemble an oak tree, with numerous roots leading to a single sturdy trunk. The metaphor for lessonware might more resemble a banyan tree, which has numerous inter-locking trunks, each with its own roots.

For open source projects, there are several major repositories where the numerous projects can be located. A principle example is SourceForge (http://sourceforge.net), which also offers development tools such as concurrent versioning software (CVS) and other relevant services to open-source projects. Nonetheless, each of the member projects is quite independent of its neighbor.

Now let us imagine an analogous repository for learnware, called LearnForge. What are its properties? We would consider such a repository to be a prototype for a digital commons for open e-learning. Contained in LearnForge would be a variety of e-learning resources (open software, open learnware, open content) as well as the associated developer (teacher) community. Our main concern here is how to nurture the growth of such a digital commons for open e-learning.

In the introduction, we presented the goal as a globally shared library of e-learning materials (software, content, scripts). However, the metaphor to a library might be too restrictive, since it tends to connote a kind of centralized control and management that we do not mean to imply. A more apt descriptor might be as a digital commons, for e-learning. A digital commons may have certain agreed upon protocols and rules, but it is not (necessarily) under any specific management authority. A commons is a repository for certain kinds of public goods. In the digital case, these may be globally dispersed, connected by means of a common index (e.g., portal), or perhaps discovered on the fly, as in various peer-to-peer models (e.g., Napster).

The original notion of a commons was a public area in a village, that all may use for grazing their cattle. In such cases of physical public goods, the more it is shared, the less its residual value. However, in the case of a digital commons, such as the sharing of information and software on the Internet, the more it is shared, the ***greater*** is the value. It is this kind of communal synergy that we seek to create in the educational domain, especially for developing countries.

E-Learning Prospects for Developing World

While we consider that an e-learning commons would be of value for industrialized countries, we believe it could have a much more profound impact for developing countries. There is broad consensus that education is a major priority for all countries, but especially for developing countries. The challenge is provide educational solutions that are cost effective — that somehow fit into the modest budgets of these countries, yet provide education that is *appropriate* — adapted to the local culture and economic conditions. Key challenges:

- Improve educational opportunities
- Improve professional employment opportunities

There is growing concern that the new e-technologies, rather than creating a "level playing field" for international competition, may actually be exacerbating economic differences, especially for developing countries. An important aspect for *positive globalization* is knowledge transfer and assimilation into local cultural contexts.

In the previous section, we summarized the growth conditions for an e-learning commons to be:

- Attaining an initial critical mass of contributions from which further contributions can evolve.
- Having sufficient incentives that users (teachers) will not only exploit the commons' top-draw materials as input for their own lessons, but will also have incentives to deposit new resources into the digital commons.

The following recommendations are speculative, but we believe worthy of exploration. We believe that an effective solution might include different roles between industrialized countries and developing countries to achieve the e-learning commons, based on different applicable incentive schemes.

The open-source projects that we have observed are primarily a phenomenon of first world countries, where there is sufficient disposable income to permit altruistic gifts of one's time and efforts to achieve a collective good that is of general value to everyone. This applies both to the creation of open-source software, open content, and potentially to open lessonware. The individual incentives for this creativity are largely intangible and non-economic. There may also be economic incentives for companies and educational institutions to contribute time and effort for common e-learning materials, e.g., for standard background subjects.

As a further incentive for teachers to contribute lessonware, a digital record might be kept of downloads, which might comprise part of the teacher's public service performance for promotion reviews. This might also include tracking of the downloads of further lessonware derived from the teacher's production (a kind of pyramid scheme of brownie points). This, we believe, might be sufficient to create the initial critical mass of lessonware and other e-learning materials to make the e-learning commons attractive.

The more significant challenge is to have the e-learning commons promulgated throughout the developing world, with frequent and detailed localization of its content in a series of evolutionary re-creations. An important part of the challenge is that teachers in developing countries do not have as much leisure time as in industrialized countries. Indeed, in most poor countries there is no such thing as "a teacher" — "not because teachers don't exist but because teachers are also farmers, shopkeepers, tradespeople, parents and have many other roles (www.unicef.org/teachers/discuss).

We believe that the best way to achieve creative contributions to the e-learning commons from the developing world is to provide financial incentives. This notion is generally regarded as taboo for open-source contributions — that programmers participate in open source for self-actualization and other, more altruistic motives. Teachers in poorer countries may also feel these motivations, but they are much more constrained in their time. They need to eke out a living for their family. We seek to make their creative contributions instrumental to this goal and potentially providing a light at the end of the tunnel of poverty.

A possible solution for doing this is a combination of micro-payments and so-called "network marketing" also known as "multi-level marketing" (MLM) (www.network-marketing-works.com). Network marketing is essentially a legalized form of pyramid scheme, where there can be an arbitrary number of sub-agents before the final sale. The notion here would be that each teacher/author would be remunerated a tiny amount for each download (or some other measure of usage) of their lessonware, as well as a fraction of this amount for each version of someone else's lessonware that was derived from theirs. The amounts paid would be accumulated as micro-payments.

In first world countries, the notion of micro-payments never gained much momentum because the amounts were too small — even for large numbers of transactions. For instance, at $.01 per transactions, a million transactions generations only $10,000. This does not make anyone rich in the first world. However, for a poor teacher in the third world this is a fantastic sum.

Concluding Remarks

In this chapter we have attempted to generalize from the successes of open-source projects to another domain, e-learning. The particular aspect that we sought to capture is the synergistic creativity observed in open source projects, which is seldom seen in other kinds of open digital media projects. The goal is to create and nurture a digital commons for e-learning that provides a rich supply of educational resources that are freely available globally, with special attention to the needs of the developing world. We concluded that achieving a digital commons for e-learning depends on two major factors: achieving an initial critical mass of contributions, and providing adequate incentives for continued participation, especially among teachers in developing countries. Industrialized countries might provide an initial critical mass through incentives similar to those for open-source projects. A network marketing scheme was suggested as a possible solution providing modest, but critical financial incentives for teachers in poor countries.

References

Bawa, A. R. (2003). Quality education is the solution to poverty. Retrieved from *http://www.mclglobal.com/History/Mar2003/17c2003/17c3n.html#h*

Cedergren, M. (2003). *Open content and value creation*. Retrieved October 4, 2004, from *http://firstmonday.org/issues/issue8_8/cedergren/*

Daniel, J. (2002, October-December). Higher education for sale. *Education Today: UNESCO Newsletter.* Retrieved from *http://www.swaraj.org/shikshantar/mceducationforall. htm*

Falstein, N. (2002). *Interactive stealth learning.* Retrieved October 4, 2004, from *http://ecolloq.gsfc.nasa.gov/archive/2002-Spring/announce.falstein.html*

Feller, J., & Fitzgerald, B. (2002). *Understanding open source software development.* Addison Wesley.

Global VSAT Forum. (2002). Report and accounts 1999-2001. Retrieved from *www.gvf.org*

Granovetter, M. (1978, May). Threshold models of collective behavior. *American Journal of Sociology, 83*, 1420-1443.

Hardin, R. (1968). The tragedy of the commons. *Science, 162*(1243).

Hardin, R. (1982). *Collective action.* WA: Resources for the Future.

Henry-Wilson, M. (2003). Inaugural speech for education program at Institute of Management Sciences. Jamaica, July 9, 2003.

Kersemaekers, S. (2002). *Why multinationals contribute ICT education to bridge the digital divide: Synergies between business benefits and socio-economic development in emerging economies.* Master's thesis, Erasmus University.

Kimbrough, S. (1997). *Whatever happened to programmed learning?* Retrieved October 4, 2004, from *http://www.upenn.edu/pennnews/features/1997/090297/Kimbrough.html*

Lessig, L. (2001). *The future of ideas: The fate of the commons in a connected world.* Vintage Books.

Markus, L. M. (1987). Toward a "critical mass" theory of interactive media. *Communication Research, 14*(5), 491-511.

Oliver, P., & Marwell, G. (2001). Whatever happened to critical mass theory? A retrospective and assesment. *Sociological Theory, 19*(3), 293-311.

Oliver, P., Marwell, G., & Teixeira, R. (1985). A theory of the critical mass. I. Interdependence, group heterogeneity and the production of collective action. *American Journal of Sociology, 91*(3), 522-556.

Ramanathan, S. (2003). Electronic market as a collective action: A research framework for studying critical mass. Paper presented at the *Research Symposium on Emerging Electronic Markets,* Bremen.

Raymond, E. S. (1999). *The magic cauldron.* Retrieved October 4, 2004, from *http://www.catb.org/~esr/writings/magic-cauldron/magic-cauldron.html*

Raymond, E. S. (2000). *Homesteading the noosphere.* Retrieved October 4, 2004, from *http://www.catb.org/~esr/writings/homesteading/homesteading/index.html*

Raymond, E. S. (2001). *The cathedral and the bazaar.* Retrieved October 4, 2004, from *http://www.catb.org/~esr/writings/cathedral-bazaar/cathedral-bazaar/*

Silverman, B., Johns, M., Weaver, R., & Mosley, J. (2003). *Authoring edutainment stories for online players (aesop): A generator for pedagogically oriented interactive dramas*. Retrieved October 4, 2004, from *http://www.acasa.upenn.edu/gameLink.htm*

SpaceDaily. (2002, April 9). PanAmSat brings VSAT Internet to upper Amazon. Retrieved from *http://www.spacedaily.com/news/vsat-02f.html*

Stallman, R. (1985/1993). *The gnu manifesto*. Retrieved October 4, 2004, from *http://www.gnu.org/gnu/manifesto.html#translations*

Stallman, R. (2003). *Speech at wsis, 16th July*. Retrieved October 4, 2004, from *http://www.gnu.org/philosophy/wsis-2003.html*

Warschauer, M. (2003, August). Demystifying the digital divide. *Scientific American*, 42-49.

Endnote

[1] We are aware of the distinctions in terminology between "open source" and "free software," as reflected by their respective organizations: Open Source Initiative (#open-source_initiative) and Free Software Foundation (#free_software_foundation). Throughout this chapter we adopt the now popular usage of "open source" in a generic sense, intended to include all software where the source code is freely available.

Appendix: SCORM Standards for E-Learning

An important aspect in support of a digital commons for e-learning is the emergence of standards for interchange and inter-operability among authorware and learning management systems more generally. The Sharable Content Object Reference Model (SCORM) is an effort in this direction.

SCORM was originally developed by the Advanced Distributed Learning (ADL) group of the US Department of Defense (DOD). The purpose was to establish standards for reusable modules of learning content to support the various training programs of the DOD. Letts (2002) elaborates:

> *"In contrast with the way that most computer-based instruction is written today, SCORM demands that content be authored to stay independent from any larger contexts. A lesson on how to apply gauze bandages, for example, must be written in a way that does not depend on it following instructions for disinfecting a wound. ...SCORM also requires content to stay independent from the software used to render it. ...SCORM can be expected to spur development of XML-based authoring tools and e-learning systems."*

The Academic ADL Co-Laboratory at the University of Wisconsin is helping nearly 40 higher-education institutions evaluate SCORM-compliant tools and technologies. Though SCORM claims to be pedagogically neutral, it does have some biases and limitations due to its origins. Rehak et al. (2002) comment that "SCORM is essentially about a single-learner, self-paced and self-directed. It has a limited pedagogical model unsuited for some environments." This is mainly a consequence of the needs of the main initiators of SCORM: the US federal government in general, and the Department of Defense in particular. Their needs are mainly in the area of training for specific systems and situations by people who are not generally in full-time education. This need is addressed very well by the spec, but "SCORM has nothing in it about collaboration. This makes it inappropriate for use in HE and K-12."

The SCORM standard continues to develop, and it is expected that future versions will have broader applicability. The next version, SCORM 2.0, will already attempt to incorporate an adaptive architecture. "The plan is to evolve SCORM to ensure that it is as useful as possible to the ADL community." Yet, respect continuity of existing investments (backward compatibility) (CETIS, 2002).

Chapter XVI

Strategic and Operational Values of E-Commerce Investments in Jordanian SMEs

Ala M. Abu-Samaha, Amman University, Jordan

Abstract

This chapter aims to articulate lessons learned from a quantitative study of Internet Technology use in Jordan. To do so, a questionnaire was devised to measure the business value and strategic advantages of being online for Jordanian organizations. The findings of the study are presented and discussed in the course of this chapter. The survey shows that an increasing number of organizations in Jordan depend heavily on e-mail to conduct their daily activities within and across organizational boundaries. And the majority of SMEs in Jordan believe that the Internet has become a fact of business life. The statistical analysis shows that the majority of surveyed organizations use the Internet for more than simply promotional reasons. They use the Internet for provisional and co-operative reasons. On the other hand, the number of transactional Web sites is very limited due to many shortcomings in the available technical, legislative and organizational infrastructures.

Introduction

The end of the 20th century has been marked with the emergence of the Internet as an alternative communication medium to the more "traditional" face-to-face, phone and fax modes. For the last few years, terms like "Digital Economy," "Digital Libraries," "Electronic Markets," "Digital Stores," and most recently "Electronic Commerce," "Electronic Business" and "Internet Business" have taken the world by storm. The promise of this "Electronic" era ("Electronism") is to replace the traditional "hard" or "physical" mode of the way we conduct trading, business or transaction by a stream of electrons. This switch from "hard," "physical" and "paper-based" style to "electronic," "paper-less" and "virtual" style is meant to be faster, easier to handle, hassle free and cheaper. Two main factors have enabled such change. Firstly, the telecommunications revolution, the advanced techniques for transmitting digital data which has enhanced both the quality and quantity of data transmission on public telephone systems cheaper than ever before. Secondly, the computer revolution, the latest advancements in the computer industry have yielded cheap, user friendly and versatile hardware and software used in almost every part of our lives (Collier & Spaul, 91/92; Parfett, 1992). "Electronism," "Digitization," or even "Virtualization" is all about connectivity, the ability to use digits to replace the existing modes of transactions like fax, post and hand-to-hand mail.

Despite the much-publicized worldwide proliferation of the Internet in the form of internal and external communication services, products and applications, the usage of such a medium in developing countries is still at its minimum. Forrester Research Institute (www.forrester.com) shows that worldwide trade over the Internet and across organizational boundaries will reach $500 billion to $3 trillion. On the other hand, eMarketer (eMarketer.com) reports that US online Shoppers exceeded 63.4 million shoppers, making revenues of $37 billion. By year 2003 eMarketer expected this figure to rise to 106 million shoppers, making revenues of $104 billion. While many sources estimated that worldwide Internet users reached 375 million by end of 2000, only a fraction of that — 1.9 million — are in the Arab World. More specifically, Inta@j, the Information Technology Association of Jordan, estimates the number of Internet Users in Jordan to be 42,000 Internet users, 50% of them served via Internet Cafes (REACH 1.0, 2000).

The purpose of this research is to reflect on emerging technologies and their digital promise, i.e., Electronic Commerce; Internet Business; and Digital Exchange, in order to study the effect of such connectivity on developing countries compared with developed countries. One of the major aims of this research is to evaluate the pervasiveness of Internet Technologies and their applications in the context of developing countries, namely Jordan. As well the chapter will attempt to evaluate and explain local trends in Internet Technology investments through a quantitative analysis of Internet use in Jordanian Small and Medium Size Enterprises (SMEs).

The chapter starts by describing the objectives of the research project, and then follows by a description of the questionnaire used, and then the findings from responses are articulated. Finally, a discussion of results and relationships is presented.

Background

One of the major challenges in evaluating any Information Technology intervention is choosing the criteria of evaluation or measures of success or lack of it (Abu-Samaha & Wood, 1998, 1999a, 1999b).

Brown (1994) and Chan (1998) distinguish between two types of measures, hard and soft measures. Hard being tangible and direct, while soft being intangible, indirect, and strategic. The latter is more difficult to obtain evidence of its expected benefits or outcomes and harder to measure using whatever agreed-to instrumentation. Chan (1998) explains the importance of bridging the gap between hard and soft measures in IT evaluation realizing that "this in turn requires the examination of a variety of qualitative and quantitative measures, and the use of individual, group, process and organisation-level measures."

Since measures or criteria of evaluation are hard to identify whether on the user side or on the organization side, a number of researchers in the realm of Internet business have identified a number of expected theoretical and practical benefits or outcomes for Internet Technology investment. The following is a summary of some of them.

Amor (2002, p. 17) identifies the following advantages of electronic business: expanding market reach (global availability, small compete with large organizations); generating visibility at very low cost; strengthening business relationships (EDI, and B2B using XML); offering new services online; cost reduction (paperless inter- and intra-business activities); shorter time to market and faster time to respond to changing market demands; and improved customer loyalty.

Amongst other advantages, Lawrence et al. (1998) reflect on companies' experiences that turned to the Internet faster than their competitors, defining the following outcomes of Internet Technology investments: e-mail; real-time training and conferencing; personalization of goods; and enabling employees to carry out tasks internally and externally. Regarding strategic importance of the Internet, Lawrence et al. (1998, p. 8) show that "the Internet and intranets give businesses the opportunities to improve their internal business processes and customer interfaces to create a sustainable competitive advantage."

Turban et al. (2000, p. 15) identify the following organizational benefits for Internet Technology investment: expanding to national and international markets; reducing cost of creating, processing, distributing, storing, and retrieving paper-based activities; reducing inventories and overheads; and customization of products and services.

Simpson and Swatman (1998) identify 13 business objectives for Internet investment. Their longitudinal survey shows that the most valuable attributes of the Internet were: to save time to look for resources; obtain useful expertise from the 'Net; savings in communication costs; and better company image. While trading in virtual environment, significant increase in sales and better supplier relationships were indicated to be significant in a prior initial exploratory survey in the same research project.

Despite the much publicized possible benefits and outcomes of Internet Technology investment, practical experience shows that these perceived benefits are more elusive to realize than previously thought. This is even made worse in case of small- and medium-size enterprises. Giaglis and Ray (1998) explain that "existing practical applications have not always been able to deliver in practice the business benefits they promise in theory." A conclusion confirmed by Lawrence and Hughes (2000), they explain that, "We discovered that the hype and speculation by business and media reports about what the Internet can do was not always right, sales increase and profitability were far lower than expected." Lawrence and Hughes's (2000) survey aimed to articulate the current picture of how SMEs in the United Kingdom are using the Internet and its technology, including the benefits, issues and the impact of use on business activities. Their survey shows that the Internet was very useful in gathering information and communicating with suppliers or customers. As well as the Internet is perceived by SMEs in the United Kingdom to be an extremely attractive method for SMEs to expand their markets and reach global audience. Ho (1997) in his study of 1,000 business Web sites to identify the primary reason for organizations to be online, points out attention to the following facts. First and foremost, to customers the values of the sites are primarily promotional, despite the fact that the marketing approach pursued by most online promoters is taken to be conventional. Secondly, provisional content is primarily in the form of published results, while the results of his survey showed that little value lays in the provisional content to customers and investors. Thirdly and most importantly, Web use for processing business transactions is largely underdeveloped. In most cases, online ordering is often simply an extension of e-mail.

Objectives of This Research

The objectives of the research project can be articulated as follows:

- To assess Internet Technology penetration and usage in Jordanian organizations (Small and Medium Size Enterprises).
- To articulate the primary purpose of Internet Technology use in Jordanian SMEs.
- To articulate the perceived benefits of using Internet Technologies within and across organizational boundaries of Jordanian SMEs.
- To articulate the challenges encountered in using Internet Technologies in Jordanian SMEs.

In order to do so, the researcher used a detailed questionnaire distributed to a sample of Jordanian SMEs over a period of two months (November 2002 till January 2003). The statistical analysis of returned questionnaires aimed to identify relationships between organizational characteristics, size and industry, and a number of anticipated Internet Technology operational and strategic benefits, mainly cost reduction and competitive position enhancement.

Figure 1. Diagrammatic representation of dependent and independent factors

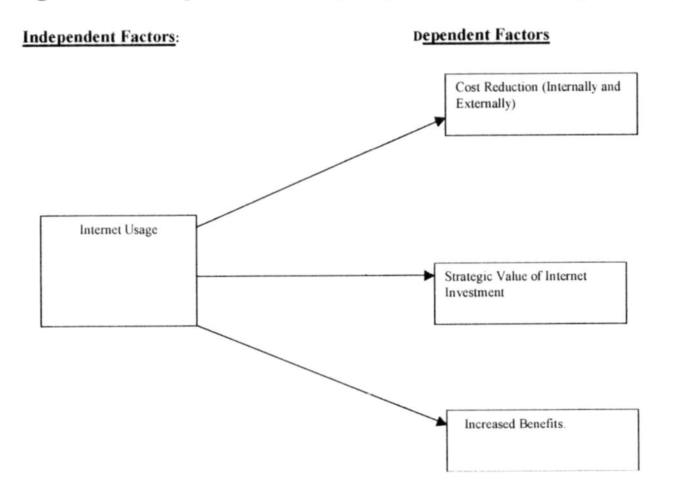

Figure 1 provides a diagrammatic representation to show dependent and independent factors of the research plan.

Questionnaire

The questionnaire consisted of two sections and a cover letter. The cover letter provided instructions of filling out the questionnaire and a guarantee that the information will be used for academic and research reasons only. The first section aimed at providing a contextual description of the surveyed organizations in terms of: type, size, speed of processors used, speed of Internet access used, whether the organization has a Web site or not, and Web site content language used. The second section aimed to provide a detailed description of Internet use, value and perceived benefits, these include: primary purpose of Web presence, strategic value of the Internet investment, and Intranet, Extranet and Internet use. The first section used a number of scales ranging from Yes and No answers to a scale of five. The second section used a scale of five answers ranging from (not available, less than 25%, 25-50%, 51-75%, and more than 75%). IT managers or IT personnel in the respective organizations mainly filled the questionnaire.

130 copies of the questionnaire were distributed to Jordanian SMEs, out of which 71 were returned, a percentage of 55%. Since Jordan lacks the existence of countrywide databases that can be used to draw out random target organizations to be surveyed, the researcher relied on purposive stratified sampling rather than fully random one. This was used to overcome the operational difficulty of sampling, and to overcome any bias in responses. 20-30 companies in the five chosen categories (Service-Based, Product-Oriented, Educational, Public, and Manufacturing) were randomly selected from the nationwide business directory and a copy of the questionnaire was delivered in hand or via post or

e-mail. As well, a question on organisation size was included in the first section to query respondents on their organizational size, from (less than 50, 50-100, 101-150, and 151-250 employees). Any organisation that did not fall within these categories was excluded from the sample. Since the sizes of organizations in Jordan are much smaller than their counterparts elsewhere in the region and worldwide, smaller size categories were used to accommodate such a difference.

Although the survey was intended to measure the effects of Internet and Web usage on Jordanian SMEs, it was impossible to draw a random sample from Internet users only. As explained earlier, Jordan lacks the provision of sophisticated databases of Jordanian organizations that can be used to draw random, stratified or even purposive samples for research. So it was virtually impossible to obtain organizational details regarding size and technology use prior to questionnaire distribution to verify the validity of the sample. On the other hand, it is realistic to say that the 45% of not-returned copies of questionnaire were so because either such enterprises did not use the Internet or were not interested in participating in such a survey. As well as, this effort was even made more difficult knowing that no such surveys were undertaken earlier, neither by research parties nor by Jordanian governmental agencies. Hence, a comparative or a longitudinal study of the effect of Internet investment is virtually impossible.

In order to generate the questionnaire the author relied on Ho's (1997) framework of Web site purposes and values and on Hughes and Lawrence (2000) Internet survey of SMEs in the UK. Despite the fact that these studies were conducted in a North American and West European context, they provide a good repository of Internet values, benefits and operational feedback. Taking the cultural, organizational, political, personal and technical differences between developed and developing countries into account, the author generated a number of benefits and values that suit the local context powered by the above global studies. This enabled the researcher to measure the benefits and effects of Internet Technology investment on local SMEs and provided a good ground to compare results with global and western studies of the same orientation.

Ho (1997) classified the business purposes of a commercial Web site into three categories: promotion of product and services (promotional value), provision of data and information (provisional value), and processing of business transactions (transactional value). A fourth dimension was added by the author based on the literature survey presented earlier of Internet Technology/Web site benefits and value. This fourth dimension is internal and external communication (co-operative), i.e., using the Internet to facilitate the exchange of data within and across organizational boundaries and for sharing organizational resources with other trading partners. This fourth dimension adds another emerging dimension to Internet use to enhance a firm's relationship with its customers, suppliers, distributors, trading partners, etc. This four-dimension framework was used in the questionnaire to query respondents regarding the expected strategic value of Internet Technology investment in Jordanian SMEs.

Many of the perceived values used in this questionnaire are adopted from the literature reviews presented earlier, mainly Lawrence and Hughes' (2000) study of Internet use in United Kingdom Small and Medium size Enterprises (SMEs) after being subjected to intense argumentative analysis to measure their relevance to SMEs in the Jordanian context. Many reasons led to such a decision, most importantly the majority of these values was subjected to and came from research based on SMEs. Lawrence and Hughes' (2000) research into Internet use yielded a number of perceived values. These include: information gathering, communication, advertising, market and product research, information distribution and dissemination, gain competitive advantage, research and development, transaction, customer service, administrative support, strategic planning, resource sharing and video-conferencing. Many of these values were used to measure the operational value of Internet Technology investment in Jordanian SMEs. In an attempt to overcome few of the many differences between British and Jordanian organizations, the questionnaire contained an open space for respondents to add their remarks regarding their organisation's use of Internet technology. This space was seldomly used by respondents.

Findings

Sector Analysis

Out of the 71 responses, the statistical analysis shows that the majority of respondents are actually from organizations in the service/product provision and educational sectors. These include schools, colleges, universities, banks, and financial institutions. Table 1 shows the number of participating organizations in different industrial sectors:

Size Analysis

A scale of four was chosen for this part of the study ranging from (less than 50, 50-100, 101-150, and 151-250 employees). Such a scale was used to correlate Internet value to size of participating organizations. The survey shows that the majority of respondents were either less than 50 employees (38% of respondents) or more than 150 employees (37% of respondents).

Table 1. Frequency/percentage

Frequency/Percentage		
Service-Based	18/ 71	25.35
Product-Oriented	14/ 71	19.72
Educational	24/ 71	33.80
Public	7/ 71	9.86
Manufacturing	8/ 71	11.27
Total	71/ 71	100

Technical Infrastructure Analysis

In terms of processor speed the majority of surveyed organizations use Pentium II and III Personal Computers (PCs). In terms of Internet introduction to surveyed organizations, the survey shows that 75% of such organizations have used Internet Technologies for more than a year. Regarding Internet speed, the survey shows that the majority of the surveyed organizations use Dial-up and Leased Lines, 37% and 39% respectively. The speed of which ranges from 56 KbPS to 512 KbPS. Table 2 provides a tabular analysis of Internet speed by respondents.

The survey shows that 49 organizations out of the 71 respondents do have a Web site, a percentage of 69%. 30 of which are hosted internally. The majority of which have English- and Arabic-based content. Table 3 provides a tabular analysis of Web site content language.

In comparison with the Hughes and Lawrence (2000) study, the number of Internet users and the percentage of Web sites are lower in Jordan compared to their counterparts in

Table 2. Internet speed of respondents

Internet Speed		Percentage
Dial-Up	26/71	36.62
Leased Line	28/71	39.44
ADSL	11/71	15.49
Via Satellite	4/71	5.63
Others	1/71	1.40
Total of Valid	70/71	98.59
Void	1/71	
Total	71/71	100

Table 3. Web site content language

Language		Percentage
English Only	15/47	21.13
Arabic Only	2/47	2.82
English and Arabic	26/47	36.62
Other languages	4/47	5.63
Total	47/47	100

the UK. While 55% (based on the number of respondents interested in taking part in this study) of surveyed organizations in Jordan are Internet users, the number rises to 79% in the UK. And while 69% of those companies have Web sites in Jordan, the number rises to 78% in the UK. These numbers would drop down sharply if the number of non-respondents were taken into account as an indicator of the limited spread of Internet usability.

Value Analysis

Regarding the main value of Internet presence, the survey shows that primarily the surveyed organizations use the Internet for more than one purpose. Table 4 provides a tabular analysis of Internet presence value for organizations.

Table 4 shows that the majority of respondents (46%) use the Internet primarily for provisional values, i.e., use the Internet to disseminate data and information within and across the organizational boundaries. A sizeable percentage of the surveyed organizations use the Internet for promotional and co-operative reasons. Thirty-four percent of respondents use the Internet for co-operating and resource sharing with organizational partners as well for exchanging organizational data within and across organizational boundaries, while the same percentage of respondents use the Internet for product and/or service promotion and advertising. This shows that SMEs in Jordan have surpassed the use of the Internet for promotional and co-operative values to provision of data and information both internally and externally. On the other hand, the survey shows that the use of the Internet for transactional purposes is very low. Only 22% of respondents indicated that they use their Web sites to sell or buy products online. This can be justified by the lack of high-bandwidth and secure connections.

Regarding the value of Internet investment, the majority of surveyed organizations feel that the Internet provides them with an important to necessary source of strategic value. Table 5 shows a statistical breakdown of Jordanian SMEs perception of the value of their investments in Internet Technology. The percentage of responses in Table 5 shows that 26% of respondents feel that the Internet provides them with a very important mean to

Table 4. Internet presence value for organizations

Value	Frequency
Promotion of products and services	24
Provision of data and information	33
Transactional (Buying and Selling)	16
Co-operative (Internal and External communication)	24
Others	4

Table 5. Perceived strategic importance of Internet presence for organizations

Internet Value Percentage		
Least Important	12/71	16.90141
Less Important	8/71	11.26761
Important	14/71	19.71831
Very Important	19/71	26.76056
Necessary	17/71	23.94366
Total of Valid	70/71	98.59155
Void	1/71	1.408451
Total	71/71	100

gain competitive advantage, while 23% of respondents feel that the Internet has become necessary for them.

This question was used to measure the organizational attitude towards the value of investment and the confidence in future direction. The table shows that the majority of respondents are confident that their investments yielded a good return on investment and such return on investment provided them with a good strategic advantage. This implies that the majority of respondents feel that their presence online and their utilization of Internet Technology provided them with strategic values beyond the mere operational benefits investigated in the next section.

Regarding operational values and uses of Internet Technology investment, Table 6 provides a tabular analysis of Internet usage/benefit by respondents using a scale of five.

Table 6 shows that an increasing number of organizations in Jordan depend heavily on e-mail to conduct their daily activities within and across organizational boundaries. While many of the respondents feel that using the Internet does lead to reducing transaction cost, the majority of them feel that the Internet does not lead to massive reduction in internal cost. As well as a small number of respondents use the Internet for product research, resource sharing, payment clearance, and human resource development. While a good number of surveyed organizations use the Internet to exchange data electronically, promote current and future products and/or services, to disseminate information (including providing customers with information), and to reach out to international markets. This is in line with Web site primary use, as the survey shows that the majority of surveyed organizations use the Internet for provision of data and information, co-operation internally and externally, and promotion of products and services respectively. On the other hand, only a fraction of the surveyed organizations use the Internet for video-conferencing purposes.

Taking the last column of Table 6, more than 75%, as in indicator of the usability of the Internet, we can rank the reasons for investing in Internet technologies for the following reasons in descending order:

Table 6. Perceived Internet benefit/usage

Usage	Not Available	Less than 25%	25-50%	51-75%	More than 75%
Percentage of Internet use by employees	5	23	12	8	23
Dependency rate using electronic mail (e-mail) to facilitate exchange of data internally and externally	9	16	8	14	21
Usage of Internet to exchange data electronically with suppliers, distributors and any other organizations (extranet)	25	8	8	12	17
Usage of Internet to lower transaction cost (the cost of offering services and/or products to customers)	14	14	11	14	14
Usage of Internet to lower internal operation cost	21	15	12	9	13
Usage of Internet to disseminate information (including providing customers with information)	17	18	15	8	13
Usage of Internet to promote current and future products and/or services	20	17	8	13	12
Usage of Internet for customer and product research (including feedback from customer)	28	12	6	13	12
Usage of Internet to reach out to international markets	24	14	12	9	11
Usage of Internet to share available resources within organizational boundaries	27	14	13	7	8
Amount of financial resources allocated for Internet development and human resources	13	26	13	2	8
Usage of Internet for transactional purposes (to buy/sell and bill suppliers, distributors or customers)	31	17	7	6	7
Usage of Internet to provide further human resources development (including online training and certification programs)	37	18	7	5	3
Usage of Internet video-conferencing (to communicate via audio/video link with suppliers, distributors, business partners, regional offices or customers)	49	11	6	4	0

- Firstly, the Internet is primarily used to facilitate exchange of data internally and externally via Electronic Mail (e-mail)
- Secondly, the Internet is used to exchange data electronically with suppliers, distributors and other organizations (extranet)
- Thirdly, the Internet is believed to reduce Transaction Cost (the cost of offering services and/or products to customers)

- Fourthly, the Internet is believed to aid in reducing Internal Operation Cost
- Fifthly, the Internet is used to disseminate information (including providing customers with information)
- Sixthly, the Internet is believed to be a valuable tool to promote current and future products and/or services of organizations
- Seventhly, the Internet is used for customer and product research (including feedback from customers)

While the rest of reasons are less important to being online, these being:

- Using the Internet to share available resources within organizational boundaries
- Using the Internet to conduct online business transactions (to buy/sell and bill suppliers, distributors or customers)
- Using the Internet to provide further human resources development (including online training and certification programs)
- Using the Internet for video-conferencing (to communicate via audio/video link with suppliers, distributors, business partners, regional offices or customers)

As mentioned earlier, the survey shows that 49 organizations out of the 71 respondents do have their own Web site. This represents a percentage of 69%. On the other hand, the responding organizations pointed out that although they may not have their own Web sites, they do use the Internet extensively. This shows that the percentage of Internet using organizations is much higher than those who use the Internet to establish their own Web presence. Such organizations use the Internet to communicate with their suppliers, distributors, sister companies, and regional offices or headquarters abroad. As well as, they use the Internet to conduct their market search online to attract trading opportunities through other companies' Web sites and to reach out to trading partners outside their geographical presence, and elicit possible markets abroad.

The statistical analysis of Table 6 shows that 33% of respondents believed that Internet usage did not impact their organizational activities. Twenty-three percent of respondents indicated that the Internet had less than a 25% effect on their organizational activities. Fourteen percent of respondents believed that the Internet had a 25-50% effect on their organizational activities. Thirteen percent of respondents showed that the Internet had a 51-75% effect on their organizations. On the other hand, 17% of respondents indicated that the Internet had more than a 75% effect on their organizational activities. These numbers show that despite the actual benefits of the Internet, a sizable percentage of respondents felt that the Internet did not impact their organizational activities positively, with 48% of respondents indicating that the Internet had less than a 50% impact on the organizational activities of their organizations.

Relationships

As explained earlier, the statistical analysis of the responses was used to measure the existence of relationships, or lack of them, between organizational characteristics, namely size and industry, and Internet Technology perceived operational benefits and strategic values. In order to assess whether there exists a relationship between Internet Technology and organizational characteristics, the author used a number of statistical correlation methods, Chi Square and One-way NOVA. To do so, the answers of particular questions in section two are correlated with organizational size and industry in section one of the questionnaire. On occasions more than one question was correlated, and on others one question was used. Then the mathematical significance is measured. If the significance was less than 0.05, that was an indicator that a relationship exists between the two independent factors (size and type of organization) and other dependent factors (like cost).

Regarding the effect of industrial sector on Internet use and value, the Chi Square analysis shows that there is not any possible link between the organization type and value of Internet investment. However, the One-way Nova analysis of the survey shows that there is a relationship between type of organization (service, product, educational, public and manufacturing) and reducing Internal cost of this Organization. Table 7/ Figure 2 and Table 8/Figure 3 of Appendix A provide a detailed description of the relationship between organizational industry and reducing transaction cost and reducing operational cost respectively. Transactional cost is taken to mean the cost incurred in the provision of products and/or services for consumption by buyers/customers. While operational cost is taken to mean the cost incurred internally. Service-based organizations believe that Internet use has yielded more positive results in terms of reducing transactional cost compared to other industries, mainly educational institutes. While product-based organizations believe that Internet use has yielded more positive results in terms of reducing operational costs compared to other industries, mainly educational institutes. This implies that cost reduction in educational institutes is at its lowest compared to other industries.

Also, there exists a relationship between organizational industry and Internet speed. Figure 4 of Appendix A provides a diagrammatic representation of this relationship. While all organizational industries use a myriad of Internet connections with different speeds, educational institutes seem to have the highest percentage of leased lines and public agencies are solely reliant on such a connection medium. As explained earlier the majority of SMEs in Jordan utilize either a dial-up or a leased line in terms of Internet speed.

On the other hand the survey indicates that no relationship exists between organizational industry and the following Internet uses/values:

- opportunity to provide employees with computer & Internet literacy
- strategic value of the Internet
- when Internet was first introduced

- whether an organization has a Web site
- language used in Web content
- where Web site is hosted
- type of organization and language used in Web site content
- dependency rate of using Internet internally

Regarding the effect of organisation size on Internet use and value, the One-way NOVA statistical analysis and correlation analysis of the survey shows that there is a relationship between organization size and opportunity to provide employees with computer and Internet literacy and training courses. As well as, there exists a relationship between size of organization and Internet speed. Figure 5 of Appendix A provides a diagrammatic representation of this relationship. Organizations of a smaller size, less than 50 employees, are reliant on dial-up connections. While organizations of a medium size, 50-100 employees, are reliant on leased lines.

On the other hand, Chi Square analysis shows that there exists a relationship between organization size and Internet usage to reduce internal costs. The Pearson Correlation indicator shows that this relationship is weak and the smaller the organization is the more likely it is to benefit from the Internet to reduce internal cost. Also, it shows that there exists a weak relationship between organization size and Internet connection speed. The smaller the organization, the more likely the faster its connection speed will be. And it shows that a weak relationship exists between organization size and Web site presence. The smaller the organization, the more likely it will have a Web site of its own.

On the other hand the survey indicates that there is no relationship between organisation size and the following Internet uses/benefits:

- strategic value of the Internet presence
- primary purpose of presence on the Internet
- when Internet was first launched

Future Trends

The outcomes of this research confirm one of the major assumptions of the research plan. A mixed approach to evaluating possible outcomes of IT Investment must be sought to mix hard, direct, and quantitative measures with soft, indirect and strategic measures of performance or criteria of evaluation. Such a mixture aids researchers in reflecting the whole picture of IT investment as perceived by IT users and by investing organizations. Based on this assumption and the outcomes of the survey, a Jordanian implementation model can be devised from responses to reflect the local attitudes towards Internet Value.

This model takes into account the unique characteristics of Jordanian SMEs and the value of Internet Technology investment organized according to importance in descend-

Jordanian Internet Technology Implementation Model

Value Category	Usage
Co-operative Value	Using the electronic mail (e-mail) to facilitate exchange of data internally and externally
	Usage of Internet to exchange data electronically with suppliers, distributors and any other organizations (extranet)
	Usage of Internet to reach out to international markets
	Usage of Internet to share available resources within organizational boundaries
Provisional Value	Usage of Internet to lower Transaction Cost (the cost of offering services and/or products to customers)
	Usage of Internet to lower Internal Operation Cost
	Usage of Internet to disseminate information (including providing customers with information)
	Usage of Internet for customer and product research (including feedback from customer)
Promotional Value	Usage of Internet to promote current and future products and/or services
Transactional Value	Usage of Internet for transactional purposes (to buy/sell and bill suppliers, distributors or customers)

ing order. In comparison with developed countries, the results of the survey show a degree of relevance between the developed and developing countries in terms of Internet values and benefits (see above model).

On the other hand the degree of Internet Technology pervasiveness is much lower in developing countries compared with that in developed countries. Nevertheless, Hughes and Lawrence (2000) show that the majority of SMEs in the UK use the Internet primarily for: information gathering, communication, and advertising. While the use of the Internet for market and product research, information distribution and dissemination is used in a lesser degree. This is to a certain degree similar to SMEs in Jordan regarding communication and promotion, but regarding information gathering and market and product research it is almost non-existent in the Jordanian context.

The survey shows a positive attitude of Jordanian SMEs towards Internet Technology investment as the majority of respondents of the survey believed the Internet provided them with a very important to necessary medium to gain competitive advantage. On the other hand, the survey shows that many of the promised benefits of Internet Technology investment are far harder to realize than previously thought. Internet use is like the tide, it ebbs and flows according to business need and availability of secure technical, legislative and organizational infrastructures. This current situation in Jordan is set to change with the many changes introduced by the Government of Jordan to induce more investments by Jordanian companies in Internet technologies. On the legislative level, the National IT plan, REACH, has called upon the Government of Jordan to build on the recent successful liberalisation and reform efforts to develop Electronic Commerce legislation and to enforce intellectual property rights (REACH 2.0, 2001). The REACH initiative identifies 25 laws, bills and articles of urgent need for amendment or ratification by the upper and lower Houses of Representatives of the Jordanian House of Parliament. The government of Jordan has recently succeeded in getting 11 of those 25 pieces of legislation ratified, amongst of which Electronic Commerce and Electronic Signature legislation. According to these laws computer print-outs and e-mails electronically signed will be held as legal evidence in a court of justice.

In terms of national infrastructure, Jordan has one telecommunication company, partly owned by the private sector and the rest by the government. As well as three mobile network operators mostly owned by the private sector and more than 11 Internet Service Providers (ISPs) all of which are owned and operated by the private sector. Twenty one more ISP licenses have been issued by the Jordanian government to meet the increasing demand for high quality and high bandwidth Internet connectivity. The continuous deregulation of the telecommunication industry, since the inception of the telecommunication act of 1995, is projected to lead to a second national telecommunication operator by the end of 2004. Currently, most network accesses are provided via dial-up connectivity, while most recently ISPs have introduced ISDN (Integrated Services Digital Network) and ASDL (Asymmetric Digital Subscriber Line) access to the Internet, which is expected to increase bandwidth, and accelerates access to the Internet. Most recently Mobile network operators have introduced Internet services over mobile phones. However, affordable prices of both hardware and telecommunication devices as well as the high cost of telephone calls are perceived as the major constraints on the proliferation of IT in Jordan. The Jordanian National IT plan (REACH) has called upon both the private and the public sectors to provide preferential access to the high-speed lines and permit private up- and down-links, as well as to provide competitive pricing on high-speed lines (REACH 1.0, 2000; REACH 2.0, 2001). The current IT infrastructure of Jordan can be described as primitive in comparison to leading countries in the IT and Telecommunication sectors. For Jordan to become a regional leader in the Middle East, Jordan needs to work out its local infrastructure both in the technical terms as well as in the economical terms.

While the Internet provides many opportunities, the challengers posed by such technologies need to be handled technologically and economically both locally and regionally.

Conclusion and Theoretical/Practical Implications

Despite the positive attitude of Jordanian SMEs towards Internet Technology investment, the survey shows that many of the promised benefits of Internet Technology investment are far harder to realize than previously thought. The majority of respondents of the survey believed the Internet provided them with a very important to necessary medium to gain competitive advantage. The majority of respondents felt that the Internet affected a smaller than expected portion of their organizational activities. Although it might be seen conflicting, but important nevertheless, it is evident to say that the outcomes of this research project confirms an earlier conclusion that publicized possible benefits and outcomes of Internet Technology Investment are more elusive to realize than previously thought (Giaglis and Ray, 1998; Lawrence and Hughes, 2000). The survey shows that the majority of SMEs in Jordan use the Internet primarily for data exchange within and across organizational boundary via e-mail rather than to sell products and/or services online. While a sizable number of the sample believed that Internet use reduces both transaction cost (the cost of offering services and/or products to customers) and internal operation cost. On the other hand many of the other publicized benefits were not materialized like product and market research, expanding into new markets, resource sharing within organizational boundaries, conducting online business transactions (to buy/sell and bill suppliers, distributors or customers), human resources development (including online training and certification programs), and video-conferencing (to communicate via audio/video link). It can be said that SMEs in Jordan concentrate on the connectivity or communicational values of the Internet. This can be attributed to the high cost of wire and wireless communications in Jordan. The Internet provides a much cheaper medium of communication when compared to other media of communications.

Despite the fact that the Internet provided smaller organisation with a strategic medium to compete with larger, more resourceful organizations, this is seldomly the case. Many SMEs in Jordan felt that they are incapable, organizationally, to compete globally, to attract a global customer or even to take advantage of the Internet to expand into new markets. This can be attributed to many reasons, like competitive product/service quality and price. But primarily it can be attributed to lack of opportunity exploitations. Many organizations still do not believe that technology can replace physical markets.

The study shows that Jordanian small and medium size enterprises concentrate on co-operative, promotional and provisional values of Internet Technology investment. While, transactional value creation is still at its minimum. In order to increase the use of the Internet for trading products and services online, more secure links, for payments processing mainly, and greater bandwidth is needed. The survey shows that the majority of respondents use dial-up connection to connect to the Internet.

Lack of advanced and secure technical infrastructure, lack of high-volume of Internet users, and limited use of credit cards in Jordanian society remain the main reasons why most organizations in Jordan refrain from using the Internet to exchange products/

services and funds online. To justify investment in online technologies and processes, high velocity of Internet traffic is needed. REACH 2.0 (2001) shows that despite the sharp fall in phone tariffs, affordability of Personal Computers remains the main hindrance to engaging in online activities for individuals. These issues need to be solved on national and regional levels rather than on national level only. Regarding payments methods, whether physical or digital, the limited spread of credit cards makes the use of online experiences to exchange digitized and physical products and services for funds a true nightmare. It is becoming evident that more creative mechanisms should be used to overcome such a hurdle, like cash exchange at delivery time or using pre-paid cash cards or even using Automatic Teller Machine (ATM)/Debit cards as an alternative to credit cards.

Regarding relationships, the statistical analysis of the responses revealed a number of relations between organizational characteristics and Internet use and value. There exist two relationships between organization industry (service, product, educational, public and manufacturing) and reducing internal operation cost and Internet speed. While two more relationships exist between organization size and opportunity to provide employees with computer and Internet literacy and Internet speed. Also, a number of weak relationships exist between organization size and Internet usage to reduce internal costs, Internet connection, and Web site presence. This limited number of relationships indicates the sporadic nature of Internet Technology investment in Jordan. This sporadic nature of investment makes it hard for the researcher to predict or forecast the future trends in Internet Technology investments. One of the most interesting findings of this research project is the demographics of Internet users amongst small and medium size enterprises in Jordan. The survey shows that 49 out of the 71 respondents have established their own Web sites, the majority of whom have done that for more than a year till the date of this survey. Further on, the second section of the questionnaire, dedicated to measure Internet usage and value, shows that more than that do actually use the Internet without setting up their own Web sites.

References

Abu-Samaha, A., & Wood, J.R.G. (1998). *Proceedings of the 11th Bled Electronic Commerce Conference. Evaluating Inter-organisational Systems the Case of (EDI) in General Practice,* Bled Slovenia, June 7-9 (pp. 174-190).

Abu-Samaha, A., & Wood, J. R. G. (1999a). Healthcare computing conference. GP/ Provider Links: Who Benefits? Harrogate UK, March 6-9, 114-120.

Abu-Samaha, A., & Wood, J. R. G. (1999b). *Proceedings of the 6th European Conference on Information Technology Evaluation. Soft Evaluation: A Performance Measures Identification Method for Post-Implementation Reviews,* Brunel University, Uxbridge, London, November 4-5 (pp. 221-228).

Amor, D. (2002). *The e-business (r)evolution: Living and working in an interconnected world.* New York: Hewlett-Packard Books.

Brown, A. (1994). *Proceeding of the 1st European Conference on IT Investment Evaluation. Appraising Intangible Benefits from IT Investment,* Henley, UK (pp. 187-199).

Chan, Y. (1998). *5th European Conference on The Evaluation of Information Technology. IT Value-The Great Divide between Qualitative and Quantitative, and Individual and Organisational, Measures,* University Management Unit, Reading, UK (pp. 35-54).

Collier, P., & Spaul, B. (1991/1992). An introduction to EDI. *Accounts Digest, 276*, 1-9.

Giaglis, G., & Paul, R. (1998). *Proceeding of the 11th Bled International Electronic Commerce Conference. Dynamic Modelling to Assess the Business Value of Electronic Commerce,* Bled, Slovenia, June 8-10 (pp. 57-73).

Ho, J. (1997). Evaluating the World Wide Web: A study of 1000 commercial sites. *Journal of Computer Mediated Communication (JCMC), 3*(1), 1-11. Retrieved from *http://www.ascusc.org/jcmc/vol3/issue1/ho.html*

Lawrence, E., Corbitt, B., Tidwell, A., Fisher, J., & Lawrence, J. (1998). *Internet commerce: Digital models for business.* Brisbane: John Wiley.

Lawrence, J., & Hughes, J. (2000). *Proceeding of the 13th Bled International Electronic Commerce Conference. Internet Usage by SMEs: A UK Perspective,* Bled, Slovenia (pp. 1-30).

Parfett, M. (1992). *What is EDI: A guide to electronic data interchange.* London: NCC Blackwell.

REACH 1.0. (2000). INTAJ. Retrieved from *http://www.reach.jo/*

REACH 2.0. (2001). INTAJ. Retrieved from *http://www.reach.jo/*

Simpson, P., & Swatman, P. (1998). *Proceeding of the 11th Bled International Electronic Commerce Conference. Small Business Internet Commerce Experience: A Longitudinal Study,* Bled, Slovenia, June 8-10 (pp. 295-309).

Turban, E., Kuy Lee, J. & King, D. (1999). *Electronic commerce: A managerial perspective.* NJ: Prentice Hall Business Publishing.

Appendix A (Supplementary Tables and Figures)

Table 7. Relationship between type of organization and reducing transaction cost

	Service Based	Product Oriented	Educational	Public	Manufacturing	Total
Not Available	2	2	8	2		14
Less Than 25%	3	1	7	3		14
25-50%	2	3	3	1	2	11
51-75%	3	5	3		3	14
More than 75%	7	3	1		3	14
Total	17	14	22	6	8	67

Figure 2. Relationship between type of organization and reducing transaction cost

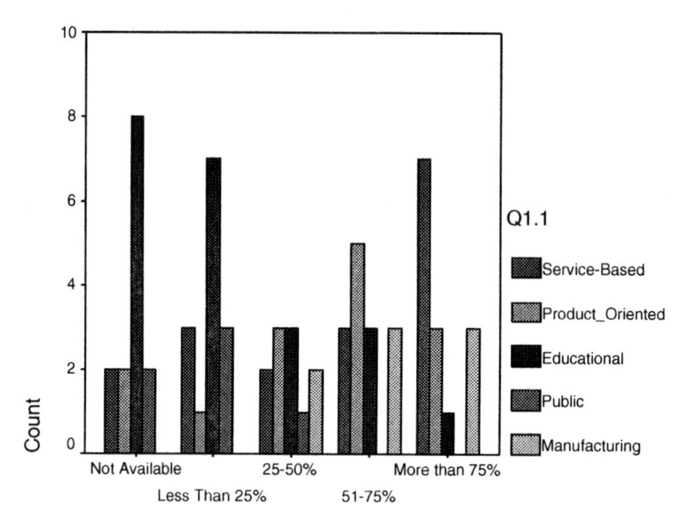

Table 8. Relationship between type of organization and reducing operational cost

	Service-Based	Product-Oriented	Educational	Public	Manufacturing	Total
Not Available	4	1	11	3	2	21
Less Than 25%	3	2	6	3	1	15
25-50%	3	3	4		2	12
51-75%	4	2	1	1	1	9
More than 75%	4	6	1		2	13
Total	18	14	23	7	8	70

Figure 3. Relationship between type of organization and reducing operational cost

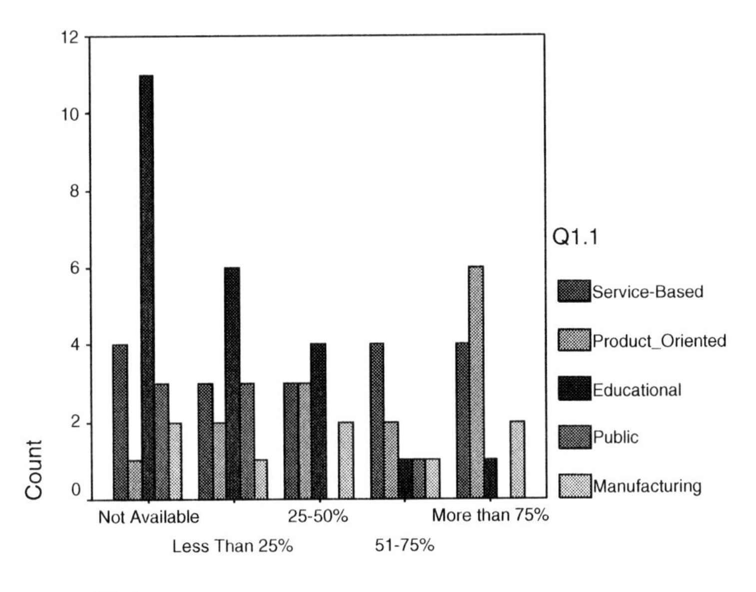

Figure 4. Relationship between type of organization and Internet speed

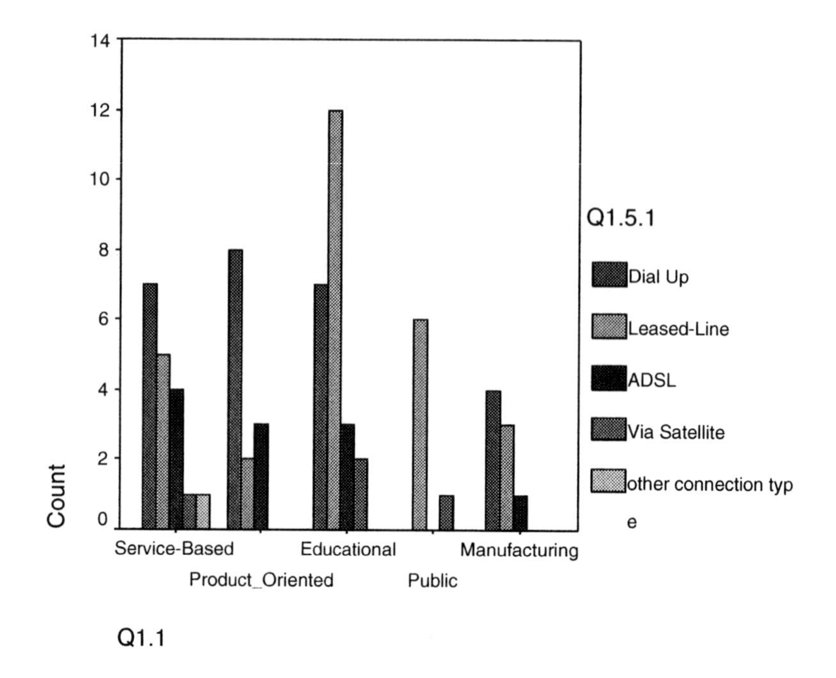

Figure 5. Relationship between size of organization and Internet speed

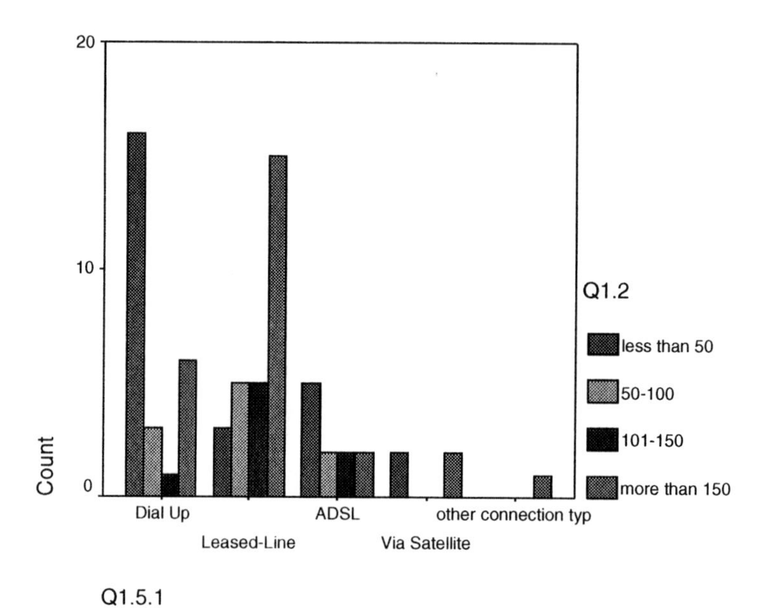

Chapter XVII

E-Commerce Infrastructure in Developing Countries

Aladdin Moawad Sleem, Trendium Inc., USA

Abstract

This chapter provides an overview of the existing e-commerce infrastructure in Developing Countries. A comprehensive survey of all the technologies used in e-commerce is introduced at the beginning of the chapter followed by a description of e-commerce infrastructure components. This chapter groups e-commerce infrastructure into three difference categories, namely: technological infrastructure, financial infrastructure, and the legislative infrastructure. An overview of the current technological, financial, as well as legislative e-commerce infrastructure in Developing Countries is provided which shows that many developing countries do not have the infrastructure required to provide e-commerce services. The author acknowledges the importance of e-commerce and its impact on the economy of any country and hopes that a clear understanding of the entire underlying infrastructure can help the Developing Countries to accelerate the growth of its e-commerce services.

Table 1. Projected B2B e-commerce by region, 2000-2004 ($Billions)

	2000	2001	2002	2003	2004	As a % of worldwide B2B commerce, 2004
North America	159.2	316.8	563.9	964.3	1,600.8	57.7
Asia/Pacific Rim	36.2	68.6	121.2	199.3	300.6	10.8
Europe	26.2	52.4	132.7	334.1	797.3	28.7
Latin America	2.9	7.9	17.4	33.6	58.4	2.1
Africa/ Middle East	1.7	3.2	5.9	10.6	17.7	0.6
Total	226.2	448.9	841.1	1,541.9	2,774.8	100.0

Source: e-Commerce and e-Business, by Zorayda Ruth Andam. Available at http://www.apdip.net/documents/eprimers/e-Commerce-and-e-Business.pdf

Introduction

As electronic commerce (e-commerce) growth becomes more and more significant, it will not be enough for developing countries to understand it, but in fact to engage in it. This engagement is not only for realizing its potential of growth for their trade and industry but also as a means of survival in the new world of e-commerce-based trade and business. Their ability to do so will depend on several factors, most important of which will be their infrastructure, both physical (the telecommunication network), as well as their financial and legal framework. It will also depend on the availability and price of hardware (computers, routers, switches, etc.) and software, as well as the human resources and education standards of the country. E-commerce activities in the developing counties are still far behind that in industrial countries. Table 1 shows the projected business-to-business (B2B) e-commerce by region.

Due to the existing development gap and differences between the developing and the industrial countries, developing countries face a distinct challenge in order to realize the promises inherent in this new technological development. The first challenge is how to equip the developing countries to benefit from and use the Internet as a tool for development, and secondly, how to ensure and manage the growth and development of the Internet as a utility that would promote development. The former is a developmental challenge and requires investments in the infrastructure of telecommunications sector to ensure the easy and affordable availability of required equipment and software, and in training and Internet literacy. The latter is more of a global regime challenge which involves the growth and regulation of the Internet and its facilities on a universal scale.

E-commerce for developing countries should not be only a vision. It is actually a fact, since the international trade and business are moving in this direction. Most developing countries are unfortunately far behind and unsure of the path forward. To appreciate the issues involved, this chapter highlights the technologies required to establish e-commerce and also provides an overview of the present situation of infrastructure development for e-commerce in developing countries.

E-Commerce Technologies

In order to provide a predefined and reliable e-commerce environment, Internet service providers and e-commerce companies need a set of technologies and integration capabilities that allow companies to group applications or users into different categories with varying levels of access and service availability. Unfortunately, developing countries have been implementing some of these technologies and their underlying infrastructures with a speed that does not match the increasing role of e-commerce in the global economy. In order to create the foundation for the rapid growth of e-commerce, effective technology policies must be adopted to embrace certain crucial areas such as:

- Reliable, scalable, and fast network connectivity
- Transaction processing and business information communication protocols
- Electronic payment mechanisms
- Building online trust through improved security and protecting privacy
- Strong intellectual property protection

The rest of this section provides an overview of some of the technologies required to establish the necessary infrastructure for e-commerce.

Networking Technologies

The future of e-commerce demands a dramatic evolution of the worldwide telecommunication infrastructure. Today's infrastructure was built to carry voice traffic and has served well for many decades. The current slow transmission speeds and congestion rates represent the legacy of an outdated system that needs to be modernized. This slow speed of accessing data over interconnected networks is one of the reasons that widen the gap between e-commerce utopia and the current reality of e-commerce in developing countries. To make matters worse, service quality and reliability vary among Internet connection providers even within the same country.

High speed constant connection to the Internet (broadband access) allows users to send and receive far larger volumes of information than traditional dial-up telephone lines allow. Some examples of broadband access technologies are:

- Modified cable television lines
- Digital Subscriber Line (DSL) which is an enhanced telephone service

- Satellites
- Fixed wires

Broadband access is necessary in order to make the vision of new Internet-based economy a reality. Although the research in broadband Internetworking technologies and services has yielded some impressive results, developing countries as well as many households in developed countries are nowhere near to having affordable and worldwide high access speeds.

New technologies for multicasting and traffic isolation are highly needed in order to support some services (i.e., voice and video) that have special transmission requirements. These technologies may be critical for "multicasting" of multimedia data across networks. Also, Internet devices like routers and switches need to perform some degree of traffic isolation among services such that one shall not impact the others.

A series of Quality of Service (QoS) and Traffic Management technologies are needed to be deployed to guarantee Web QoS to customers. While several vendors have unveiled products in this area, it is likely to take a while before they can be effectively put to use and before interoperability issues are resolved. To gain high-speed data transfers in wide area networks, Asynchronous Transfer Mode (ATM) has been widely used as the access method for the carrier-based, high-speed Synchronous Optical Network/Synchronous Digital Hierarchy (SONET/SDH) infrastructure. While ATM is one of the most scalable and flexible technologies available, it is still not very affordable in may parts of the world. At the same time, companies are reluctant to invest into technologies that may be superseded by newer ones such as Wave Division Multiplexing (WDM) or Internet Protocol (IP) over SONET.

In addition to the wire-line infrastructure, recent advances in wireless telecommunication technologies have made commercial deployments of wireless networks possible, opening up a potentially huge opportunity for wireless e-commerce or what is sometimes called mobile-commerce. Wireless connectivity offers benefits such as faster access speeds, user mobility, flexibility in use, and easy scalability of networks that in turn improves productivity and enhances competitiveness. The wireless networking domain has seen the emergence of a number of standards. The alphabet soup of 802.11 standards seems to be ever widening. The term Wi-Fi, a common term used for the 802.11b standard, has become synonymous with Wireless Local Area Networks (WLANs). WLANs were initially used by enterprises in various verticals (warehousing, factory, health care, retail, etc.), but now they are increasingly finding applications in the general-purpose public space (cafes, airports, hotels, etc.) and homes. This transition represents a tremendous growth opportunity for establishing a wireless e-commerce infrastructure.

Business Information Exchange Protocols

Until very recently, the primary technology for B2B electronic commerce of large firms has been electronic data interchange (EDI). EDI uses a store and forward messaging technology, a process which parallels the exchange of electronic mail (e-mail) with the addition of structured message content and functionality to assure audit ability and assurance of delivery. Because EDI enables communications between the business systems of trading partners, messages are translated into highly structured formats, allowing both the sending and the receiving organizations to share a common and identical understanding of the content and context of the messages. Private, value-added networks (VANs) have been the primary carriers of EDI communications, using a variety of communications protocols. EDI technology has been available for more than two decades, but has only been implemented by a very small segment of the business community worldwide (UNESCAP- Development Research and Policy Analysis Division, 2003). Today, EDI is undergoing a transformation spurred by the rapidly increasing connectivity provided by the Internet and is itself an integral part of the sweeping wave of Internet commerce automation. The Internet is eroding the value-added attributes of private networks, shifting the balance of supplier value-added from the network itself to software and services. In fact, the failure of EDI in its traditional form to achieve greater penetration reflects the complexity and expense of integration with existing business systems owing to the lack of agreement on standards, combined with the continuing costs of an EDI server and VAN network support resources.

As previously explained, with the rapid growth of the Internet, organizations are increasingly using the Web to conduct business with greater speed, reach, and efficiency. This transformation is especially prevalent in B2B commerce. A large number of companies worldwide have adopted e-procurement systems. Each procurement system uses different B2B protocols for interaction with seller systems. Many of these protocols are proprietary and specific to the system. Standard bodies are defining protocols and message formats for B2B processes. The electronic business XML (ebXML) framework, sponsored by the United Nations Center for the Facilitation of Procedures and Practices for Administration Commerce and Transport (UN/CEFACT) and the Organization for Advancement of Structured Information Standards (OASIS), includes a messaging service, a Collaborative-Protocol Agreement (CPA) specification, and a business process specification scheme (Business Software Alliance, 2003). These are used for enabling the interaction between business processes.

The Web services approach defines a messaging and a remote procedure call mechanism using Simple Object Access Protocol (SOAP). On top of SOAP, the Web Services Description Language (WSDL) defines an interface for Web-based B2B remote procedure calls similar to the Common Object Request Broker Architecture (CORBA) interface definition language (IDL).

Online Trust Technologies: Ensuring Security and Protecting Privacy

The future of electronic commerce ultimately depends on how much the transacting parties trust the security of the transmission and content of communications. It also rests on the faith that these communications will be granted enough recognition in any domestic or foreign jurisdiction. In e-commerce transactions, agreement and settlement are transferred in open network environment, such as the Internet, by exchanging digital information, which raises many security concerns. Security is also an issue in regard to the creation of the electronic data message, data protection techniques applied to the data message, and its transmission to the recipient. Different levels of security have to be selected between different trading partners according to the vulnerability of their business processes. The following are the most important threats for messaging:

- a message may be duplicated
- a message may be lost or replayed
- a message may be intercepted and modified
- a third party may pretend to be a valid message sender
- the sender may claim she/he never sent a particular message
- the recipient may claim she/he never received a particular message
- a message may be read by a third party (unauthorized disclosure of message content)

Providing enterprises with tools to address access control, confidentiality and integrity, and audit and monitoring is not enough, even though it takes care of many of their security needs at the enterprise level. The real concern for the policy makers is to provide the required foundation for trust by making sure that there are trust management products and services for wide use on the Internet.

To build trust in an open network environment, such as the Internet, the trading parties need to feel that their business transaction have certain attributes. Among these attributes are confidentiality, integrity, and authenticity.

Confidentiality

This is to ensure that the information in the network remains private and no one other than the trading parties has access to the information being exchanged among the parties. This is typically accomplished through encryption. There are two major encryption schemes. The first one is the secret-key scheme. In this scheme, a common key is used by the sending party for encrypting the message and used again by the receiver to decrypt the message. This common key is held secretly between the sender and the receiver. The other encryption scheme is the public key where different keys are used for encryption and decryption. In this scheme, one key is kept secret and the other key is made public. This scheme makes it very difficult to figure out the secret key from public key using calculations.

In the typical e-commerce transaction, both of two schemes are used combining the advantages of both schemes. At first, to establish a session, a common key is transferred using public-key encryption. Next, commercial information is sent using common-key encryption with the common key, which is transferred formerly. The most famous secure communication protocol is the Secure Socket Layer (SSL) protocol. SSL is normally set in the browser software for searching Web information in the Internet. SSL realizes not only confidentiality but also integrity and authenticity, using the two types of cryptographic schemes.

Integrity

This is to ensure that a message has not been modified in transit. There are two ways for ensuring message integrity. The first is message digest with secret information. The second is digital signature with public-key cryptography. It is needed to register the public key before transaction.

Authenticity

This ensures that the writer of a document or the sender of a message is who he claims to be. The methods of personal authentication fall into three types. The first is biometrics type of fingerprint and handwriting. The second is token type of driver's license and credit card. The third is secret information type of password and digital signature. Certification authority (CA) organization is essential to register public keys, in case of using a digital signature with public-key cryptography.

The following is an overview of some of the technologies used to establish trust in the new electronic marketplace.

Encryption Technologies

Secret-Key Encryption Scheme

The early cryptosystems were character-based systems which used mathematical functions for encryption and decryption and either substituted characters for one another, transposed characters with one another, or did both operations. The security of the cryptographic algorithm was based on keeping the algorithm secret, which didn't seem to provide enough security since a person intercepting the message can try different letter transformation and use tables of the frequency of letters and words in the language

Figure 1. Encrypting a message with secret-key scheme

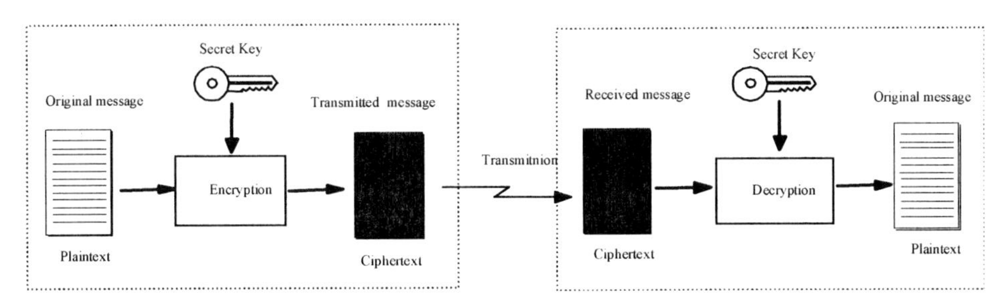

until the correct transformation is found. Modern encryption algorithms are key-based algorithms. Unlike classical cryptography, key-based cryptography encryption and decryption algorithms are known to everyone, but the encryption and decryption operation can not be done unless a certain key is passed to the algorithm. The security of the system is based on keeping the key hidden (Schneier, 1996).

In secret key cryptography, a secret key is established and shared between two individuals or parties and the same key is used to encrypt or decrypt messages. Therefore, it is also referred to as symmetric key cryptography.

Figure 1 illustrates the secret-key encryption scheme. The worldwide standard algorithm for symmetric key cryptography is the Data Encryption Standard (DES).

Public Key Encryption Scheme

In 1976, Whitfield Diffie and Martin Hellman, introduced what is called public key cryptography which laid the foundation of many schemes necessary for electronic commerce (Stinson, 1995). In public key cryptography, each participant runs a program that generates a "public key" and a "secret key." The secret key is never shared with anyone, while the public key is announced. Public key systems are set up so that anyone with the public key can encrypt a message but not decrypt it, only the person who has the corresponding secret key can decrypt it. This means that for two persons; Alice and Bob, to communicate safely using public key cryptography, Bob can encrypt a message using Alice's public key and no one (even Bob himself) but Alice can decode it using her secret key as shown in Figure 2.

As shown in Figure 3, when Alice encrypts a message with her secret key, anyone who has the corresponding public key can decrypt it. The pairing between the public and the secret key ensures that if someone is able to decrypt a message using Alice's public key, then this is evidence that the message was encrypted by Alice (who holds her secret key). This is very important since it allows a user to make "digital signatures." There are many secure modern public key algorithms. RSA, named after its inventors Rivest, Schamir, and Adleman, is the most widely used.

Figure 2. Encrypting messages with public key cryptography

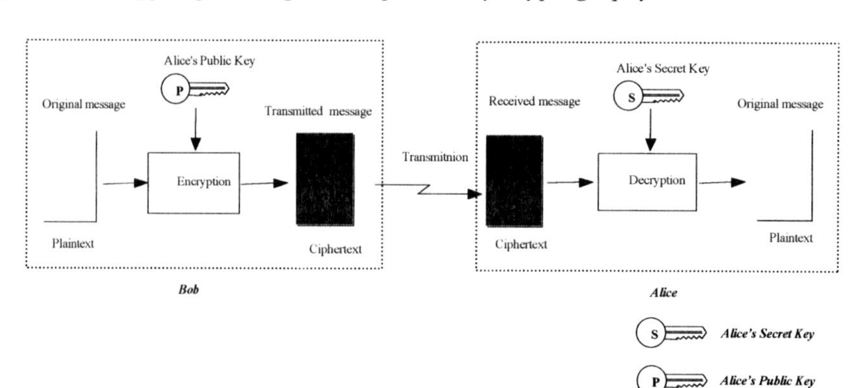

Figure 3. Decrypting messages with public key

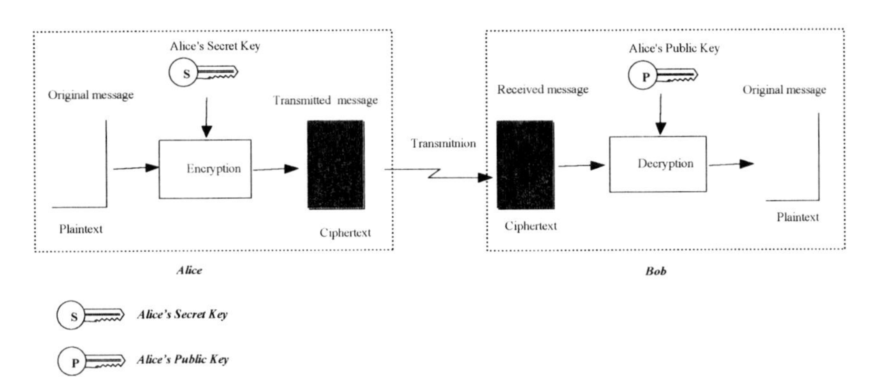

Hybrid Encryption Scheme

Public key algorithms are significantly slower than secret (symmetric) algorithms. Symmetric algorithms are generally at least 1000 times faster than public key algorithms. In most practical implementations, public key cryptography is used to secure and distribute session keys, and those session keys are used with symmetric algorithms to secure message traffic. This is sometimes called a hybrid cryptosystem.

Message Digest

Although Alice may encrypt her message to make it private, there is still a concern that someone might modify her original message or substitute it with a different one. One way of guaranteeing the integrity of Alice's message is to create a concise summary of her message and send this to Bob as well. Upon receipt of the message, Bob creates its own summary and compares it with the one Alice sent. If they agree, then the message was received intact.

A summary such as this is called a message digest, one-way function or hash function. Message digests are used to create short, fixed-length representations of longer, variable-length messages. Digest algorithms are designed to produce unique digests for different messages. Message digests are designed to make it too difficult to determine the message from the digest, and also impossible to find two different messages which create the same digest, thus eliminating the possibility of substituting one message for another while maintaining the same digest.

Digital Signature

Digital signature technology has laid the foundation for several mechanisms that are currently used in e-commerce. It is an implementation of public key cryptography that enables the creation of tamper-proof documents, which unequivocally associate their owners. Digital signatures are used to authenticate both the sender and message (i.e., to provide proof to the recipient that the message stems from the sender and that the message's contents have not been altered since leaving the signatory). A digital signature is based on the actual contents of the message itself. It is a small amount of data that is recorded on an electronic medium. To digitally sign a message, the sender produces his digital signature by applying certain calculations to the message. This process is called the "*signature function.*" The resulting signature, which looks like random data, only has meaning when read in conjunction with the message used to create it. The recipient of the message checks the digital signature by performing another set of calculations on the signature and the message. This is called the "*verification function.*" The result of these calculations reveals whether or not the signature is a genuine authentication of both the sender and the message.

A signature is only useful if it can be performed exclusively by the intended authorizer. This is why the method for creating any individual signature must be kept secret by its owner. Since it is impractical to construct new signature and verification functions for each user, a common method with a large number of unique keys is used, each of which produces a unique signature. This means that the function, which processes the message as input, needs another additional parameter (for both the signature and verification functions). This simplifies both signature and verification processes because only this one parameter needs to be unique for each user. These parameters, called keys, allow different signatories to use the same digital signature method but produce different results from their own unique keys.

There are many methods for producing and checking digital signatures. These methods can be classified into two categories according to whether they use secret key or public key techniques. In *secret key signature methods*, the same key is used to both create and verify signatures, and the signature and verification functions are alike. Secret key signature methods are therefore called symmetrical signature methods.

In *public key signature methods*, a secret key is used to sign messages, but a public key is used to verify the signatures. The secret key and public key always form a key pair,

Figure 4. Public key signature

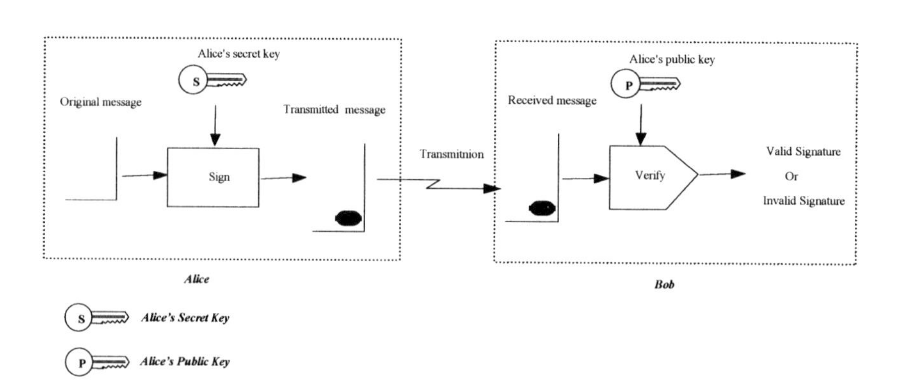

i.e., a public key can only be used to verify signatures and does not need to be kept secret. Its special design means that the recipient of a message can verify the authenticity of the signature, but cannot use it to determine how the signature was created during the secret key process. The public key cannot be used to create unauthorized signatures.

Digital signature with public-key encryption is popular in e-commerce. Legal understanding of a digital signature and certificate of public key is necessary for their effectiveness in business activities.

Blind Signature

The blind signature technique was first introduced by Chaum, Fiat and Naor in 1988 (Chaum, Fiat, & Naor, 1990). It is a cryptographic invention that allows the production of digital money by permitting numbers to serve as electronic cash. It solves the problem of authenticating an electronic bill while preserving anonymity. Chaum, Fait and Naor introduced a protocol by which a virtual bank, or any other authorizing agency, can authenticate a number so that it can act as a unit of currency, yet the bank itself does not know who has the bill, and therefore cannot trace it. Chaum managed to come up with a proof that this sort of anonymity could be provided unconditionally, with all the assurance of mathematical proof that no one could violate it. Chaum continued to build upon those ideas, addressing problems like preventing double-spending while preserving anonymity. In a particular clever mathematical twist, he came up with a scheme whereby one's anonymity would always be preserved, with a single exception: when someone attempted to double-spend a unit that he or she had already spent somewhere else. The blind signature is an extension of the digital signature technique.

Digital Certificates

A digital certificate is an electronic file that is used to uniquely verify the identity of the holder or sender of the certificate. It associates the name of an entity that participates in a secured transaction with the public key that is issued to sign the communication with that entity in a cryptographic system. Typically the issuer of the certificate is a "trusted third party" or "certification authority" (CA). All participants who use such certificates agree that the CA is a point of secure storage and management of the associated private signing key. The CA issues, creates, and signs certificates. The following is an example of the contents of a digital certificate:

- Version
- Serial number
- Signature algorithm ID
- Certificate issuer
- Certificate holder
- Validity period (the certificate is not valid before or after this period)

- Attributes, known as *certificate extensions,* that contain additional information for this certificate
- Digital signature from the CA to ensure that the certificate has not been altered and to indicate the identity of the issuer
- Public key of the owner of the certificate

Using a digital certificate simplifies the problem of trusting that a particular public key is in fact associated with a participating party, effectively reducing it to the problem of trusting the associated CA service.

Secure Communication Protocols

Due to its convenience, the most widely used secure communication protocol is the Secure Sockets Layer (SSL) protocol. The SSL protocol is a protocol layer which may be placed between a reliable connection-oriented network layer protocol (e.g., TCP/IP) and the application protocol layer (e.g., HTTP). SSL provides for secure communication between client and server by allowing mutual authentication, the use of digital signatures for integrity, and encryption for privacy.

The protocol is designed to support a range of choices for specific algorithms used for cryptography, digests, and signatures. This allows algorithm selection for specific servers to be made based on legal, export or other concerns, and also enables the protocol to take advantage of new algorithms. Choices are negotiated between client and server at the start of establishing a protocol session.

A special version of the HTTP protocol (HTTPS) is needed to allow secure communication between the client and the server using the SSL protocol.

Defense Against Unauthorized Access

Guarding the enterprise network from external intrusions is a critical task in protecting the enterprise's e-commerce assets. This can be accomplished cost-effectively through a standard firewall-safe network configuration. The term firewall has been around for quite some time and originally was used to define a barrier constructed to prevent the spread of fire from one part of a building or structure to another. Network firewalls provide a barrier between networks, which prevents or denies unwanted or unauthorized traffic. When implemented correctly, firewalls can control access both to and from a network. They can be configured to keep unauthorized or outside users from gaining access to internal or private networks and services. They can also be configured to prevent internal users from gaining access to outside or unauthorized networks and services. Firewalls can be configured to require user authentication. This allows network administrators to control access by specific users to specific services and resources. Authentication also allows network administrators to track specific user activity and unauthorized attempts to gain access to protected networks or services. Firewalls can provide auditing and logging capabilities. By configuring a firewall to log and audit activity, information may be kept and analyzed at a later date. Firewalls can generate statistics based on the information they collect. These statistics can be useful in making policy decisions that relate to network access and utilization. Some firewalls function in a way that can hide internal or trusted networks from external or untrusted networks. This additional layer of security can help shield services from unwanted scans. Firewalls can also provide a central point for security management. This can be very beneficial when an organization's human resources and financial resources are limited.

Electronic Payment Technologies

To enable e-commerce, electronic payment mechanisms have to be available. Online payment processing requires coordinating the flow of transactions among a complex network of financial institutions and processors. Technology has simplified this process so that, with the right solution, payment processing is easy, secure, and seamless for the trading parties.

Credit Card Payments

Credit card payment processing in the online world is similar to payment processing in the offline or "brick-and-mortar" world, with one significant exception. In the online world, the card is "not present" at the transaction. This means that the merchant must take additional steps to verify that the card information is being submitted by the actual owner of the card. Payment processing can be divided into two major steps: authorization and settlement. The authorization step verifies that the card is active and that the customer has sufficient credit available to make the transaction. Settlement phase involves transferring money from the customer's account to the merchant's account.

Smart Cards

Identical in size and feel to credit cards, smart cards store information on an integrated microprocessor chip located within the body of the card. These chips hold a variety of information, from stored (monetary) value used for retail and vending machines, to secure information and applications for higher-end operations such as medical/healthcare records. New information and/or applications can be added depending on the chip capabilities.

Different types of cards in use today include contact, contactless, and combination cards. Contact smart cards must be inserted into a smart card reader. These cards have a contact plate on the face which makes an electrical connector for reads and writes to and from the chip. Contactless smart cards have an antenna coil, as well as a chip embedded within the card. The internal antenna allows for communication and power with a receiving antenna at the transaction point to transfer information. Close proximity is required for such transactions, which can decrease transaction time while increasing convenience. A combination card functions as both a contact and contactless smart card.

Payments Service Providers

As more B2B trading partners conduct business and provide customer service over the Web, it makes sense to handle invoicing, billing and payment processing in the same fashion. For many small e-commerce shops, online payment processing might not be a feasible solution. As an alternative to processing the payment online, small e-shops use a third party to process the payment. This third party is usually called an "electronic payment service provider." There are a number of different types of electronic payment service providers in business, including biller-focused, commerce-focused, and payer-focused providers. In the biller-focused area, CheckFree is the leader (Loshin & Vacca, 2003). The company processes 49 million electronic payments per month. Another player in this area is Princeton eCom.

In the commerce-focused area, the major players include CyberCash, CyberSource, and VeriSign. For the payer-focused providers, the service is provided for the payer. Examples of payer-focused providers are PayPlace.com and ProPay.com.

Digital Cash

This is the digital equivalent of cash. Users withdraw digital coins from their Internet bank accounts and store them on their hard disks. Whenever someone wants to make a payment, he or she uses these coins. The main design parameter that differentiates systems in this category is the degree of anonymity provided by the system. Some systems provide unconditionally untraceable electronic cash as long as there is no double spending. Other systems provide less anonymous cash but with a high degree of scalability (the ability to add new users to the system without affecting the performance of the system). The main advantages of this mechanism are:

- *Privacy:* Electronic cash is untraceable. The bank does not link the numbers to a particular person, therefore, it is impossible to link payment to payer. The customer does not have to worry about being added to dozens of mailing lists. Unless of course, the customer has ordered actual merchandise that needs to be sent to their home, instead of information that can be sent over the Internet.
- *Limited liability:* A customer can only use as much money as they are carrying on their hard disks. People may be more willing to deal with electronic cash and only risk the $20 in their electronic wallet than to send their $5,000 credit card number across the Internet.

Intellectual Property Protection Technologies

With advancements in Internet technologies and increasing demands on online multimedia businesses, digital copyright has become a major concern for businesses that engage in online content distribution through various business models, such as pay-per-view, subscription, trading, and so on. In electronic commerce Web sites or applications, digital contents can be categorized into four basic types of media data: image, audio, video and text. Digital watermarking is viewed as an enabling technology to protect these media data from re-use without giving adequate credit to the source or in an unauthorized way. A "watermark" is a signal added to digital data that can be detected or extracted later to make an assertion about the data. Currently, no universal watermarking technique exists that satisfies all requirements of all applications. Instead, the specific requirements of each watermarking application depend on the protection objectives, the kind of object and its digital size, and possibly the kind of distribution channel.

Digital watermarks are embedded in digital objects (such as images) so that owners and perhaps end users can detect illegitimate copying or alteration. Digital watermarks can be made either "perceptible" (by people) or "imperceptible." A "fragile" watermark is damaged by image distortions and thus serves to detect alterations made after the watermark is applied. A "robust" watermark survives distortions such as trimming away most of the image and thus can serve as evidence of provenance. Both kinds can be embedded in most varieties of digital objects. Watermarks are currently of most interest for images, audio signals, and video signals.

A fragile watermarking scheme has two procedures, one for watermark insertion and one for watermark extraction. The input to the insertion procedure consists of the unmarked object, the watermark, and a key associated with the creator of the object (or another authorized party in the distribution chain). The output is a watermarked object. The input to the extraction procedure consists of the watermarked object and the key used during insertion. If the object has not been altered since it was marked and the correct key is used, the output of the extraction procedure is the watermark. If the object has been altered or the wrong key is used, the extraction procedure outputs an error message. Some fragile watermarking schemes can identify the unauthorized alteration. Others detect only that an alteration has occurred.

In a robust watermarking scheme, it is assumed that the marked object may be altered in the course of its normal use. For example, robustly watermarked images may undergo compression and decompression, filtering, scaling, and cropping. The inputs to and outputs from a robust watermarking insertion procedure are the same as in the case of fragile watermarking. The inputs to the detection procedure are the watermarked object (which may have been legitimately altered in the course of normal use), the watermark that was inserted into the object, and the key. The detection procedure then indicates whether the object contains a mark that is "close to" the original watermark. The meaning of "close" depends on the type of alterations that a marked object might undergo in the course of normal use.

Another scheme for watermarking is the fingerprinting scheme. In this scheme, there is an additional input to the insertion procedure that depends on the recipient of the specific copy. The output is a marked object in which the mark (the fingerprint) identifies the recipient. Two different customers purchasing the same work would receive objects that appeared the same to human perception but contained different watermarks. If unauthorized copies were later found, the fingerprint could be extracted from those copies, indicating whose copy had been replicated.

E-Commerce Infrastructure in Developing Countries

E-Commerce Infrastructure Components

The current evolution of e-commerce requires three types of infrastructure: technological infrastructure, financial infrastructure, and legislative infrastructure.

Technological infrastructure is required to establish an Internet marketplace. This relies on a variety of technologies, the development of which are proceeding at breakneck speeds (e.g., interconnectivity among telecommunications, cable, satellite, or other Internet backbone, Internet service providers (ISPs) to connect market participants to that backbone, and end-user devices such as PCs, TVs, or mobile telephones). The financial infrastructure is needed to provide e-commerce services with the necessary trusted and secure financial transactions. This infrastructure makes payment over the Internet possible (through credit, debit, or Smartcards, or through online currencies). The third infrastructure component is the set of protocols, laws and regulations that affects the conduct of those businesses engaging in and impacted by electronic commerce, as well as the relationships between businesses, consumers, and government.

Technological Infrastructure

All the consumer confidence and legal support in the world won't boost e-commerce if there's no way to deliver electronic content to consumers efficiently and quickly. The success of the e-commerce revolution from the point of view of the developing countries is dependent on several key preconditions. The first (though obvious) is the widespread availability of the Internet. For developing countries, access to modern telecommunication systems is therefore perhaps *the* defining element of electronic commerce. A well–functioning, modern telecommunication infrastructure and a satisfactory distribution of electricity; along with access to computer hardware, software and servers are the basic requirements for electronic transactions and therefore e-commerce.

Many developing countries lag far behind developed country markets in the availability of the technical prerequisites for conducting electronic commerce. The gaps in the two main requirements for Internet (i.e., telephone and computer availability) highlight the difference (Singh, n.d.). In addition to the limited technological infrastructure in developing countries, telecommunications services are often unreliable, high cost or both. There are also enormous differences in access to telecommunications both between and within developing countries. For instance, while in developing countries a considerable proportion and sometimes the majority of the population lives in rural areas, a high percentage of the main telephone lines are located in urban areas.

Many developing countries may not be able to generate or attract the large investments needed for the telecommunications infrastructure in their entire country. Perhaps, therefore, a more strategic option would be to formulate and implement appropriate

Figure 5. E-commerce infrastructure components

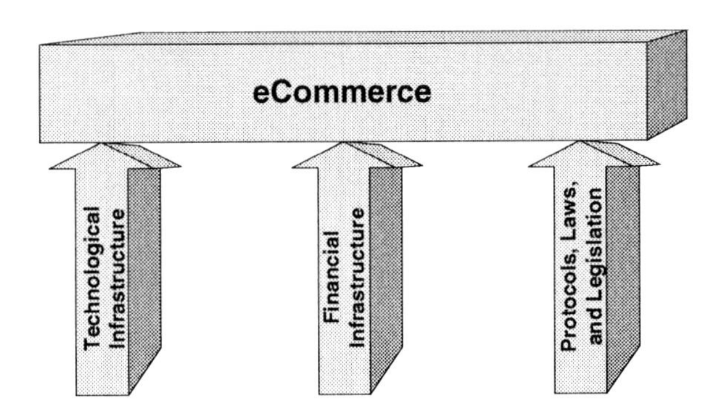

policies by concentrating on the areas in which electronic commerce is most likely to bring the highest benefits to their respective national economies: business-to-business, business-to-consumer and business-to-government transactions. This can be both geographic as well as sectoral. If developing countries wait to take initiatives in the area of e-commerce until a complete countrywide modern communications infrastructure for electronic commerce is in place, the gap with the developed world may grow even wider. What is important to bear in mind is that massive investments and flawless technological solutions are not always necessary or possible. Even existing networks can be reengineered and global services for the Internet and e-commerce sites can be utilized, especially in urban concentrations, where the maximum potential for trade and commerce presently exists. Several developing countries have followed this strategy of providing Internet access initially to only important commercial centers, especially where there is a concentration of export-intensive industry. Others have set up business centers in far flung places, linked to a common network and service that assists not only in access but also business promotion. Such models provide an immediate and workable solution that can be emulated by other developing countries.

Network Backbone and International Bandwidth

The availability of international bandwidth is very critical for developing countries because a large part of their Internet traffic tends to be international. The UNCTAD e-Commerce and Development Report for year 2003 (UNCTAD, 2003) highlights the fact that most of the limited international links that exist in developing countries tend to connect to the United States or Europe.

A handful of African countries have links with their neighbors. This means that a high percentage of intra-African Internet traffic flows through expensive intercontinental circuits. The situation in Asia and Latin America, although somewhat better, also limits their participation in the global information economy. Improvements in bandwidth availability and costs tend to be associated with a regulatory environment that promotes competition. Bandwidth scarcity may reflect the high cost of connecting small, low-

Table 2. International Internet bandwidth (Mbps) by region, 2000-2002

Region	2002	2001	2000
Africa	**2,118**	**1,231**	**649**
Asia	**78,584**	**51,044**	**22,965**
Europe	**909,159**	**675,348**	**232,317**
Latin America	**26,287**	**15,893**	**2,785**
North America	**381,904**	**272,187**	**112,222**

Source: UNCTAD e-commerce and development report, 2003

income (and therefore low-usage) or landlocked markets to the Internet backbone (UNCTAD,2003).

High international bandwidth costs can also be determined by international charging practices whereby developing-country Internet service providers (ISPs) must pay the full cost of an Internet connection with a developed country. Since for many ISPs in developing countries the cost of international bandwidth represents quite a sizeable proportion of their total costs, they have to pass it on to their subscribers. In the end, such arrangements imply that developing-country users subsidize developed-country users' access to information hosted in the developing countries.

Table 2 shows the Internet bandwidth connected across international borders grouped by region.

It can be seen from the table that despite the relatively fast growth of the last three years, the average African Internet user still has about 20 times less capacity than the average European user, and 8.4 times less than a North American one. Even these rather grim overall figures hide the virtual digital isolation of some of the poorest African countries, where the international bandwidth available can be measured in terms of kilobits per second (Kbps) and may correspond to the needs of a mid-size European or US enterprise.

Network Access

The cost of Internet access depends on several factors, including the price of routers and other computing facilities and to a lesser degree on the wages of operators. Such charges may be significantly higher in developing countries than in developed ones, reflecting the high cost of capital as well as possible inefficiency and monopoly profits of telecommunications. Internet access cost is a major factor in the growth of electronic commerce. In many parts of the world the cost of telephony, the availability of telephone lines, and Internet Service Providers' access charges are key determinants of the rate at which electronic commerce can expand. The state or private monopolies that run telecommunications in most of the world are unable or unwilling to cut the cost of local access.

Inexpensive bandwidth is crucial to the development of economies based on electronic commerce, but will only come about if markets are privatized, deregulated and competitive.

Broadband networks are now accessible to mostly, but not exclusively, in high-income market economies. UNCTAD estimates of the number of broadband users worldwide to vary between 55 and 100 million people, more than 75% of whom reside in just six countries. The Republic of Korea leads the world in per capita terms, with more than 21 subscribers per 100 inhabitants. Broadband is progressing fast in several other countries as well. In most countries, where broadband penetration remains below the 10% mark and prices are still fairly high, UNCTAD is predicting that the short-term impact of this technology on most business operations will remain limited. However, significant penetration levels (above 40%) could be achieved in several markets provided that the regulatory environment ensures sufficient competition among providers, both within and across the various technologies used to deliver broadband. Particularly in developing countries, where DSL may not face competition from cable, fiber optic or wireless technology, it may be advisable for regulators to issue licenses for alternative modes of delivery, such as fixed wireless.

Some developing countries have adopted different strategies to facilitate Internet access in an attempt to increase Internet usage. For example, the Egyptian government outlined a broad agenda that touched on everything from improving land and mobile phone availability and ensuring fair competition in the telecoms and IT market, to developing an IT-based economy and bringing the state bureaucracy online. One of the pioneering initiatives the Egyptian government introduced was the "free Internet" initiative, which boosted Internet users from 400,000 in 1999 to 3.5 million in June 2004 (Palmgren, 2002). The "free Internet" initiative allows the end-users to enjoy the Internet free of subscription fee because users are still paying the nominal cost of 20 cents per hour for their Internet calls. That's good news for the ISPs, which collect 70% of the call revenues from the phone company. This agreement was brokered by the Ministry of Communication and Information Technology (MCIT), which figured the state-owned operator, Telecom Egypt (TE), ought to play a part in a national drive to increase Egypt's online presence. In the old model, ISPs had a hard time making ends meet by selling monthly subscriptions priced at $4 for unlimited hours. Since salaries are low, on average $100 per month, many users were sharing one account. Now, when users no longer take turns, traffic has increased. The revenue-sharing model has enabled the ISPs to upgrade Egypt's Internet infrastructure by placing their own equipment in the operator's exchanges.

Internet Usage

A certain level of Internet usage is required to have a reasonable rate of e-business activities. One of the metrics used to measure Internet usage is the number of Internet users. While a high number of Internet users does not necessarily mean a high rate of e-business activity, it can be argued that if individuals find it difficult to use the Internet due to limited access, then the technological conditions for establishing an e-business environment are probably not being met. In addition, access is by no means the only bottleneck in the development of a digital economy.

Table 3. Internet users (thousands) in selected countries, 2000-2002

	2002	2001	2000	% Change 2001-2002	% Change 2000-2001
Africa	**7,943**	**6,510**	**4,559**	**22.01**	**42.81**
Algeria	500	200	150	150.00	33.33
Morocco	500	400	200	25.00	100.00
Nigeria	200	115	80	73.91	43.75
Togo	200	150	100	33.33	50.00
Tunisia	200	150	100	33.33	60.00
Zimbabwe	500	100	50	400.00	100.00
Latin America & Caribbean	**35,459**	**26,163**	**17,673**	**35.53**	**48.04**
Argentina	4,100	3,650	2,600	12.33	40.38
Brazil	14,300	8,000	5,000	78.75	60.00
Colombia	1,982	1,154	878	71.75	31.44
Mexico	4,663	3,636	2,712	28.27	34.04
Asia	**201,079**	**150,472**	**109,257**	**33.63**	**37.72**
India	16,580	7,000	5,500	136.86	27.27
Thailand	4,800	3,536	2,300	35.75	53.74

Source: UNCTAD e-commerce and development report, 2003

In the absence of sufficiently comparable and truly global information about the intensity of Internet use, as opposed to the mere absolute number of people with access, estimates of the number of users provide a straightforward, objective — even if imperfect — indication of whether the foundations of a "digital economy" — for example, awareness, access, experience and trust — are present in a society.

Using data from the International Telecommunication Union (ITU), the UNCTAD presents the global number of Internet users and the distribution of them among regions of the world (UNCTAD, 2003). The report shows that the number of Internet users in the world reached 591 million in 2002. At the end of 2002, developing countries had 32% of the world Internet users, up from 28% in 2001. The report indicates that if current trends continue, Internet users in developing countries could constitute 50% of the world total in the next five years. Table 3 shows the growth of Internet users in selected countries from 2000 until 2002.

The factors that stimulate Internet usage vary from one country to another. The following are a few examples from southeast Europe (IDC CEMA, 2002).

Bulgaria

The number of Internet users has been increasing rapidly in Bulgaria, due to the increasing popularity of the Internet and computer gaming clubs. To ensure continued growth in Internet usage, the government has launched several initiatives. B2C Internet commerce in Bulgaria has been stimulated by the existence of a number of local online retailers as well as by the setting up of two online payment systems. However, much online consumer spending is still made via foreign sites.

Romania

In Romania, Internet usage from public places such as Internet cafés and education facilities already accounts for the majority of Web users. This proportion is likely to increase as a major government initiative to connect schools to the Internet gets underway. There are more inhibitors to the development of both B2C and B2B Internet commerce in Romania than in the other southeast European countries. These include the lack of a legal background to enable online payment systems. True Internet commerce is still at negligible levels, even in the consumer sector. Web sites are used mainly for marketing purposes, with any real business taking place off-line. On the positive side, Romania's relatively large population and more sizeable economy should make it easier for online retailers to reach critical mass in the medium term.

Slovenia

Slovenia is one of the front runners in Internet adoption in Central and Eastern Europe. Factors including relatively high disposable incomes and government initiatives have contributed to the high rate of Internet usage. Most home PCs are now connected to the Internet, the vast majority of companies have Internet access and most schools can provide Internet facilities to all of their pupils. New government measures to implement computer literacy programs, establish Internet kiosks in public places and make school terminals available to the general public will help to further increase Internet usage. Yet, despite high Internet penetration and widespread usage of more advanced Internet applications such as online banking services, e-commerce has not taken off in Slovenia. Although consumers are increasingly buying products online, they do so mainly from foreign sites due to a lack of choice among Slovenian online retailers.

Table 4. Internet hosts (thousands) by region, 2000-2002

	2002	2001	2000	% Change 2001-2002	% Change 2000-2001
Africa	**281**	**274**	**217**	**2.68**	**25.92**
Latin America & Caribbean*	**3,413**	**3,413**	**1,968**	--	**73.40**
Europe	**18,363**	**15,325**	**12,533**	**19.83**	**22,27**
North America*	**109,084**	**109,084**	**82,931**	--	**31.54**
Asia	**10,803**	**10,809**	**7,172**	**-0.05**	**50.70**
Oceania	**3,305**	**2,732**	**1,973**	**11.09**	**38.48**
Developing countries	**7,279**	**7,212**	**12,392**	**0.93**	**-41.81**
Developed countries	**137,700**	**134,424**	**94,402**	**2.44**	**42.39**
World	**144,979**	**141,636**	**106,795**	**2.36**	**32.62**

*2001

Source: UNCTAD e-commerce and development report, 2003

Internet Hosts

Internet hosts are more markedly concentrated in the developed world than are users (UNCTAD, 2003). Table 4 shows the distribution of Internet hosts among world regions. North America and Europe account for as much as 89% of all of the Internet hosts in the world. Contrary to the trend in the number of users, the number of Internet hosts is growing faster in the developed countries than in the developing world.

As indicated in the UNCTAD report, these numbers point out that the relatively few people who use the Internet in developing countries compete among themselves for access to a proportionally much smaller number of computers connected to the Internet, and they have access to little locally hosted Internet content. It must be noted, however, that hosting content in a server located in a developed country may be the best option for some enterprises in developing countries. For example, it may be preferable to host information about a tourist destination on a server located in or near the countries where the potential tourists reside.

Financial Infrastructure

This infrastructure is needed to provide e-commerce services with trusted and secured transaction as well as payment components. The faceless nature of e-business requires that transactions between two parties be secured. The poor banking services, inadequate Information and Communication Technology (ICT) infrastructures and absence of legislative and regulatory framework for e-commerce in developing countries require a slightly different approach for implementing e-commerce. Liberalization of trade in financial services helps strengthen the financial system and the economy in general by bringing about improvements to the financial infrastructure. The presence of foreign financial services firms encourages a competitive financial services market which in turn leads to increased efficiency, greater innovation, lower product pricing and increased consumer choice. Developing countries must recognize that the elimination of obstacles to trade in financial services will play a critical role in developing the financial services infrastructures required for e-commerce. Developed countries should increase technical support to developing countries to help them develop financial services and the legislative, regulatory and human resource requirements, which are key components of efficient electronic marketplaces.

Several countries have taken steps, although at an uneven pace, to reduce barriers to foreign financial services firms. As economies increasingly turn to global financial markets to meet a portion of their capital needs, foreign financial firms are helping to channel both foreign and domestic savings to these markets. Non-resident financial services firms face many barriers such as:

- limitations on cross-border access, including the right to buy and sell financial products across borders and to participate in and structure transactions
- lengthy and difficult approval for new products

- restrictions on foreign exchange
- limitations to trading in domestic stocks
- considering the significant contribution of foreign financial services firms to development, nations should reduce and eventually eliminate these barriers by establishing clear and non-discriminatory approval procedures for the activities of non-resident firms

Protocols, Laws, and Legislations

Electronic commerce is attracting increasing attention from policy makers at the national, regional and international levels. It presents a number of challenges that are highly distinctive. First, the speed of development in electronic commerce and the speed of the changes that it brings strain the processes of traditional policy formulation. Second, the issues cut across a broad range of technical, legal, economic and institutional questions that have often been treated organizationally in separate ways by different entities. Third, policy makers are confronted with acutely different participation in electronic commerce and with participants having different levels of awareness of its implications and consequences. Both within and between countries, there are striking disparities in infrastructure development, as well as technical and economic access to the Internet. Fourth, electronic commerce is conducted on a global medium, which requires international coordination and uniformity of approach in order for it to be exploited effectively and to its full potential. Finally, special attention is needed to ensure that all developing countries have the potential to benefit from these new processes.

The developments in technology and the rapid acceptance of electronic commerce have, to a certain extent, outstripped the laws governing conventional business trading. Several regulations and laws need to be modified or created in order to accommodate new and changing technology. However, the process of enacting new laws or legislation is generally a long and slow process. The increased use of electronic means of communication, such as EDI, e-mail and the Internet, has raised concerns about their legal effect, validity and enforceability. In most of the developing countries, the existing national laws do not contemplate the use of modern means of communication. National and international laws impose restrictions on the use of electronic communication techniques by requiring "written," "signed" or "original" documents. As electronic commerce is not restricted by national boundaries, its adoption requires that the legal ramifications are considered by all those interested in international trade and development.

To realize the potential of electronic commerce, however, governments and the private sector must work together to create a predictable legal framework, to ensure that the Internet is a safe business environment and to create human resource policies that endow workers and the populations at large with the skills necessary for jobs in the new digital economy.

E-Commerce Future Trends

The e-commerce revolution is far away from being complete. The fact is, the revolution has hardly started. A recent survey by PricewaterhouseCoopers and The Conference Board stated that large enterprises were moving into e-business at a much slower pace than previously expected. Nearly 78% of the large enterprises surveyed were not yet processing transactions online. For 83% of the companies, e-business was generating less than 8% of revenue.

One of the driving changes that will expand the digital marketplace is the next-generation Internet, which provides very high bandwidth at very low cost. The result will not only be vast number of new users, but users who will be logging on with an array of new devices. What that means is that over the next few years, you will see a marketplace that is defined by explosion and convergence: an explosion of new devices, new users, new media and transactions, and a convergence of standards to bring it all together.

Improved Performance and Information Availability

Delivering information in a way that doesn't keep customers waiting requires much more than fast servers. It will involve a whole new level of connectivity supporting an unprecedented level of integration across the virtual marketplace so that customer-critical information is available whenever and wherever needed. Developing countries still have a long way to go to establish broadband network access infrastructure that can support not only faster online business transactions but also better integration in the electronic marketplace, allowing the exchange of information as needed.

E-Commerce Service Providers

Many online retailers and manufactures have realized that owning and operating e-commerce infrastructure does not make economic or operational sense. This has suggested that next generation e-commerce merchants should not own and operate the e-commerce infrastructure. An emerging sector within the e-commerce arena is the E-commerce Service Providers (ESPs). They leverage the Internet itself to deliver a complete online channel solution with guaranteed levels of performance quality. Companies contract for a fully branded online store, all of the applications and services required to manage it, and a partner committed to their ongoing performance improvement. This approach will allow retailers and manufacturers to deploy a complete e-commerce solution in less time and less cost than before. From a profitability and reliability standpoint, businesses will be able to justify e-commerce to their shareholders and customers. By enabling companies to focus on their core business, ESPs unlock the full potential of online sales channels. ESPs will provide the sustainable e-commerce solution that manufacturers and retailers have been seeking.

Wireless and Mobile E-Commerce

The fast-growing wireless segment of the Internet market has the potential to boost e-commerce and elevate it to a totally new level. The integration between wireless voice and data service has opened the door for fast and easy delivery of information to users wherever they are. The future of e-commerce will be greatly impacted by the growth in short message service (SMS) as well as the multimedia messaging service (MMS).

Wireless Internet access will continue to play a major role in changing the way people are using the Internet. Wireless local area network access, known as WiFi, is already radically changing the way in which the Internet is accessed in the home, in the office, in public locations and on the move. However, wireless local area networks (LANs) are only one part of a much bigger wireless broadband landscape that is now unfolding before us.

WiMAX and 802.20 threaten to have just as big of an impact on outdoor broadband wireless service delivery as WiFi has done in indoor environments. The combination of Wi-Fi, WiMAX and/or 802.20 threatens to radically change the way in which broadband services are provisioned, creating new business models and challenging those of the established providers. ADSL, cable modem and 3G service providers all have something to fear from WiMAX. But at the same time, each could derive direct benefits by harnessing this new technology to increase service coverage and lower costs. The next few years will largely see small-scale trials and proof-of-concept technology pilots for fixed broadband access. Initial deployments will address the fixed access market, as WiMAX PCMCIA cards or WiMAX-enabled laptops are unlikely to ship until 2007/8. The ratification of 802.16 Rev E will pave the way for support of mobile access, but it is still too early to know whether WiMAX will be able to dominate the mobile market.

Conclusion

Perhaps the clearest indication of the growing importance of e-commerce in the global economy is the rapidity with which Internet use has grown and spread during the last decade. The boom in e-commerce also includes increased use of other media for trade, such as the telephone, television, fax, and electronic payment. To what extent e-commerce can have an impact on the economy of a country is greatly facilitated by services that enable payments to be made and that facilitate the delivery of the product.

Many developing countries do not have the infrastructure required to provide these services. The e-commerce revolution requires three types of infrastructures:

- technological infrastructure
- financial infrastructure
- legislative infrastructure

To establish an Internet marketplace, a technological infrastructure that relies on a variety of technologies needs to be in place. This technological infrastructure includes the Internet backbone, Internet service providers (ISPs) to connect market participants to that backbone, and end-user devices such as PCs, TVs, or mobile telephones. The financial infrastructure is needed to provide e-commerce services with the necessary trusted and secure financial transactions. This infrastructure makes payment over the Internet possible (through credit, debit, or Smartcards, or through online currencies). The third infrastructure component is the set of protocols, laws and regulations that affects the conduct of those businesses engaging in and impacted by electronic commerce, as well as the relationships between businesses, consumers, and government.

A crucial issue for the success of e-commerce is how to establish trust. People purchasing products online need assurances about the companies they are dealing with and about the products they are buying. Firms selling products online need to be confident that payment will be made. Trust-enhancing processes to support the use of e-marketplaces are very important and these are very weakly developed in many developing countries.

The widespread availability of the Internet is one of the most important factors that helps e-commerce transactions to increase. For developing countries, access to modern telecommunication systems is perhaps the defining element of e-commerce. A well functioning, modern telecommunication infrastructure and a satisfactory distribution of electricity, along with access to computer hardware, software and servers are the three basic requirements for electronic transactions and e-commerce.

References

Business Software Alliance. (2003). Necessary elements for technology growth. Business Alliance Software.

Chaum, D., Fiat, A., & Naor, M. (1989). Untraceable electronic cash. *Advances in Cryptology- CRYPTO '88 Proceedings.*

Goepfert, J., Kusnetzky, D., & Whalen, M. (2003). *Finding financial freedom: Using technology to address business challenges within financial services.* InformationWeek White Papers.

Harris, R. (2004). Information and communication technologies for poverty alleviation. UNDP-APDIP.

IDC CEMA. (2002). Internet usage and eCommerce in Southeast Europe, 2000-2005. IDC.

Loshin, P., & Vacca, J. (2003). *Electronic commerce* (4th ed). Charles River Media.

Mann, C. (2000). *Electronic commerce in developing countries: Issues for domestic policy and WTO negotiations.* Retrieved April 13, 2003, from *http://www.iie.com/publications/wp/2000/00-3.pdf*

Palmgren, M. (2002). *Egyptian flock to new net plan.* Retrieved August 12, 2004, from *http://www.wired.com/news/print/0,1294,52993,00.html*

Schneier, B. (1996). *Applied cryptography* (2nd ed). John Wiley.

Singh, D. (n.d.). *Electronic commerce: Issues for the south.* Retrieved April 24, 2004, from *http://www.southcentre.org/publications/e-Commerce/toc.htm*

Stinson, D. (1995). *Cryptography theory and practice.* CRC Press.

UNCTAD. (2000). *Building confidence: e-Commerce and development.* UNCTAD.

UNCTAD. (2003). *E-commerce and development report 2003.* United Nations Publications.

UNESCAP- Development Research and Policy Analysis Division. (2003). *Economic and social survey of Asia and the Pacific.* United Nations Publications.

Chapter XVIII

An E-Commerce Longitudinal Case Study from Ukraine

Murray E. Jennex, San Diego State University, USA

Donald L. Amoroso, Appalachian State University, USA

Abstract

The need to develop a strategy for e-business applications is an important facet of doing business, especially where organizations can provide value-added services to the customer base. In this chapter, we identify key value-added e-business applications and focus on success factors that clearly support small enterprise goals. Performance factors are presented that show the impact of e-business applications on the organization's bottom line. The case study in this chapter uses a methodology of direct observation, unstructured interviews and document review to gather data for identifying issues in starting a small business using the Internet. The case organization provides services around office and business support, energy consulting, and Web development. The formation of International Business Solutions has been difficult due to the banking system, work culture, and infrastructure of Ukraine. IT issues investigated include telecommunications, energy, hardware, software, and the availability of technical skill sets. E-business issues that were found to be important in this study include: (1) difficulty of the user interface, (2) a lack of a planning process for e-business applications, (3) development and testing concerns, (4) finding the Web site and a lack of branding, and (5) the lack of evidence to support a formal budgeting process.

Introduction

Information and Communication Technology (ICT) can provide a small enterprise an opportunity to conduct business anywhere. Use of the Internet allows small businesses to project virtual storefronts to the world. Heeks and Duncombe (2001) discuss how IT can be used in developing countries to build businesses. This chapter discusses the case of a small startup company in Kyiv, Ukraine, and its use of ICT to help it succeed. The subject company was founded in early 2000 with the initial goal of providing business service support to companies wanting to do business in Ukraine. ICT was seen as a method of marketing and contacting clients and potential clients. Offering ICT services was also seen as a potential service for clients. To date the company has succeeded in surviving and growing, however, it has not been as successful as initially expected. This chapter looks at where ICT has been successful and where ICT limitations raised issues.

The chapter first presents an overview of how ICT can be expected to support companies in developing or emerging economies. This is followed by a description of the subject company and an analysis of how ICT has supported the company's goals. The chapter ends with conclusions on the impact of this case on perspective uses of ICT in emerging economies. Discussions on methodology and limitations are provided to aid readers in assessing the validity of the chapter.

Background

E-Business for Small Companies

Developing strategies to adopt and market e-business technologies and services requires an organization to make significant investments. Deciding to make the initial and ongoing investment is contingent on the organization's perception that the future benefits will outweigh the costs involved. Mitra and Chaya (1996) propose that there is a need to quantify the benefits from the investments in e-business systems and that building quality e-business systems will require solid evidence of added value to customers. The added value for customers will result in additional profit for the organization, as they are able to maintain current customer relationships and develop new ones based on the attractive offerings a new e-business presence affords. Later in this chapter, we discuss the results of the case study with our impressions of the Web site for the case study.

Developing an e-business niche will allow an organization to provide the best possible deliverable to the customer, even if this means passing part of the deal to a competitor who specializes in another aspect of the e-business system. Bakos (1991, 1997) found that ultimately all e-business systems reduce buyers' search costs and increase the efficiency of e-commerce transactions, and therefore create numerous possibilities for the strategic uses of these systems. Operating within an e-business framework offers a more cost-effective model, with feasible elimination of several steps of the traditional sales process.

Powell and Dent-Micallef (1995) identified an emerging trend from single-source sales channels to electronic markets, lowering coordination costs for producers and retailers and resulting in fewer distribution costs. Smaller businesses ultimately benefit from lowered sales costs and gain access to larger markets. Executives are "sold" on technology spending based upon the strategic value-added nature of the opportunities, as found by Jarillo (1988). Given an adequate availability of talent for development of e-business systems, as reported later in this chapter, we feel that it is imperative for e-business applications to have strong consideration in small organizations.

Amoroso and Sutton (2002) found that small organizations need to focus on providing their customers with a set of Web applications/services that best serve the customer, rather than using a hit-and-miss approach. They found evidence to support that the greater the degree of clarity of e-business service offerings, the more the need to have these Web-based applications/services developed by partner firms. The decision to outsource e-business applications/services is crucial to building a quality Internet presence, and especially important for smaller organizations. Organizations that focus on key online applications/services will have a greater degree of success. These applications/ services will need to be planned in advance considering the Ukrainian marketplace, for example, in order to yield bottom-line, value-added results. The e-business applications/ services found to have the greatest impact on the success of small organizations, along with percentage of added value from the Amoroso study, are presented in Table 1. With the many e-business applications/services available to development and the careful discretion of resource allocation, these applications/ services were found to be the most beneficial for generating downstream revenues and cost reduction for small businesses. Due to the span of components in an e-business system, many organizations find difficulty in fulfilling all customers' needs. Therefore, a focused strategic planning session around value-add will yield IT investment suc-

Table 1. Value-added e-business applications/services

Applications:	
• Electronic catalogs	71.4%
• Workflow systems	71.4%
• Online customer service	71.4%
• Order management	57.1%
• Electronic billing systems	42.9%
• Online auctions	28.6%
• B2B exchanges	28.6%
Services:	
• E-database integration	57.1%
• Web monitoring	57.1%
• Supply chain automation	42.9%
• Online communities	42.9%
• Sales force automation	42.9%
• Electronic marketplaces	28.6%
• XML solutions	28.6%

cesses (Brynjolfsson & Hitt, 1996). Applications and services in the e-business space that were found to provide a lesser degree of value-add included electronic software distribution, purchasing cares, data warehouses for e-business, electronic accounting interfaces, and intranet systems.

The authors examined the factors that would lead to the success of a small organization's e-business applications/services. Clearly, several mega-categories of e-business success factors emerged including: (1) understanding the customer base needs, (2) support of substantial e-business initiatives in an ongoing basis, (3) developing e-business applications using a proven development methodology, (4) branding Web sites for competitive advantage, and (5) reshaping the organization's corporate culture (Amoroso, 2002). How the organization manages the knowledge gathering process for their customer base in Ukraine will have strong impacts on their ability to meet their needs and their needs for specific Web site features, such as multi-lingual support, click-through capabilities, and customer profiling. The degree of Internet application/service maturity will determine the way that customers interact with the Web sites and how integrated the applications will support customer-side requests. The development of small organization applications/services will depend upon acquiring development methodologies that extend the company's technical talent and enable the company to successfully manage e-business projects. Having a set of strong e-business applications to deliver to company's customers via the Web will not be successful if they are not accessible to the customer base, thus facilitating the need for strong product branding efforts. Finally, the corporate culture will need to support conducting business in a new way given new e-business initiatives and ultimately new corporate processes for promoting and delivering products and services.

Amoroso (2001) found a set of corporate performance factors that small organizations would need to consider and eventually quantify with respect to e-business initiatives. Table 2 lists the key performance factors that are crucial. E-business initiatives, like other corporate investments, will need to provide the needed payback to the organizational bottom line. Clearly, e-business applications have been found to have a greater emphasis on speed-to-market than traditional IT applications (Hart & Saunders, 1998).

The Asian-Pacific Economic Cooperative (APEC) Readiness Initiative, in conjunction with APEC economies for developing countries, developed an e-commerce readiness

Table 2. Corporate performance factors impacting e-business

- Customer attraction
- Customer retention
- Customer satisfaction
- Incremental revenue growth
- Market cap positioning
- Lower cost of sales
- Cost control
- Increased market share
- Business process streamlining
- Decreased cycle time

Table 3. E-commerce readiness assessment factors

- Basic IT Infrastructure
  - Access to basic infrastructure
  - Speed and functionality of the infrastructure
  - Price of the infrastructure
  - Reliability of the infrastructure
  - Availability of terminal equipment
  - Infrastructure market conditions
  - Interconnection and interoperability
- Access to Necessary Services
  - Internet service providers
  - Non-IT services and distribution channels
  - Financial institutional involvement
- Current Level and Type of Use of the Internet
  - Access type to the Internet
  - Number of Internet hosts
  - Government use of the Internet
- Promotion and Facilitation Activities
  - Economic policies with regard to standardization
  - Effects of e-commerce on employment
  - Government support of adaptive technologies
- Skills and Human Resources
  - Education and training policy
  - Schools' access to the Internet
  - Facilitation of internationalization of business
- Positioning for the Digital Economy
  - Taxation policies
  - Legal framework
  - Electronic authentication
  - Security and encryption
  - Copyright
  - Privacy laws
  - Consumer confidence

assessment guide that targets a blueprint for assessing and analyzing electronic commerce penetration into economies that are underdeveloped (APEC, 2000). This research helps countries identify impediments within their country borders to successfully deploying e-commerce initiatives. Based upon this research, several key IS, non-IS, and e-commerce specific categories were determined for the distribution of e-commerce throughout a developing country, as shown in Table 3.

One impediment to the development of e-commerce in emerging economies is the information environment. Many developing countries do not have a history of sharing data. The ability to pool data for statistical analyses is necessary for many business processes and organizations. The absence of shared data can result in a lack of effective information systems due to the lack of reliable and consolidated marketing, customer, and economic data. This also usually results in low data quality and trust in the data that is available. Chepaitis (2002), using Russia as a model, identified 12 factors that affect the information environment in an emerging economy. These factors focus on the business culture of the economy and will limit the emergence and scope of e-commerce in these

Table 4. Factors affecting the information environment in an emerging economy

- Unsuccessful and intrusive government planning and regulation
- Barriers to entry and dictated pricing in distribution, supply, and regulation
- Informal entrepreneurship: black markets, barter
- Ineffective methods for managerial accounting
- Unanticipated shortages and other factors inhibiting demand
- Political fear and widespread avoidance of information sharing
- Unstable currency, nascent financial regulations, and a dearth of financial services
- A reluctance to divulge information without compensation or reciprocity
- Proprietary attitudes towards data ownership
- Rigid, hierarchical management styles with a reluctance to share information or empower employees
- Communication behaviors that rely on oral tradition or more than one language
- An emphasis on price and availability to the exclusion of quality

economies. Since Ukraine and Russia were part of the former Soviet Union and shared a common infrastructure and business culture these factors are likely to affect businesses in Ukraine (Table 4). Missing from these factors are the integrated networks and systems and integrating organizations that companies in the developed countries use to share and consolidate data. Infrastructure will be discussed in the analysis section of this chapter.

Economy of Ukraine

A description of the economy of Ukraine is provided by the Central Intelligence Agency's World Fact Book Ukraine page:

> After Russia, the Ukrainian republic was far and away the most important economic component of the former Soviet Union, USSR, producing about four times the output of the next-ranking republic. Its fertile black soil generated more than one-fourth of Soviet agricultural output, and its farms provided substantial quantities of meat, milk, grain, and vegetables to other republics. Likewise, its diversified heavy industry supplied the unique equipment (for example, large diameter pipes) and raw materials to industrial and mining sites (vertical drilling apparatus) in other regions of the former USSR. Ukraine depends on imports of energy, especially natural gas, to meet some 85% of its annual energy requirements. Shortly after independence in late 1991, the Ukrainian Government liberalized most prices and erected a legal framework for privatization, but widespread resistance to reform within the government and the legislature soon stalled reform efforts and led to some backtracking. Output in 1992-99 fell to less than 40% the 1991 level. Loose monetary policies pushed inflation to hyperinflationary levels in late 1993. Ukraine's dependence on Russia for energy supplies and the lack of significant structural reform has made the Ukrainian economy vulnerable to external shocks. Now in his second term,

> President Kuchma has pledged to reduce the number of government agencies and streamline the regulation process, create a legal environment to encourage entrepreneurs and protect ownership rights, and enact a comprehensive tax overhaul. Reforms in the more politically sensitive areas of structural reform and land privatization are still lagging. Outside institutions — particularly the International Monetary Fund, IMF — have encouraged Ukraine to quicken the pace and scope of reforms and have threatened to withdraw financial support. Gross Domestic Product, GDP, in 2000 showed strong export-based growth of 6% — the first growth since independence — and industrial production grew 12.9%. As the capacity for further export-based economic expansion diminishes, GDP growth in 2001 is likely to decline to around 3%.

Ukraine's Washington, D.C.-based embassy to the United States' Web site describes Ukraine's banking system as a system in the process of development. There are approximately 200 commercial banks in Ukraine of which 2 are wholly owned subsidiaries of non-Ukrainian banks. The number of foreign-owned banks is not expected to increase due to opposition within the banking regulatory agency. Weak bank regulation enforcement and loose lending policies led to a banking crisis in 2000-2001. Many banks failed and, as Chervachidze (2001) reports, there is little confidence among Ukrainians in the banking system. The result is that the majority of Ukrainians do not have bank accounts nor use banking services. Ukraine's United States Embassy advises that companies doing business in Ukraine may confront delays in transferring funds domestically and internationally, converting currency, and in repatriating in foreign currency. Additionally, state authorities such as the tax directorate have wide-ranging powers to freeze bank accounts or to withdraw funds for payment of taxes or fines without the need to obtain a court order or authorization.

Methodology

This chapter examines the use of e-commerce by International Business Solutions, IBS, over a period of three years. Two data points were taken. The first data point used direct observation, unstructured interviews, and document review to gather data. Interviews and observation were conducted in Kyiv, Ukraine, during the second week of August 2001. Follow-up e-mail interviews with the founder of the subject company continued as needed for the next six months. Analysis of the data was done using models from the previously discussed literature and DeLone and McLean's IS Success Model (Molla & Licker, 2001). Analysis resulted in a set of recommendation for improving the e-commerce site.

The second data point also used direct observation, unstructured interviews, and document review to gather data. Interviews and observation were conducted in Kyiv, Ukraine, during the third week of January 2004. The purpose of the second data point was to determine how well IBS implemented the recommendations and how effective the recommendations were.

Description of the Subject Company

The subject company, International Business Solutions (IBS), is located in Kyiv, Ukraine. Dave Sears, an American expatriate living in Kyiv, formed IBS in March of 2000. The company has an affiliated United States company, Energy Solutions, also formed in March 2000. Energy Solutions is a Nevada-based corporation. IBS is a small enterprise. In December 2001 it had three full-time and five part-time employees, all Ukrainian, in addition to its founder. To minimize costs and increase flexibility, IBS utilizes an extensive network of independent contractors to provide its services. IBS has three areas in which it provides services:

- Office and business support to business people traveling to the Ukraine
- Energy consulting
- Web development

IBS was originally formed to broker the knowledge and contacts that the founder had made managing logistics during a contract assignment into a business that offered business people everything they would need when doing business in Ukraine. This includes basic services such as translation, interpretation, transportation, business office support, and escorts. It also includes some not-so-common services such as cell phones, introductions to government and business leaders, customs, airport entry, and currency exchange assistance, and message center services. These services are offered due to the large difference in culture, language, and regulatory and business practices between Ukraine and the United States. To provide these services, IBS owns an office near the center of Kyiv. The office is equipped with phone, fax, computer, and Internet. IBS has six computers connected in an Ethernet LAN using a Windows 2000 server. Additionally, IBS has several cell phones that are available for rent. IBS maintains contracts for services with a large number of drivers, translators, interpreters, guides, apartment brokers, and business support staff. IBS assesses potential contract workers to ensure they meet the standards that IBS advertises.

Energy consulting services are offered as a by-product of the founder's expertise in running and managing power plants and his previous work performing a Y2K risk assessment on the energy sectors of Ukraine, Armenia, and Georgia. Demand for these services fluctuates greatly and is dependent upon United States policy.

Web development services are offered, as it was discovered during the Y2K risk assessment that there is a large Web development talent pool available in Kyiv that is very reasonably priced and motivated to work. Given the weak business climate in Ukraine these services are targeted to United States and other developed countries' companies. Originally IBS offered fixed fee contracts for development projects. Problems with managing projects remotely and incomplete user requirements drove IBS to switch to providing developers on an hourly basis working under direct control of the contracting client. IBS is able to offer developers at a very attractive rate, approximately $15.00 per hour United States Dollars, USD. The new approach has been more successful and IBS has several developers under contract to outside companies.

IBS markets itself through word of mouth and its Web sites. Word of mouth is through business gatherings in Kyiv and a good word from clients to potential clients. Web marketing is through three Web sites, one for each of the major business areas. The Web sites are:

- www.ibs-websolutions.com
- www.ibsukraine.com
- www.energy_solutions_llc.com

IBS uses the Virtual Face, E-Shop online business model as described by Tetteh and Burn (2002). The Web sites are interactive and provide a good overview of IBS services, examples of the skills of the IBS Web developers, and a means of contacting IBS. A United States Internet Service Provider, ISP, hosts the sites. A United States ISP was chosen because they were considerably less expensive than Ukrainian ISPs, approximately $7.00 USD per month for the United States ISP versus approximately $40.00 USD per month for the Ukrainian based ISP. Another issue was bandwidth, 48 kbps is becoming common, and DSL is being introduced but is very expensive, greater than $100.00 USD per month. Reliability is still an issue with Ukrainian ISPs as power quality is poor and phone lines are degraded.

Analysis

Non-IT Issues

It has taken IBS over 18 months to become operational and to make money; and IBS is still not fully operational. This is primarily due to the business climate in Ukraine. Things take time to complete. Paying expedited fees is normal and it is rare to pay the standard fee and have something done. Americans pay more for everything, up to 10 times what a Ukrainian would pay. IBS avoids this by the founder living in Kyiv and by having Ukrainian directors and partners. Regulations and tax requirements are difficult to understand and comply with. Getting paid is difficult. Transfer payments are expensive with an approximate $350.00 USD fee applied, and take a long time, approximately five weeks. IBS is set up to accept off-shore payments through a non-Ukrainian bank. This allows IBS to accept credit card payments. Since Ukraine is a cash economy, checks aren't accepted. Non-Ukrainian customers must pay in cash or via credit card. This is difficult for many customers to understand and work with. The author faced this during contract work in Ukraine. Drivers, interpreters, and translators all expected to be paid in cash at the time of the service. Invoicing and paying by check were not accepted. Receipts were informal unless written by the author. The author's parent organization, an American electric utility, did not do business in this manner and had no business process in place to support cash on delivery payments. Fortunately it was worked out and a method established that was acceptable to the client, the United States government. IBS has the same issues with large customers.

Networking within the business community is vital. Business within Ukraine is done face-to-face and usually over meals or social gatherings. Reputation is important and whom you know is vital. Doing a good job supports continued business but does not get the initial job. IBS discovered this after several months. Once the founder realized this mode of business he spent a majority of his time mixing with and meeting people in Kyiv.

Learning to work within the Ukrainian culture has also been difficult. Ukraine has 33 holidays compared to approximately ten in the United States. Punctuality is not a standard practice. Break times are whenever. Language has been an ongoing issue since the founder did not speak Russian when he first arrived in Kyiv and still is not fluent.

The information environment affecting IBS within Ukraine are summarized by the factors listed in Table 3. Essentially all the factors affecting the emergence and scope of e-commerce in Russia are present in Ukraine as stated and they have had an affect on IBS. The preceding paragraphs illustrate how these factors have impacted IBS. In particular the need to socialize to gain business, the lack of a stable banking system, and the inability to figure out business rules demonstrate the weak information environment.

IT Issues

Phone communications for calls outside Ukraine are very expensive. Calling cards are available that significantly reduce costs, to about $0.45 USD per minute, but require accessing special numbers and switches (these cards are typically available for about $0.25 USD per minute in the United States). File transfers are slow. Bandwidth is not readily available due to degraded communication lines. When lines are new or have been upgraded, bandwidth is available for a reasonable cost from a United States perspective. One of the authors was able to find an Internet café that offered 128 kbps connections for a cost of about $4.00 USD per hour. It should be noted that this is considered expensive in Ukraine. The clientele being primarily non-Ukrainians confirmed that it is expensive. Another negative is the poor condition of the telecommunications infrastructure. Jennex et al. (1999) found no digital/IT equipment used for energy management and plant communications. Analog phone switches and equipment were normal. Telecommunication service between dispatch centers and plants or other dispatch centers were very unreliable. Ukraine has a wide area network for monitoring the power system that is based on SM1420 and SM2 computers (these are DEC1000 and PDP11 clones). This system was observed to be frequently out of service requiring system operators to rely on voice communications for dispatch functions. These also were frequently out of service requiring system and plant operators to use analog radios or to simply load follow resulting in poor power quality with little frequency stability with the previously discussed effects on IT. Observations of phone lines in hotels catering to westerners found that dial up connections of greater than 9800 bps were difficult to impossible to sustain for more than a few minutes due to line noise and errors. Ultimately this reduces the effectiveness of the Internet, e-mail, and fax processes, raising the cost for these services. Additional analysis of the telephone system from the Central Intelligence

Agency's World Fact Page confirms the antiquated state of the system:

> Ukraine's telecommunication development plan, running through 2005, emphasizes improving domestic trunk lines, international connections, and the mobile cellular system. At independence in December 1991, Ukraine inherited a telephone system that was antiquated, inefficient, and in disrepair; more than 3.5 million applications for telephones could not be satisfied; telephone density is now rising slowly and the domestic trunk system is being improved; the mobile cellular telephone system is expanding at a high rate. Two new domestic trunk lines are a part of the fiber-optic Trans-Asia-Europe (TAE) system and three Ukrainian links have been installed in the fiber-optic Trans-European Lines (TEL) project which connects 18 countries; additional international service is provided by the Italy-Turkey-Ukraine-Russia (ITUR) fiber-optic submarine cable and by earth stations in the Intelsat, Inmarsat, and Intersputnik satellite systems.

Ukraine has poor power quality. Jennex (2001) looked at IT in the energy sector of Ukraine and found it to be at a 1960s or 70s technological level. Frequency oscillations of 0.5 hertz or more are routine and power outages common. Most critical building and hotels, as well as many residences, keep and maintain backup generators. North American standards have frequency oscillations controlled to 0.05 hertz or less. Digital equipment does not function well nor last long with the large observed frequency swings. Digital clocks in Ukraine routinely lose about 20 minutes a day. The average house or offices in Ukraine (including IBS offices) have a 45-amp fuse box while the average house or office in Southern California has 150-200 amps. It should also be pointed out that Ukrainian fuses are the old type that actually fuses and not modern, re-settable trip breakers. What this means is that the electrical infrastructure in Kyiv and Ukraine does not readily support a modern office's IT electrical needs. Large companies compensate by installing their own power equipment. Small companies make do with what they have with the result that they have less reliable IT. Houses and consumers have to choose between running the computer and running the house appliances, limiting the availability of local consumers to participate in e-commerce.

Availability of hardware and software can be issues. Leading edge hardware such as personal computers, digital cameras, printers, and communications equipment are very expensive and hard to get. Additionally, many companies differentiate between hardware sold in Europe and that sold in the United States. As an example Jennex, et al. (1999) took an inexpensive Epson printer to Kyiv. When the ink cartridges expired, replacements were found in Kyiv. However, the printer refused to recognize the replacements even though they were the correct product per the model number. It was later learned that the coding used on the cartridge was different for models sold outside the United States than for those sold in the United States. Another issue is incompatible character sets. Ukraine uses the Cyrillic alphabet. The character set used to display this alphabet on computers in Ukraine makes the generated files unreadable on computers running English character sets.

Software is readily available. The issue is the great abundance of pirated software. Virtually any software package can be purchased in the local markets for approximately

$2.00 USD per compact disc, CD. Authentic software is available but costs as much or more than it would in the United States. This makes buying and using authentic software unattractive. IBS uses only authentic software since it has a United States affiliate and the founder and owner is a United States citizen. Competing companies in Kyiv do not have this constraint so this gives IBS an unattractive cost differential. Fortunately this isn't a significant issue as IBS does not provide software to its developer contractors and only needs software for its own business use. Software costs can be a significant issue to other companies in Ukraine, especially if they are trying to use "legal" versions and the competition does not care.

Jennex (2001) reported an adequate availability of technical talent in Ukraine. Olearchyk (2001) in the Kyiv Post English language newspaper reports that there is a growing shortage of talent. It is stated that approximately 2500 IT specialists are leaving Ukraine each year. Additionally, the schools are not producing usable IT professionals due to their focus on theory and not practical education. The net impact is that Ukraine software developers are turning down contracts due to lack of work force. Two issues exist. The first is that IBS will not be able to keep Web developers under contract for potential work, other companies offering steady work will get them first. This is mitigated by IBS being willing to offer part-time work to developers to do in their spare time. Given the low wages, there may be an adequate number of developers willing to do this. The second issue is the upward pressure on wages. As developers become shorter in supply, companies will have to pay more to retain them. This will force IBS to pay more for developer services and that will result in higher prices that IBS must charge. This may make IBS a less competitive player in the outsourcing market.

E-Business Issues

The Web sites used by IBS are distinctive, sophisticated, interactive sites. They provide audio and image information as well as text information. They work best with high-speed connections and higher end personal computers. Technically the sites are very good. They appear easy to use and navigate, although this is the authors' impression and not verified through any usability testing. These issues in context of the success factors discussed earlier include:

Understanding Customer Base Needs:

- The sites constantly play music. While the music changes with the different sites, there are no controls other than the user's computer's audio controls for turning it off. After a few minutes, this can get annoying for visitors to the site.
- Needing to scroll down through information. Molla and Licker (2001) consider scrolling a detriment to content quality.

Support of Substantial E-Business Initiatives:

- There is no indication that IBS has an ongoing planning process for e-business applications.

Developing the Web Site:

- Testing the sites on multiple computers yielded various results. When accessed with a new, top-end personal computer connected to a high-speed network the site worked fine. When accessed using a mid-level personal computer with a 115 kbps dial up connection using America Online, AOL, as the ISP, the music did not work with the audio controls at full volume and the introduction screen partially failed while loading very slowly, taking approximately 1 minute. After that, the site worked well with all sites loading within 10 seconds although the music never did work. A low-end computer or connecting at 56 kbps was not tested but are not expected to work well.

Branding the Web Site:

- Finding the sites is difficult. Searches were conducted using the AOL, MSN, and Yahoo search engines. Searching on "Ukraine" did not locate the sites in the first 100 hits. Searching on "IBS" did not locate the sites in the first 100 hits. Searching on "Ukraine" AND "IBS" found the site with the first hit on Yahoo but did not locate the sites in the first 100 hits of AOL or MSN. However, the first hit, Estate2000, an apartment-renting firm, contained a link to the sites. This indicates the sites are not registered well with search engines. Further evidence of this is in the site visit count observed by the author. The author was visitor 1345 to the business services site and 1109 to the Web development site, counts that are less than expected.

Reshaping the Organization's Corporate Culture:

- Evidence to support a budgeting process in future years toward e-business initiatives was not present. Even though it was found that a technical staff was available for development, no formal e-business planning process was observed.

E-Commerce Readiness Assessment (Table 3):

- Evidence to support the basic IT infrastructure being ready to support e-commerce was not present. Even though it was found that a basic infrastructure with respect to availability of hardware and software is present, the speed, reliability, price, and interoperability/interconnection of the basic infrastructure is not sufficient for e-commerce.
- Since IBS customers are primarily located in Western Europe and the United States, ample evidence of sufficient Internet usage was found. However, it is observed that Internet usage within Ukraine is insufficient to support e-commerce.
- Evidence to support Ukraine being positioned for the digital economy was not present. All the bullets listed in Table 3 for this heading are issues, with most being discussed in the non-IT issues section of the analysis and Economy of Ukraine section of the Background.

Ultimately the value of the site is in the business it generates. By the company's own admission, the sites have generated many inquiries but little to no business. While Molla and Licker (2001) do not include organizational impact in their modification of DeLone and McLean's IS Success Model (1992), this analysis considers a lack of organizational impact in the form of sales as a key indicator that the sites are not successful. Molla and Licker (2001) do include customer trust and return visits as indicators of success. The sites do use secure transactions for credit cards and their state on the site that they respect the privacy of their clients. However, no statement is made on the privacy of client information collected on the sites and the site counter indicated the author was visitor number 1300, this indicates potential issues with regard to trust and return visits by clients.

Conclusion

IT can help make small enterprises in developing countries successful. However, these enterprises face many technical and non-technical issues that impact the ability of these enterprises to take full advantage of IT. These issues for Ukraine can be summarized as:

- Regulatory environments that don't support business development by restricting the flow of funds and information
- Cultural issues which make business and communication with developed economies difficult
- Inadequate and unreliable telecommunications systems limiting communications and e-commerce and causing high communication costs
- Inadequate power quality degrading IT equipment
- Lack of personnel with IT abilities
- High costs of leading-edge and high-end software compounded by the ready availability of inexpensive, illegal versions of this software
- High costs of leading-edge and high-end hardware that may not be compatible with their United States' counterparts

These issues can be overcome by managing IT well and by persistence. However, care has to be taken with e-commerce. The tendency is to rush to be in the world market place. The risk is that the enterprise will actually alienate customers with unsuccessful Web site designs. IBS has technically good sites but is not successful with them. This could be corrected with some relatively simple fixes such as planning, registering with more search engines, and removing some annoying features. Molla and Licker's (2001) e-commerce modification of DeLone and McLean's IS Success Model provides some insight into how a successful e-commerce site should be designed.

Epilogue

IBS has used the results of this case study to revamp its Web sites. The music has been removed and registration has resulted in IBS showing up in the top five results when a user searches on "International Business Solutions" and "Ukraine" or "Ukraine" and "IBS." However, searching for Energy Solutions or Web Development or Web Solutions still does not readily find the company. Additionally, changes to the Web sites based on the above analysis are improving the look and feel of the site but changing to a Ukrainian ISP has slowed page loading and continual changes to the site are keeping the site from being highly reliable and from developing branding. Ultimately, it is expected that the changes being implemented will greatly improve the traffic and return of the IBS Web site.

Additionally, issues discussed in the conclusion section are still observed to be valid although power and communications systems are becoming more reliable.

Areas for Future Research

This chapter focuses on a single company. The methodology used to assess IT and e-commerce in this company provides a framework that can be used with other companies and in other countries. The next logical step is to replicate this research in a number of companies located in various countries. The results from these assessments can then be used to generate generic models and a survey/assessment instrument that can be used for any assessment.

References

Amoroso, D. (2001). *e-Business success factors.* Working Paper, San Diego State University.

Amoroso, D. (2002). *Successful business models for e-Business: An exploratory case analysis of two organizations.* Working Paper, San Diego State University.

Amoroso, D., & Sutton, H. (2002). Identifying e-Business readiness factors contributing to IT distribution channel reseller success: A case study analysis of two organizations. *Proceedings of the 35th Hawaii International Conference for System Sciences.* IEEE.

APEC Readiness Initiative. E-Commerce Readiness Assessment Guide. Retrieved February 2002 from *http://www.apecsec.org.sg/*

Bakos, J., (1991). A strategic analysis of electronic marketplaces. *Management Information Systems Quarterly, 15*(3), 295-310.

Bakos, Y. (1997). Reducing buyer search costs: Implications for electronic marketplaces. *Management Science, 43*(12), 1613-1630.

Brynjolfsson, E., & Hitt, L. (1996). Paradox lost? Firm-level evidence on the returns to information systems spending. *Management Science, 42*(4), 541-558.

Central Intelligence Agency World Fact Page for Ukraine. (2001). Retrieved December 2001 from *http://www.odci.gov/cia/publications/factbook/geos/up.html*

Chervachidze, S. (2001, November 29). Safe deposit? *Kyiv Post.* Retrieved from *http://www.kpnews.com/main/10155*

Chipaitis, E.V. (2002). E-commerce and the information environment in an emerging economy: Russia at the turn of the century. In P.C. Palvia, S.C.J. Palvia, & E.M. Roche (Eds.), *Global information technology and electronic commerce: Issues for the new millennium* (pp. 53-72). Ivy League Publishing.

DeLone, W.H., & McLean, E.R. (1992). Information systems success: The quest for the dependent variable. *Information Systems Research, 3*, 60-95.

Hart, P., & Saunders, C. (1997). Power and trust: Critical factors in the adoption and use of electronic data interchange. *Organization Science, 8*(1), 23-42.

Heeks, R., & Duncombe, R. (2001). Information, technology and small enterprise: A handbook for enterprise support agencies in developing countries. Institute for Development Policy and Management, University of Manchester.

Jarillo, J.C. (1988). On strategic networks. *Strategic Management Journal, 9*, 31-41.

Jennex, M.E. (2001). IT in the energy sectors of Ukraine, Armenia, and Georgia. *Proceedings of the Global Information Technology Management Conference, GITMA* (pp. 164-168).

Jennex et al. (1999). Ukraine Y2K risk assessment final report. U.S. Aid Project.

Mitra, S., & Chaya, A. (1996). Analyzing cost-effectiveness of organizations: The impact of information technology spending. *Journal of Management Information Systems, 13*(2), 29-57.

Molla, A., & Licker, P.S. (2001). E-commerce systems success: An attempt to extend and respecify the DeLone and McLean Model of IS success. *Journal of Electronic Commerce Research, 2*(4), 1-11.

Olearchyk, R. (2001, November 29). Software development workforce slipping. *Kyiv Post.* Retrieved from *http://www.kpnews.com/main/10154*

Powell, T., & Micallef, A.D. (1997). Information technology as competitive advantage: The role of human, business, and technology resources. *Strategic Management Journal, 18*(5), 375-405.

Tetteh, E.O., & Burn, J.M. (2002). A framework for the management of global e-Business in small and medium-sized enterprises. In P.C. Palvia, S.C.J. Palvia, & E.M. Roche (Eds.), *Global information technology and electronic commerce: Issues for the new millennium* (pp. 275-293). Ivy League Publishing.

Ukraine Embassy to the United States. (2001). Web site. Retrieved December 2001 from *http://www.ukremb.com*

Chapter XIX

E-Business in India: Early Evidence from Indian Manufacturing Industry

Rajeev Dwivedi
Dept. of Management, Indian Institute of Technology Delhi, India

Sushil
Dept. of Management, Indian Institute of Technology Delhi, India

K. Momaya
Dept. of Management, Indian Institute of Technology Delhi, India

Abstract

Business and industries have faced several changes from the agriculture society to information society. The recent change is due to Information Technology (IT) affecting many businesses and industries. It is changing the nature of business from the traditional way of doing business. The complete change in traditional business is due to IT. This is known as e-business transformation. The Indian manufacturing industry is undergoing this IT-enabled change and is still under process of click and brick system. Indian automobile companies are stressing the importance of e-business in the domestic automotive industry. The main aim of the chapter is to explain how the manufacturing and especially the automobile industry business has changed from traditional brick and mortar business to click and brick e-business. This chapter provides a study of e-business transformation in manufacturing industry in India using Flexible Systems (SAP-LAP) Methodology. The SAP-LAP stands for Situation-Actors-Process and Learning-Action-Performance. This methodology helps for understanding systematic nature of e-business transformation. The explanation of stakeholder flexibility due to e-business transformation is Industry will be explained.

Introduction

As e-business is reshaping industries, it is hard to ignore or hard to survive without participating in IT-enabled business. Indian companies are doing well in the e-enabled environment, especially in its manufacturing industry. The aim of this chapter is to address the e-business transformation issues in an Indian context through a real-life case study of an Indian automobile organization. The scope of the chapter covers systematic understanding of e-business transformation through Flexible Systems Methodology. The case of the automobile industry is selected purposively. The reason behind this is that automobile companies are the first users of the newest and latest technologies. Hence, it has touched the day-to-day life of Indian customers as they adopt new technology. Most important reason is that manufacturing accounts for a significant contribution in the Indian GDP system and especially the automobile sector. The big question is whether e-business can help in the manufacturing sector for enhancing productivity and efficiency. The chapter highlighted and covered current and emerging issues related to e-business strategy. The information used in SAP-LAP analysis and in the chapter, has come from the doctoral work carried out by the author on "e-business transformation and stakeholder flexibility: a study of manufacturing industry in India." It is very difficult and beyond the scope of the chapter to provide all data and information. Some concise points and information is highlighted from the doctoral work.

Chapter Organization

The chapter is organized into four levels in order for understanding the e-business in India. The levels are: e-business in India, Indian manufacturing industry, Indian automobile sector and a case of Indian premier automobile company. The Flexible Systems Methodology is used for understanding the e-business in manufacturing industry, while SWOT analysis is done with automobile sector, and the case study of Maruti Udyog Limited is explained. The case of Maruti is taken for briefly explaining the e-business in automobile company. It reflects from J.D. Power sales satisfaction and customer satisfaction index 2004, in India, Maruti received 813 points on a 1000-point scale and the industry average is 758 points in customer satisfaction, carried out with 3,600 customers across nation. The sales satisfaction index is 784 and industry average is 760 on a 1000-point scale. The flow of the chapter is shown in chain given below. Maruti is the highest scorer in both surveys and above average, which is the reason for choosing the case for knowing where e-business is helping them.

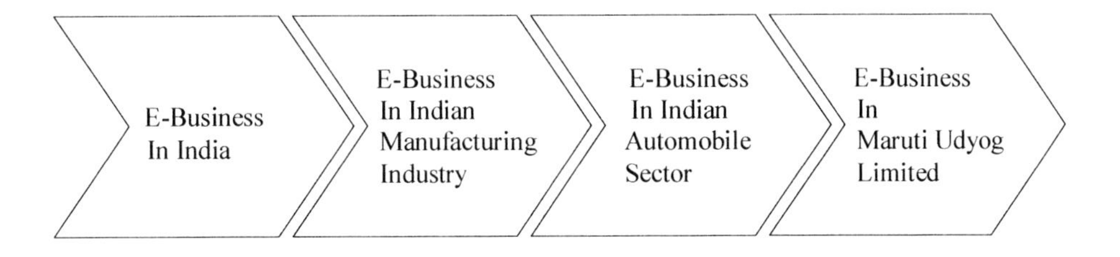

Background (Literature Review)

The literature review reveals that developed countries are using the full potential of e-business while developing countries are in the early phase of e-business. The literature available in this regard is limited and needs a lot of attention in order to understand the scope of e-business in developing countries.

- **Why e-business?:** The change comes only when the previous phase faces the saturation of S shape growth curve. Previous change was MIS. E-business is to exploit the opportunities created by the Internet and information technology for business to create virtual marketplace. E-business is concerned specially about B2B segment because of the failure of the B2C segment (dot-com failure). The failure of e-commerce brought e-business to create new opportunities for business, especially for B2B. The transaction of money is much more in B2B as compare to B2C e-business.
- **For whom is e-business?:** E-business is for all stakeholders of the business, customers, employees, and shareholders, every one. In all cases it gives flexibility and JIT services at place. From a business point of view, it's for Big MNCs to exploit the resources and opportunities available globally. They can afford the cost infrastructure of e-business with the help of world-class technology partners. The e-business is also for semi and medium-size enterprises to adopt and transform e-business step by step as Sawhny and Zabin (2001) described the ladder from inform to transform.

E-Business in Developing Countries

The Information and Communication technology is playing a vital role in the development of growing economies such as India, Brazil, Mexico, and some south Asian countries like Thailand, Malaysia, etc. IT has the potential to enhance the economies in a social, cultural and technological manner in order to enhance competitiveness of the country. India has done well so far in the right direction while understanding the need of IT for Indian industries. Some evidences have come from the two studies done by Kamel and Hussein (2001, 2002), which address the development of e-commerce in Egypt. They describe about the evolution of the Internet, profile and challenges along with opportunities. Further, they highlighted that Information and Communication Technology has become vital as a platform for business and socio-economic development. The poor and expansive ICT infrastructure, lack of awareness on e-business issues, inadequate legal and regulatory framework, absence of trust, network payment and secure transaction services are just some of the challenges faced by developing countries in their attempt to make the transition into the networked economy (*NTOKO)*.

Manufacturing Industry and E-Business

The world of business is being changed to an e-economy by new forces of global competition, increased information availability, educated consumers, changing relationships, rapid innovations, and increasingly complex products. No industry is left or untouched by e-business (Grieger, 2003). E-business transformation does not come easily in manufacturing companies. Though, a successful transformation in an e-business model, manufacturing companies can cost effectively evolve into remodeled organizations that attract, retain, and serve their customers better, while improving ROI on their technology investment in creating shareholder value (Scott, 2002). E-Business represents an evolutionary step, not the end of traditional retail. Retailers must incorporate the Web's best with good business practices. Traditional companies that viewed the Web as a business opportunity and not a threat are the ones standing strong after the dot-com fallout.

Manufacturing industry worldwide has been facing unprecedented challenges brought by ever changing, global and competitive market conditions as well as changing social demands, regional, governmental and environmental regulations. *E*-commerce and Internet technologies injected "velocity" into the front *business* activities and enabled companies to shift their *manufacturing* operations from the traditional factory integration philosophy to a supply chain-based *e*-factory philosophy. It transforms companies from a local factory focus to a global enterprise and *business* focus (Lee, 2003). Further authors explained how *e-manufacturing* as a new concept to answer the needs of *business* strategies for complete integration of all *business* elements including suppliers, customer service network, and *manufacturing* units by leveraging the Internet, Web-enabling, tether-free technologies and computational tools. Enabling tools that will be introduced to support *e-manufacturing* include the ability to monitor the plant floor assets, and predict the variation and performance loss for dynamic rescheduling of production and maintenance operations, and synchronize with other related *business* actions to achieve a complete integration between *manufacturing* systems and upper-level enterprise applications. Finally, information technologies for next-generation *e-manufacturing* transformation are discussed. Growing number of programs such as TQM, JIT Production, DFM (design for manufacturing) is not to mention lean manufacturing, reengaging, benchmarking, and the ubiquitous team approach (Hayes & Pisano, 1994). The key issues of e-business and manufacturing are shown in Table 1.

E-Business and SMEs

Lal (2002) has defined the role of e-business in India in his paper "E-Business and Manufacturing Sector: A Study of small and Medium Seized enterprises in India." He links the competitiveness, bandwidth, searching of new markets, efficiency, international orientation, size of operation conduct variables, and reduced cost in his analytical framework of e-business in India.

According to Tettech and Burn (2001), Web-based business can be an external attractive option for most SMEs to extend their customer base into a global market without vast

Table 1. Key issues of e-business and manufacturing

Year	Author (s)	Key Issue
2003	Lee	E-commerce and Internet technologies injected "velocity" into the front *business* activities and enabled companies to shift their *manufacturing* operations from the traditional factory integration philosophy to a supply chain-based *e*-factory philosophy.
2003	Tigre	For e-commerce diffusion, a factor more important than the sector itself is the scale of information flow among various agents within the economy.
2002	Oakham	The truth may be that e-business works for some sectors, like retail and wholesale, where significant improvements in logistical cycle times have been achieved, but for many sectors it is not realistic to expect a truly integrated supply chain function with benefits to all parties.
2001	Computergram Weekly	The key driver for manufacturing was the additional sales opportunities afforded by the Web, with improvements to internal efficiency in second place.
2001	Soliman and Youssef	Recent developments in information technology such as ERP, Internet, Knowledge Management systems necessitate the use of these technologies in order for the next generation manufacturers to co-evolve and survive on the new business landscape.
1997	Brynjolfsson	Flexibility in manufacturing relies not only on powerful new information technologies, as is commonly emphasized, but also on mutually reinforcing practices.

expenses, but there are many hurdles and hidden costs. SMEs can achieve global competitiveness without necessarily increasing their actual size, but rather by building on their virtual or soft assets in order to expand. These virtual assets include information skills, digital resources, and competencies for managing inter firm relation and collaborative engagement with other firms. The World Wide Web (WWW) offers exciting new opportunities for small and medium-sized enterprises (SMEs) to extend their customer base into the global marketplace.

Economic development is the main goal of most governments. Globally it is recognized that approximately 80% of economic growth comes from the SME sector. A total of 99% of businesses in North America and Europe are SMEs. It is therefore likely that government strategies, aimed at facilitating SME adaptation of Internet-enabled business processes and practices help increase national GDP (Jutala et al., 2002).

Among the SMEs, adoption of *e-business* is still low due to lack of awareness and costs constraints. Local companies should realize that *e-business* is a new way of doing *business*. Local companies, especially small and medium-sized enterprises (SMEs), can bank on *e- business* to face growing competition from other companies in the region and from around the world. It's a question of survival and the ability for everyone to compete on the same level against the backdrop of trade liberalization and globalization. Companies can understand that apart from efficiency, *e-business* can improve on revenues. Through *e-business*, a company will also enhance customer service and with daily *business* processes carried out electronically, *business* can be conducted during or after normal *business* hours. By operating electronically, *e*-businesses can help to reduce paperwork and other processing costs. In addition, companies can improve on internal logistics through faster (real-time) communications with suppliers and customers. The potential impact of e-business is great, especially with lowering of regional and global trade barriers.

E-Business in India

Rapid globalization of business, economic deregulation and the increased role of IT in shaping corporate strategy in developing nations have given rise to a need for understanding the strategic and environmental drivers for Electronic commerce. Tarafdar and Vaidya (2004) pointed out that:

> *In India, economic liberalization, initiated in the early 1990s, resulted in increased Competition in many industry sectors, as foreign companies were allowed to set up manufacturing and assembly plants. This led to intensified efforts at overall modernization and IT adoption across industries, as domestic companies felt the need to respond to the requirements of a de-regulated economic environment. In many cases, the increased emphasis on IT investments has involved adoption of EC. It is important to understand how organizations have responded to these environmental changes by adopting IT in general and EC in particular.* The study considered internal and external drivers of the electronic commerce adaptation.

According to NASSCOM (2002), major factors driving growth of the e-Business applications market in India are: Need for streamlining business process, Maturing information systems, Extending global strategies, Availability of better solutions, and Maturing first tier applications.

The study carried out by NASSCOM (Table 2) also reflects that the organizations large and within the SME segment are in different stages of deployment of e-business applications. The manufacturing sector, particularly high-tech companies are aggressively deploying e-business applications. The traditional manufacturing companies, however, have been slow movers. Banking and Finance sectors along with telecom are the exploiting full capabilities of e-business, while education and government are on first phase of adoption. High-tech companies are using CRM, SCM and enterprise portals.

Table 2. Total e-business applications market: enterprise application adoption landscape across verticals in India over a scale of 1 to 10, 2001

	ERP	SCM	BI	EP	E-Com	CRM	C-com	Others
Banking and Finance		5.0	7.5	7.5	7.5	10.0	5.0	10.0
Chemical	5.0	5.0	5.0	2.5	5.0	5.0	2.5	
Government & Education	2.5		2.5	2.5	2.5	2.5		
Healthcare	5.0	5.0	5.0	2.5	5.0	5.0	5.0	5.0
High Technology	5.0	5.0	7.5	7.5	7.5	7.5	5.0	
Hospitality and Travel	5.0	7.5	5.0	5.0	5.0	7.5	5.0	
ITES	5.0	5.0	7.5	7.5	5.0	7.5	2.5	
Manufacturing	5.0	5.0	2.5	2.5	2.5	5.0	5.0	
Public Sector	5.0	5.0	2.5	5.0	2.5	5.0	2.5	
Telecom			7.5	7.5	7.5	10.0		10.0

The study reflects the future of e-business in health care and government. Big businesses in the manufacturing industry are using existing IT infrastructure, while SME firms made first-time-purchases of ERP and SCM solutions during 2003-04. The statistics show that Internet users and broadband subscribers are increasing in multiple orders. This is helpful for enhancement of e-business adoption in India.

- **Who has adopted e-business in India?:** The pattern of adoption of e-business in India is virtually identical to that elsewhere in the world, with automotive manufacturers, FMCG and oil companies and financial institutions (banks and stock exchanges) being the early adopters. The roster of companies already entrenched on the Indian e-business landscape not surprisingly includes names such as Hindustan Lever, Pepsi, Coke, BPCL, IOCL, Bombay Stock Exchange, HDFC Bank, Maruti, TELCO, and Ford, though a full list includes a sprinkling of virtually every kind of firm.
- **What are the benefits of e-business?:** Companies are racing toward e-business to gain financial and competitive advantages in the marketplace. The following are examples of typical objectives:

- Agility
- Convenience to external parties
- Eliminate intermediaries and redundancy
- Flexibility
- Growth
- Improved perception of company and products
- Lower transaction costs
- Personalized and improved customer service
- Profitability
- Reaching new markets
- Reduced delivery times
- Scalability

Indian Manufacturing Industry and E-Business

Indian manufacturing industry is the third largest contributor in Indian GDP, apart from service and agriculture, while China is ahead in manufacturing as compare to agriculture and service sectors. The growth of Indian manufacturing is on steady pace over the past three to four years. The reason is new environment of doing business by leveraging operational excellence through e-business and Information Technology enabled business. Five years ago, India had very few globally competitive manufacturing companies such as Reliance, Hero Cycles, Sundram Fasteners, Hindustan Lever and Ranbaxy. After the transformation, a greater number of Indian manufacturing companies are making their presence known worldwide. Some of these companies are like Tisco, Telco, Bajaj Auto, TVS Motors, Sundaram-Clayton, Gujarat Ambuja, Dr. Reddy's Laboratories, Asian Paints, Hindalco Ballarpur Industries, Arvind Mills, Vardhaman Spinning, Zodiac, Balrampur Sugar, Bharat Forge, Moser Baer, Hindustan Inks, Sigma Corporation, automotive manufacturers, FMCG and oil companies and financial institutions (banks and stock exchanges) being the early adopters. The companies that have already made significant impact in manufacturing through e-business landscape are such as Hindustan Lever, Pepsi, Coke, BPCL, IOCL, Bombay Stock Exchange, HDFC Bank, Maruti, TELCO, and Ford, though a full list includes a sprinkling of virtually every kind of firm.

In addition, foreign companies like GE, Tecumseh and Hyundai have started using India as a manufacturing base for their worldwide operations. Public-sector companies like BHEL, Bharat Electronics and Hindustan Aeronautics are enhancing their manufacturing capabilities by e-business.

Flexible Systems Methodology

This phase uses a powerful tool of Flexible Systems Methodology — SAP-LAP (Sushil, 1994, 1997, 1999, 2000). The SAP part deals with analysis of the case. The LAP part deals with synthesis and learning issues, suggested actions, and their impact in terms of performance of the industry or organization. The brief SAP-LAP model for change has been applied to understand the Situation, Actors, Process and Learning, Action and Performance of e-business in Manufacturing Industry of India. The SAP-LAP Model for Change is given in Table 3.

Situation

- The Manufacturing industry has taken a lead in implementing and absorbing state-of-the-art IT technologies. Currently, manufacturing industry accounts for 15% of IT usage across sectors in the domestic market. Some key trends in IT usage by the sectors are highlighted below in Table 4.
- India's Internet user base is growing at a rapid pace. India's Internet population has grown to 29 million in March of 2003 from 10.7. million in 2002.
- The 91% of Indian corporate are having Web presence.
- China's manufacturing industry is the big threat for Indian manufacturing industry and required full e-business potential.

Table 3. Flexible Systems Methodology (SAP-LAP)

Situation
What are the major opportunities?
What are the major threats?
What are the seeds of change?

Actors
Has a guiding coalition been formed?
What are the values and beliefs?
At what levels has the vision of change been communicated?

Process
How the strategy coalition been formed?
What are the values and beliefs?
At what levels has the vision of change been communicated?

Learning
What is the understanding about the change?
Where is the consonance or dissonance?
What are potentials to change the process?

Action
What all is to be done to prepare for a cultural change?
How to initiate a structural change?
What should be done to be ahead of the change?

Performance
How will the change affect performance?
Are we finding an improvement in key performance indicators?
Where have we had set backs?

Table 4. IT trend in India

S. No.	Industry/Sector	IT Usage Percentage
1	IT and Telecom	22%
2	Banking and Service	21%
3	Manufacturing	15%
4	Government	14%
5	Education	11%
6	Energy	06%
7	Small Offices/Home/ Others	11%

Source: NASSCOM, 2003

Actor

- Companies, those that are less than five years old show low maturity, possibly because they are still consolidating their markets before implementation of complex and expensive IT system. The companies more than 20 years old also show low IT maturity. The maturity has been defined on different levels for exploiting of e-business, such as information, interaction/communication, transaction, and transformation. Resistance to change is the reason in old business. This is followed by standardized legacy systems. Hardware companies are using e-business in the age group of 5-10 years, because they have been consolidated in their operations and are becoming IT savvy (Table 5).

Table 5. IT and maturity index

Age	Maturity index
Less then 5 Years	17%
5-10 years	44%
10-20 years	29%
More Then 20 years	19%

Source: ETIG-PwC SCM Survey 2002

- Indian manufacturing companies are aware the value of knowledge management in the organization for IT Strategy.
- The stakeholders are required to join e-business model of manufacturing industry. The key stakeholders are the major actors of e-business because they get flexibility and also provide flexibility to business.

Processes

- Information Technology is an inseparable part of supply chain management, which is harnessing information flows for decision making. The ETIG-PWC survey revels that ERP systems have been implemented widely, Followed by warehousing management system and customer relationship management, while only 15% had any transport management IT system in place (Table 6).
- In most of the rest of the world, the business-to-business (B2B) e-commerce has been a multiple of the consumer sector.
- The current application in many manufacturing enterprises is centered around enterprise applications such as ERP, CAD, SCM, CRM, SRM, etc., including plant-level execution systems for shop floor automation, HR functions, quality management systems, WIP tracking systems, and job scheduling applications.

Table 6. Status of IT systems in manufacturing industry of India

System	Already Implemented %	Plan to Implemented %
ERP	67	24
Warehousing Management System	34	24
Data warehousing and data mining	32	34
Bar coding	31	38
EIS/DSS/PMS	31	37
Electronic Procurement	26	50
Advanced planning and Scheduling	23	35
Manufacturing execution system	15	22
Transportation management system	15	24
CRM	11	15

Source: ETIG-PwC SCM Survey 2002

Learning

- If India wants to achieve the 7-8% growth rates in GDP then the manufacturing sector has to contribute significantly around 15% growth (Web site of ministry of commerce). This growth rate of manufacturing industry is possible with the use of Information Technology in manufacturing sector across the entire value chain.
- The learning is from the e-business models and stakeholder participation in the models that all value chain partners should be incorporated in e-business model for cutting cost, reducing time, better services and gaining advantage.

Action

- E-business transformation should be across the value chain in every sector of the manufacturing industry for taking full advantage.
- IT and Telecom infrastructure with government support has to grow for better e-business facilities.
- The fast growth of Internet users in the country to take advantage of e-business.

Performance

- The e-business transformation will lead to the higher growth rate and sustainability by reducing cost, streamlining operations and improving customer's satisfaction due to available services and products at the right place and just in time.
- E-business will bring stakeholder flexibility in manufacturing industry of India which leads to business performance. This reflects from the study conducted in the manufacturing industry for seamless integration.
- E-business may help to India for becoming to global sourcing hub and operational base for MNCs.
- E-business helps directly or indirectly in foreign direct investment (FDI) in India, based on data collected in the last 10 years or so.

Indian Automobile Industry: Some Facts and Figures

The journey of the Indian automobile industry was started with the set up of Hindustan Motar Corporation. Ghanshyamdas Birla in 1942 incorporated the first HM plant at Okha, Gujarat. Now, the automobile industry is one of the core industries in India's economy and contributing significantly. Since the liberalization of the Indian economy in 1991,

Table 7. Global players and sourcing from India

- Fiat Plans To Source US $200 mn. Worth Of Components From India Per Annum
- Mercedes Benz (Daimler Chrysler) Has Set Up 7 Component JVs In India For Global Sourcing Of Parts
- Cummins USA Is Already Sourcing Engine Parts From India For Cummins Global Operations
- Multi National such as DELPHI and VISTEON have started exporting components made in India to their various other plants around the world-Investing further to make India as a manufacturing base

Table 8. Key points

- India is the 2nd largest two wheeler manufacturer in the world
- Second largest tractor manufacturer in the world
- 5th largest commercial manufacturer in the world
- 3rd largest car market in Asia
- Automobile industry Contributes 17% of the total indirect taxes

India has become the hub of global automobile majors (see Table 7). The automobile industry in India is gradually evolving to replicate those of developed countries.

The Indian automobile industry is broadly divided into passenger cars/multi-utility vehicles, commercial vehicles, two-wheelers and tractors. The key facts about the segments are mentioned in Table 8.

The performance of the Indian automobile industry has gone well gradually as shown by the tables in the appendix.

Indian Automobile Industry and E-Business Change

There is now growing evidence that enterprises gain substantially from e-business. Fast productivity growth in industrialized countries has been largely attributed to the widespread application of ICT, creating millions of new jobs and billions of dollars in savings.

The Indian Automobile Industry has witnessed significant changes. The first change from globalization and second change is from e-business. Most of the global automobile companies are investing in India over the last ten years. Along with the entry of multinational auto companies, the industry is changing by IT-enabled business transformation called e-business transformation. The Indian Automobile Industry is second after the textile industry in the manufacturing sector of India and contributes significantly to the national GDP. The processes of embracing the New Economy model with retaining its traditional strengths and transforming itself into an e-business enterprise.

The paradigm of e-commerce will create more opportunities for this industry. The auto industry being one of the key drivers of the economy is focused on envisioning its growth through the new economy.

E-Business Functions and Stakeholders of the Automobile Industry in India

Information Technology has simplified production processes in automobile companies. Manufacturing units demand huge periodic investments, as raw materials need to be continuously allotted to specific components, assembled on the shop floor and later placed on an IT backbone. The process is painstaking, expensive and calls for precision planning at every step. IT has revolutionized manufacturing processes. In most cases, the use of IT in the manufacturing sector is restricted to rudimentary practices. Manufacturing companies are adopting technologies like ERP, CRM and SCM to improve market competitiveness. The Table 9 demonstrated the significant use of E-business in automobile industry. The related stakeholder of e-business functions are shown in Figure 1.

The e-business is not only transforming the companies of automobile industry but it is reshaping the value chain containing major stakeholders of the business such as: employees, suppliers, distributors, dealers, retailers, technology partners, shareholders, call centers, IT partners, and ultimately customers. All major stakeholders are connected

Figure 1. E-business functions and related stakeholders

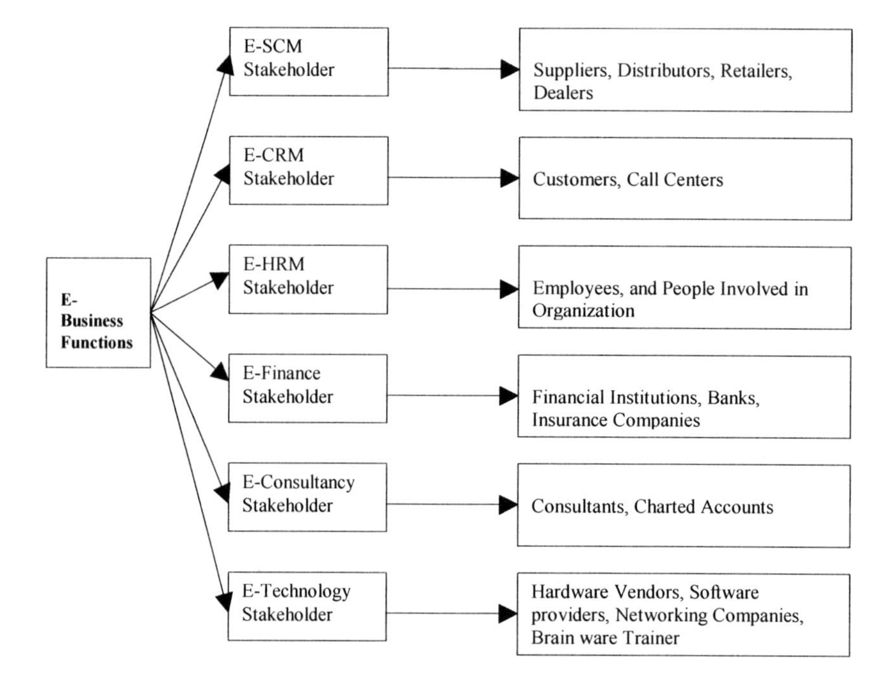

Table 9. The function of automobile and e-business

e-Function	E-Business Benefits	Example of e-Business
e-Transactional	E-business can transform unstructured processes into routines transactions	Artificial intelligence, dedicated software, groupware technology, multimedia
e-Global strategy	E-business allows an easy and fast information transfer over a great distance, making processes independent of geography	EDI, Internet (Intranet/Extranet), video-conferencing, tele-work
e-Automation	E-business can replace or reduce human labor in a process	Artificial intelligence, CAD/CAM, CNC, EDI, PLC, search engine
e-Analysis	E-business can bring complex analytical methods to bear on a process	Dedicated software, work stations
e-Informational	E-business can deal with and deliver a great amount of information expressed in several ways: text, graphics, sound, video	CAD/CAM, databases, video-conferencing, audio-conferencing, search engine
e-System	E-business can enable changes in the sequence of tasks in a process, often allowing multiple tasks to be worked on simultaneously	Groupware technology, shared databases, Internet (Intranet/Extranet),
e-Knowledge management	E-business allow the capture and dissemination of knowledge and expertise	Artificial intelligence, groupware shared databases, data mining, groupware, video-conferencing, chat systems
E-Order Tracking	E-business allows the detailed tracking of task status, inputs, and outputs	Artificial intelligence, CAI, CAM, PLC, sensors, Internet-based applications
E-interactivity	E-business can be used to connect, either synchronously and asynchronously, two parties that would otherwise communicate through an intermediary	CAD/CAM, EDI, Internet Intranet/Extranet)

Source: Adopted from N. Carbonara.

via a seamless network (Internet, extranet and intranet) for free flow of information, communication, transaction and process and operations among stakeholders of the automobile industry.

SWOT Analysis of Indian Automobile Industry

The SWOT analysis on Indian automobile industry explains the strength, weakness, opportunities and threat of industry due to e-business. The Table 10 is the SWOT analysis and Table 11 is the E-SWOT metrics for the automobile industry of India.

The first box indicates how e-business can help strengthen the opportunities. The second has been explained as how opportunities help to overcome the weaknesses of the business. Third box indicates the strength of e-business to compete against threats. The last one explains how to survive due to weaknesses in the threat environment.

Table 10. SWOT of automobile industry of India

Strengths	**Weaknesses**
• All automobile companies have their separate Information Technology department and application development teams. • Global automobile players are using India as an operational base and e-business is cutting cost of operational processes. • Cheap labors and e-material management • Strategic position of the India with the opportunity to expand to SAARC countries • Growing number of domestic suppliers which in turn decreases the dependency upon imports and seamless supply chain using e-business. • Associated sector development and business process outsourcing. • Establishment of new industrial sector dealing with waste recycling and processing of automobile industry products such as e-reverse auction • IT-educated employees	• Dependent on technology alliances from global IT players such as Microsoft, Oracle, HP, Lucent, and SAP R3. • The lack of operational efficiency due to basic infrastructure and lack of e-business in SMEs • Manual processes of small suppliers • Lack of IT awareness in SMEs employees • Unable to compete in foreign markets • Low export rate of SMEs
Opportunities • Telecom infrastructure • ROI • Seamless and JIT process • Global reach • Market research • Financial transaction • Alliances • Foreign investment • Outsourcing hub for many world-class players	**Threats** • Security • Trust • Large competition

Table 11. E-SWOT metrics of Indian automobile industry

	Strengths (S)	Weakness (s)
Opportunity(s)	Use of e-business to gain competitive advantage in the industry	Use e-business to reduce overhead cost of processes
Threat (s)	Established strong network among stakeholders through e-business	Provide flexibility to the stakeholders and innovative products using e-business

E-Business at Maruti Udyog Limited and Stakeholders Flexibility

Maruti Udyog limited (MUL) is the largest car manufacturer of India. In the year 1982, the government signed a joint venture with Suzuki Motor Corporation of Japan and set up the first joint venture of the Government of India in car manufacturing. One of the biggest weapons in Maruti's arsenal is information technology. Currently, almost 75% of Maruti's Rs 9,672 crore in yearly business transactions is conducted online and the

Table 12. E-business performance results at MUL

Improvement in Services by E-Business
• Helpful in maintaining Maruti's core competency to provide cheapest car • Customer Information Management, customer support • ERP system available to external partners and 200 dealers • Centralized application deployment and easy system maintenance • Improved IT support across entire network • Improved customer service — Maruti employees can easily access the information they need to respond more quickly to customer issues • Improved information sharing—document collaboration and publishing features enable users to work together more easily and publish documents more efficiently • Faster time to benefits
Financial Results by E-business
• Reduce 50% costs of procurement processes • Cost reduction of inventory and other overhead costs • 75% cost saving for paper and other stationary • 72 Laks Rs by online reverse auction • Online transaction of 75% of total revenue • 20% reduction in Total Cost of Operation (TCO)

company is poised to take it to 80% by 2003. On an average, Maruti invests between Rs 15-20 crore every year on IT. The Table (Appendix A) shows the key stakeholders of Maruti Udyog Limited. E-business brings more options for doing business, to change the way to do business, and freedom of choice for doing business is called e-business flexibility. All the Key stakeholders (Appendix A) are seamlessly connected with Internet, Intranet, and Extranet. All stakeholders are getting the flexibility from e-business in terms of information, communication/interaction, transaction and processes and operations to conduct business for Maruti. The reason behind the success of Maruti is e-business and provided flexibility to stakeholders due to e-business. Table 12 indicates the key performance indicators of e-business in Maruti. It has 370 showrooms, 5,000 service centers connected through its network, 6,000 trained and knowledgeable executives working in its IT-enabled environment for providing insurance, finance, accessories and almost any other car-related services, product 24*7 in one place.

Concluding Remarks

E-business has brought the change in the Indian automobile manufacturing industry. Progressive large companies have considerably changed the way of doing business. Many small and medium size companies are planning to change and have a Web presence. The Indian manufacturing industry needs to innovate to reach and remain at the frontier of "world-class" competitiveness using e-business. E-business models and technologies are becoming important for Indian Manufacturing Industry producers as they are integrated into the global value chains. The question is whether e-business delivers the value for Indian manufacturing industry to gain sustainable competitive advantage.

References

Brynjolfsson, E., Renshaw, A. A., & Alstyne, H.V. (1997, Winter). The matrix of change. *Sloan Management Review*, 37-54.

Carbonara, N. (2003). Information and communication technology and geographical clusters: opportunities and spread. *Technovation.*

Computergram Weekly. (2001, May 14). UK manufacturers reluctant to adopt e-business.

Dwivedi, R., & Momaya, K. (2003). Stakeholder flexibility in e-business environment: a case of an automobile company. *Global Journal of Flexible Systems Management*, *4*(3), 21-32.

Grieger, M. (2003). Electronic marketplaces: a literature review and a call for supply chain management research. *European Journal of Operational Research, 144*, 280-294.

Jutla, D., Bodorik, P., & Dhaliwal, J. (2002). Supporting the e-business readiness of small and medium sized enterprises: Approach and metrics, internet research. *Electronic Networking Applications and Policy, 12*, 139-164.

Kamel, S., & Hussein, M. (2001). The development of e-commerce: The emerging virtual context with in Egypt. *Logistic Information Management, 14*(2), 119-126.

Kamel, S., & Hussein, M. (2002). The emergence of e-commerce in a developing nation: case of Egypt. *Benchmarking: An International Journal, 9*(2), 146-153.

Lal, K. (2002). E-business and manufacturing sector: a study of small and medium seized enterprises in India. *Research Policy, 31*, 1199-1211.

Lee, J. (2003). E-manufacturing—Fundamental, tools, and transformation. *Robotics & Computer-Integrated Manufacturing, 19*(6), 501-507.

NTOKO, A. (n.d.). Basic e-commerce training for Pakistan: An e-business technology strategy for developing countries. Retrieved from *www.itu.int/ITU-D/e-strategy/ecdc/Seminars/ pakistan/PakistanPaper+Bio.pdf*

Oakham, M. (2002). The engines of e-business. *Metalworking Production, 146*(8), 40-41.

Sawhney, M., & Zabin, J. (2001). *The seven steps to nirvana.* Delhi: Tata McGraw Hill.

Scott, W G., & Howard, S R. (2001, Fall). Determinants of governance structure for the electronic value chain: Resource dependency and transaction cost perspective. *Journal of Business Strategies.*

Soliman, F., & Youssef, M. (2001). The impact of some recent developments in e-business on the management of next generation manufacturing. *International Journal of Operations & Production Management, 21*(5/6), 538-564.

Sushil. (1994). Flexible systems methodology. *Systems Practice, 7*(6), 633-651.

Sushil. (1997). Flexible systems management: an evolving paradigm. *Systems Research & Behavioral Science, 14*(4), 259-275.

Sushil. (1999). *Flexibility in management: Global institute of flexible system management*. New Delhi: Vikas.

Sushil (2000). *E-business: Creating flexible enterprise with new business models, in cornerstones of enterprise flexibility*. New Delhi: Global institute of flexible systems management/Vikas..

Sushil. (2000). SAP-LAP models of inquiry. *Management Decision, 38*(5), 347-353.

Tarafdar, M., & Vaidya, S. (2004). Adoption of electronic commerce by organizations in India: strategic and environmental imperatives. *The Electronic Journal on Information Systems in Developing Countries (EJISDC), 17*(2), 1-25. Retrieved from *http://www.ejisdc.org*

Tettech, E., & Burn, J. (2001). Global strategies for SMEs business: Applying the small framework. *Logistics Information Management, 14*.

Tigre, P. B. (2003). E-commerce readiness and diffusion: the case of Brazil. *I-WAYS, Digest of Electronic Commerce Policy and Regulation, 26*, 173-183.

Web Sites

http://www.automobileindia.com/general-info/auto-sector-india.htm

http://www.nasscom.org/newsline/dec02/feature.asp

www.accenture.com

www.autoindia-junction.com

www.computer-today.com

www.expresscomputer.com

www.marutiudyog.com

www.pwcglobal.com

Appendix A: Key Stakeholders of Maruti

Stakeholders	Description of Stakeholders
Customers	• All customers of all segments (around 5 million vehicles are sold till 2002 and 3,50,000 are sold every year)
Employees	• Around 2,000 employee are IT savvy out of around 5,700 employee
Suppliers/vendors	• Around 500 supplier and all are connected through Extranet
Dealers/ Retailers/ workshops/	• Around 260 dealers (show room) all are connected through WAN. • Around 277 dealers workshops • Around 1,329 Maruti authorized service stations
Technology Provider	• Suzuki Motors Japan with strong communication system
Call center	• GE Gurgaon for customer Information center • i2i enterprise across all 25 cities of India
Financial Service Providers (Bankers)	• ABN AMRO • American Express Bank Ltd • Bank of Tokyo-Mitsubishi Ltd. • Banque Nationale De Paris • Citibank N A. • Citicorp Maruti finance ltd • Corporation Bank • Credit Lyonnais • HDFC Bank • ICICI • Kotak Mahindra • Maruti Countrywide Finance • Punjab National Bank • Sanwa Bank Ltd. • SBI Maruti car loans • Standard Chartered Grindlays Bank Ltd. [Merged] • State Bank of Indore • State bank of Mysore • State Bank Of Travancore • Sundaram Finance • Union Bank Of India
Insurance	• Maruti Insurance Distribution Services (MIDS), MIDS was set up to sell insurance as a corporate agent of **Bajaj Allianz** General Insurance • Maruti Insurance Brokers (MIBL). MIBL was constituted to sell insurance products of the state-owned **National Insurance**.
IT partners	• Compaq • HP • Oracle • Microsoft
Consultants	• PWC (Auditor of the Maruti) • AT Kearney for E-Business solution • MB Athreya for HR (employees performance system)
Shareholders	• Suzuki Japan • Government of India • Recently Issued Public Shares
Media Partners	• Yahoo India • FM Radio • Indiatimes.com • Business magazines, Newspapers • TV

Source: Dwivedi and Momaya (2003)

Chapter XX

Challenges and Opportunities for Information Brokers in Brazil: A Study of Informational Needs of Southern-Brazilian Enterprises When Expanding Their Businesses

Eusebio Scornavacca, Jr., Victoria University of Wellington, New Zealand

Joao Luiz Becker, Federal University of Rio Grande do Sul, Brazil

Stuart J. Barnes, Victoria University of Wellington, New Zealand

Abstract

Information is no longer just a control tool for managers — it is a basic resource as important as raw materials or the human resource. Consequently, it is a significant managerial responsibility to decide which kind of information should support the organization's decision-making process. This is particularly the case when expanding the business. This chapter presents the results of a study examining the information needs of 796 Southern-Brazilian enterprises contemplating business growth. The survey is part of a program of research into information intermediation via the Internet. The results show a marked convergence of the sample's informational needs, but also some significant differences among specific business groups. In addition, the research has surfaced some clear "core" of informational needs, supported by some "peripheral" needs. The chapter concludes with a summary, reflections and some future research directions.

Introduction

The paradigm shift from an industrial-based economy to an information-based economy (Tapscott & Caston, 1993) becomes all the more evident when observing the expansion of the telecommunication infrastructure and role of information systems in today's organizations (Albertin, 1999; Evans & Wurster, 2000). Information is no longer just a management control tool, but a key organizational resource as important as raw materials or the human resource, and a growing number of enterprises are specializing in information management (Barnes & Hunt, 2001; Davidow & Malone, 1993; Mowshowitz, 1997; Freitas, Becker, Kladis & Hoppen, 1997).

With the popularization of the Internet, a tremendous amount of information has become available and easily accessible to the general public (Shapiro & Varian, 1999). However, in this environment, it is becoming clear that quantity of information does not necessarily mean or reflect on quality. The differentiation in quality of information is achieved by how an organization uses the information that it owns (Murdick & Munson, 1988). Having the right information resources imposes an immense managerial responsibility, since it will directly impact on the support of an organization's decision-making process.

Decision-making is a crucial activity for any organization's daily life (Freitas, Becker, Kladis & Hoppen, 1997). Indeed, Simon (1947) points out that the main activities in organizations are, essentially, activities of decision-making and problem solving. In particular, it is observed that one of the most critical moments in the life of an organization is the expansion of its businesses (Porter, 1997; Kotler, 1999). At this juncture it is prudent to get some information to support this important decision-making process. However, which information would managers and entrepreneurs typically demand to support such a decision? This is an important question for both companies specialized in information management, for researchers, and for the IS field.

This study was developed as a part of a program of research into information intermediation via the Internet, focusing on the informational needs of Southern-Brazilian enterprises when expanding their businesses. The research was conducted with the support of SEBRAE-RSi. The goal of this chapter is to present the results of this study of informational needs and identify the challenges and opportunities for information brokers in Brazil. The chapter is organized as follows. The next section provides a brief overview of e-commerce in Brazil. The following provides a summary of the research methodology, and the subsequent section examines the results of the data analysis. Finally, the chapter ends with a summary and conclusions.

Overview of E-Commerce in Brazil

According to Emarketer (2002), the number of Brazilian Internet users is projected to grow from 5.8 million in 1999 to 29 million in 2005. DeGolveia and Kassicieh (2002) point out that in 1997, Brazil was the world's 13th largest e-commerce market, and in 1999 it had already moved to 7th place. E-commerce in Brazil is expected to increase from US$300 million in 2000 to about US$2.6 billion by 2003 (DeGolveia & Kassicieh, 2002). Undoubtedly, Brazil plays a major role in the Latin American e-commerce scene and the Brazilian e-commerce industry has evolved rapidly in the past six years (Albertin, 1999; DeGolveia & Kassicieh, 2002).

London (2000) indicates that Brazilian companies which utilize the Web are still facing several challenges due to the country's social-economic structure. It is important to keep in mind that concentration of Web users in the upper economic classes limits the impact of e-commerce and a large majority of Brazilians do not have access to a phone line or computer. Therefore, the relatively small number of online users, 5 million, limits expansion of new Web sites. One example is the relatively small amount of online advertising, US$72 million for the year 2000, which imposes limitations on the number of online companies that rely on advertising as their main source of revenue (London, 2000).

In a recent survey, Albertin (2003) found that Brazilian companies are progressively using the Internet as a tool to define and support their competitive strategies. Mainly, companies use the Internet to scan the environment and identify threats and opportunities. As a result, it is easy to identify in the Brazilian market a growing number of enterprises that are specializing in information management and brokerage. On the other hand, little is known about the information needs of Brazilian enterprises when expanding their businesses.

Research Methodology

Based on studies on information intermediation via the Internet and the Intelligence Cycle model of Montgomery and Weinberg (1998), the research outline was developed (Figure 1). As shown in Figure 1, the research was divided into two phases: the first phase consisted of the development of a data collection system. The second phase is composed of the data analysis in order to qualify the existent informational demand. The informational niche — information needs of Southern-Brazilian enterprises when expanding their businesses — was defined based on perception and needs of partner companies taking part in this research.

The research established two key aims, relating both to the process and outcomes of the research:

1. To understand how informational demands can be captured through a semi-automatic data collection system (The survey system used in the research is fully described in Scornavacca et al., 2003.).

Figure 1. Research outline

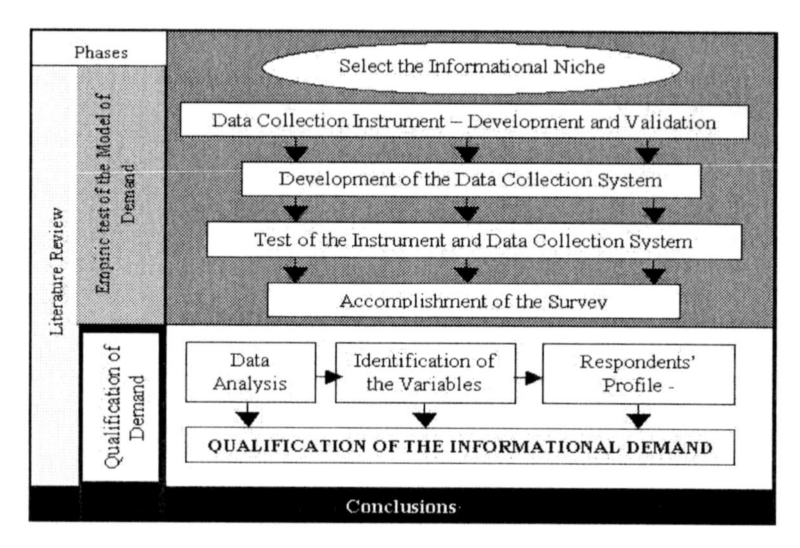

2. To discover the informational needs of the target group.

In order to achieve this goal, it was defined — supported by review of the literature and by the specialists involved in this work — that data collection would be accomplished via a survey. The next section describes this method.

Development and Application of the E-Survey

The use of surveys in business research is commonplace, supported by an abundant literature (Pinsonneault & Kraemer, 1993; Babbie, 1999; Churchill, 2001). Lately, the Internet has become an additional medium for survey work (Churchill, 2001; Bradley, 1999; Simsek, 1999; Taylor, 2000; Epstein, Klikerberg, Wiley & McKinley, 2001; Boyer, Olson, Calantone & Jackson, 2002). Comparisons between results of surveys using traditional and electronic methods of data collection have suggested evidence of equivalence, supporting methodological validity (Epstein, Klikerberg, Wiley & McKinley, 2001; Boyer, Olson, Calantone & Jackson, 2002).

Surveys conducted via the Internet can provide substantial benefits over traditional methods, including: cost reduction; the possibility of working with enormous samples; fast turnaround; and the use of images, sound and hypertext in the construction of questionnaires (Taylor, 2000). Simsek (1999) points out that the Internet also facilitates the verification of message delivery and its respective reading, reduces paper consumption, and minimizes potential errors of interpretation in the respondent's calligraphy.

Boyer et al. (2002) also suggest a reduction of non-answers and larger flexibility of presentation and codification of the questionnaires.

Notwithstanding, there are also disadvantages in surveys through the Internet. One of the most obvious is the definition of the sample. A significant portion of the general population still doesn't possess an electronic address and e-mail lists are often poorly structured (Bradley, 1999; Simsek, 1999). However, as Simsek (1999) points out, in cases where the target population is recognized as being Internet-enabled, e.g., consumers of Internet products or services, these samples tend to be quite worthy.

The e-mail list used as a sampling frame in this research was taken from the database of SEBRAE-RS. This database is very representative of the population, with a high degree of structure for sample selection. The sampling method used in the research was non-probabilistic, based on convenience. The sample was chosen were companies located in the State of Rio Grande do Sul, registered at SEBRAE-RS, and possessing an e-mail address.

Potential respondents were informed about the survey by e-mail, and invited to take part. The e-mail text also received face and content validations. These e-mail messages were generated automatically, and the code used was the company identification code of the SEBRAE-RS database. The use of individual links in the body of the e-mails had the intention of restricting and controlling access to the questionnaire (Bradley, 1999; Simsek, 1999). This ensures that a company answers the survey once only. Also, it is possible to generate follow-up e-mails for those companies that failed to answer or only partially answered the research instrument.

The URL provided to the potential respondents led them to the questionnaire home page, which provided information on the research objectives, partners involved, confidentiality, and about the issue of spontaneity in response. The data collection instrument used in the survey was developed by a group of IT specialists involved with the research program. The questionnaire focused on the information needs of growth enterprises. Specifically, the questionnaire opened with the following statement:

> *Before expanding a business, it is wise to obtain some information to support this important decision. Imagine that your company is thinking about expanding its businesses and that you will be the person entrusted making this strategic decision. What information do you consider important to have in hand to support your decision making process? Please, describe it in the fields below using keywords.*

The decision to request the respondents to use keywords instead of open text to express their needs was motivated by three considerations:

1. The evaluation of the possible use of an automatic information search system guided by keywords.
2. The verification of the capacity for respondents' synthesis.

3. The facilitation of the analysis process, since the number of respondents was close to 1,000.

After keyword information had been collected, the respondent is asked to confirm whether they wish to move to the next section, where some demographic data is collected. For this part of the questionnaire, the system consulted the database and identified the respondent's company data, which helped to automate the survey process. The use of ASP (active server pages) also allowed the capture of information - related to the operating system, manufacturer and version of the browser used by the respondent. At the end of the survey, if the respondent pressed "continue" and the data was incomplete, a message appeared informing the respondent of the problem and associated field, and asked for confirmation that the respondent wished to continue. Subsequently, the respondents were asked if they wished to receive a brief report with the research results. Concluding the process, a message of gratitude was shown.

One of the greatest challenges of this work was the development of the system to support data collection. The core of the e-survey system consisted of the ASPs, Web pages and a database hosted on the web server of EA/UFRGS. The database, in Microsoft Access format, contained the information on the sample (identification code, electronic address, name, and so on) and also recorded the new data. The Web pages that the respondents accessed were developed in HTML and ASP. The scripts, developed in ASP, created an interface between the system administrator and the system (e.g., to send e-mails, control the process, and so on). Also, a support service was put in place, by which the respondents could contact the researchers through e-mail, telephone or fax.

After online data collection had been completed, an additional phase of data collection by telephone was carried out with 30 companies that had not answered the survey (randomly selected). This procedure was used to check for the existence of a non-respondent bias. Such a bias would be verified if there are statistically significant differences among the results obtained in the previous collection (through the Internet) and the new sample (by telephone). The sample was considered representative (Churchill, 2001; Taylor, 2000).

The survey started by sending an e-mail notification to 9,730 companies registered in the SEBRAE-RS database. A window of 21 days was given for completing the questionnaire. Of the 9,730 e-mail addresses, 7,963 companies had valid e-mails. Of this group, 939 (11.80%) answered to the questionnaire. During the organization of the data, 143 questionnaires (15.20% of the 939) were excluded (104 did not present any keywords and 39 didn't present keywords pertinent to the theme). Thus, 766 (9.62% of the valid e-mails) answered the online questionnaire correctly, and 30 (0.38%) answered by telephone during the non-respondents' analysis, totaling 796 valid questionnaires (10.00% of the total of valid e-mails). The response rate obtained is excellent when compared with other studies previously conducted in the same region using the Internet as the communication media. Recent studies carried out in Rio Grande do Sul obtained 5.48% (Vieira, Viana & Echeveste, 1998) and 3.23% (Vieira, 1999) of valid answers. This high level of response can be attributed to the very rigorous control system, which was lacking in the previous investigations. The control system facilitated the sending of follow-ups to the non-respondents and also ensured that no company answered the questionnaire more than once.

Data Analysis

According to Babbie (1999), the core of a survey's data analysis is composed of two factors: description and explanation. In this section, we describe the sample and the measures of variables, before proceeding to analyze the associations among them.

Social Demographic Profile

The socio-demographic profile was accomplished by: (a) measuring the absolute frequencies of the multiple-choice questions; and (b) averaging the responses to numeric questions. The "χ^2" test was used to establish if the observed frequencies of a variable differed, in a significant way, from the expected ones, under a hypothesis of independence (Mendenhall, 1990). A test and analysis of variance (ANOVA) was used to find if there was significant difference in the averages of the studied sample (Mendenhall, 1990). Most respondents were male (79.56%), and the average age was 37.03 years (Min. = 18, Max. = 67 and $\sigma = 9.6$). Interestingly, the males averaged of 38.11 years ($\sigma = 9{,}43$) and the women 32.77 years ($\sigma = 8{,}34$), and the t-test found female respondents significantly younger ($p < 0.05\%$). Turning to the organizational positions of the respondents, 52.80% are entrepreneurs or business owners, 18.96% directors, 18.55% managers, and 9.69% occupy some other position. It is important to remember that, in the data collection process, it was emphasized strongly that an individual responsible for the company must answer the survey. The level of the respondents' education didn't present statistically significant differences, although it is interesting to observe that the total number of people who completed higher education was 60.85%, while the number of people who failed to complete higher education was also quite high at 27.15%.

A categorization of the companies' size was accomplished according to its sector of activity (SEBRAE-RS, 2000). The large and medium companies were grouped, because of their low frequencies, to facilitate analysis. Table 1 presents a comparison between the size and the sector of the participant companies, indicating frequencies. A χ^2 test revealed a significant correlation between size and sector of respondents ($p < 0.05$). The significance of specific cells (*) is also shown at the 5% level (all positive unless otherwise indicated). As can be observed, there are a large number of small firms in the

Table 1. Representation of the sectors and size of the respondent companies

Size	Industry	Trade	Services	TOTAL
Micro business	17.62% (74)*(-)	23.33% (98)	59.05% (248)*	**100% (420)**
Small Business	36.97% (88)*	25.63% (61)	37.39% (89)*(-)	**100% (238)**
Medium/Large	52.00% (39)*	13.33% (10)	34.67% (26)	**100% (75)**
TOTAL	**27.42% (201)**	**23.06% (169)**	**49.52% (363)**	**100% (733)**

Table 2. Average age of the companies by sector and size

Size/Sector	Industry		Trade		Services		General AVER.	
	Average	N	Average	N	Average	N.	Average	N
Small business	15.00	74	12.87* (-)	98*	9.67* (-)	248* (-)	11.35* (-)	420*
Small business	23.66*	88*	21.43*	61*	13.37* (-)	89* (-)	19.24*	238*
Medium / Large	33.31*	39*	32.50*	10*	26.31*	26*	30.77*	75*
Total	22.34*	201*	17.14	169	11.77* (-)	363* (-)	15.91	(733)

service sector. The number of medium and large industries in the sample is also quite expressive. In particular, this supports the assertion by Megginson et al. (1998) that industry, by its own nature, tends to possess an organizational structure of medium or large size.

The average age of companies by sector and size is presented in Table 2. Statistically significant differences were detected through ANOVA. The significant cells are marked (*), where $p < 0.05$. It is interesting to observe that the age of firms in the industry sector is superior to the other sectors. This can be explained by the degree of dynamism found mainly in the service, but also the trade, sectors. Among the three sectors, industry, in a certain way, can be characterized as a more traditional and less dynamic sector (Drucker, 1998).

Keyword Analysis

The keywords given by the respondents were analyzed using lexical analysis (Freitas, 2000). A total of 3,902 keywords were collected, of which 1,037 were unique. Freitas and Jenissek (2000) affirm that any text analysis begins with the complete organization of the used vocabulary. This organization consists of isolating each graphic form delimited by two characters. Most of the time, it is delimitated by the spaces between words, or punctuation marks. Each appearance is called an occurrence. These are, in turn, counted to form the lexicon.

In addition, data organization also included verification of spelling, pertinence, and separation of the words. Keywords composed of two or more words (such as "purchase price") were transformed in a single graphic form ("purchase_price"). The next steps were to group synonymous keywords and create keyword categories. During this process, dictionaries were created to control which words composed each group. To ensure domain relevance, significance and validity, an IT consultant with more than 30 years of market experience validated these. The researchers and industry professionals debated each disagreement to arrive at a common perception, trying to reduce the subjectivity of the analysis. The result of this time-consuming process was the creation of 92 categories corresponding to 3,726 citations (176 citations were considered inappropriate and therefore excluded). The average number of keywords for each respondent was 4.90, with a standard deviation of 2.97 (Min. = 1 and Max. = 25).

The categories were ranked according to the total number of citations in each group. Figure 2 provides a curve representing the cumulative percentage of citations according

Figure 2. Percentile accumulated in each category

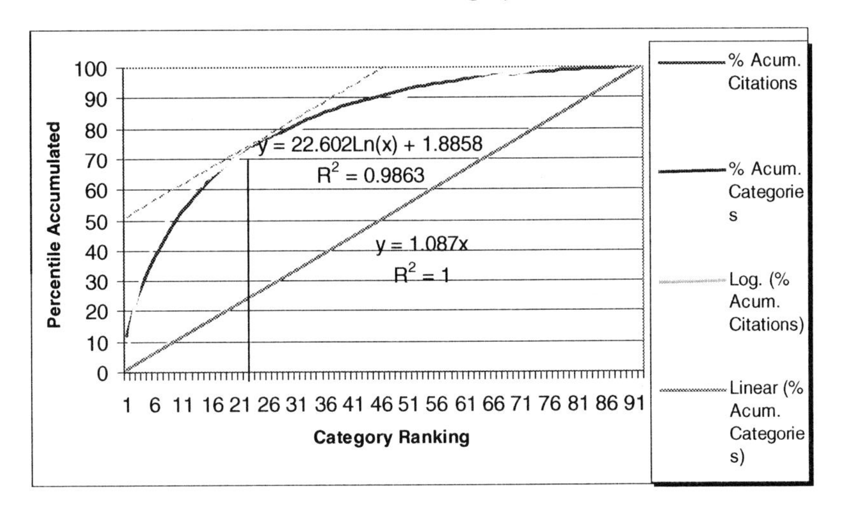

to their ranked categories. To observe an interesting contrast, Figure 2 also has a straight line representing a linear function. If the degree of importance were distributed evenly, this linear function would indicate the cumulative percentage of citations. Notice that the first ten categories are responsible for 50.21% of the analyzed citations, demonstrating a clear convergence in the informational needs declared by the respondents.

To accomplish the analysis, it was necessary to establish a division criterion for the categories. First, a tendency line (Log. (% Acum. Citations)) of the curve obtained in the graph (% Acum. Citations) was generated. After that, a line parallel to the linear function (% Acum. Categories) was drawn, and the tangency point with the curve was calculated (ii). This operation facilitated the selection of a group of categories for which the percentile of representativeness of each category is bigger than the average. The result was the formation of two groups. The first, labeled "Primary Needs," contains the 20 most important categories according to the respondents (21.74% of all categories) and represents 2,598 citations (69.73%). The second group, called "Secondary Needs," consisted of 72 subsequent categories (78.26%), representing only 1,128 citations (30.27%).

In the analysis of the informational needs expressed by the categories, "Primary Needs" represents the information that the majority of respondents consider fundamental to the decision-making process for business growth. "Secondary Needs" represents less relevant information needs, being characterized as the respondents' individualized requirements. Thus, it is necessary to give a special attention to the categories contained in the first group of informational needs. Table 3 presents the "Primary Needs" categories in rank order (according to frequency and percentage upon the total of citations). It is noticeable that the informational needs with respect to market information ("Market"), represents quite a significant portion of the obtained citations (12.31%). "Competitors" (6.30%) and "Payback" (6.17%) also stand out among the others.

Table 3. "Primary Needs" (ranked by number of citations on each category)

Rank	Categories	Freq.	%	Rank	Categories	Freq.	%
1	Market	459	12.31	11	Market Research	104	2.79
2	Competitors	235	6.30	12	Capital	94	2.52
3	Payback	230	6.17	13	Demand	82	2.20
4	Human Resources	186	4.99	14	Logistics	77	2.07
5	Costs	151	4.05	15	Personal Attributes	74	1.99
6	Products	144	3.86	16	Product Attributes	70	1.88
7	Clients	123	3.30	17	Technology	62	1.66
8	Investments	120	3.22	18	Place	60	1.61
9	Price	116	3.11	19	Infra-structure	53	1.42
10	Planning	107	2.87	20	Knowledge	51	1.37
					TOTAL	**2598**	**69.73**

Factor Analysis

Factor analysis was used to discover the existent interdependence among the categories belonging to the group "Primary Needs." Thus, the goal of this analysis was to generate a better understanding of the fundamental structure of the 20 categories, and to combine them to develop new concepts.

The first step was to organize the data in a spreadsheet. Each row represented a respondent and each column a category, creating a 796x20 matrix. For each category mentioned by a respondent a "1" was entered in the corresponding cell, while "0" was entered for those categories that the respondent failed to mention. Next, the data were transferred to SPSS for principal components' analysis. The result was the creation of eight factors, explaining 51.85% of variance. (The factor analysis, a Varimax rotation with

Table 4. Results of principal component analysis

Component	Categories	Quant.	Proposed nomenclature
1	Products, Product attributes, Price	3	Product
2	Personal attributes, Knowledge	2	Managerial Capability
3	Payback, Investment	2	Income
4	Human resources, Infrastructure	2	Personal and Physical structure
5	Planning, Market research	2	Projection
6	Competitive, Clients	2	Market Environment
7	Demand, Location	2	Place
8	Costs	1	Costs
Load < 0,5	Market, Logistics, Capital, Technology	4	Distributed among the components

Kaiser normalization, converged in 61 iterations.) In the rotated matrix, only items with a load higher than 0.5 were considered (Aaker & Day, 1989). Table 4 presents the eight resulting components of these interactions and the respective categories. It also presents a nomenclature suggestion for the identified components.

The categories belonging to first component are clearly related to products, while component 2 indicates the managerial capacity and business knowledge. For the third component, payback and investment demonstrate the need for information related to income (and profit), which the business expansion process can propitiate to the company. When analyzing the keywords contained in the categories "Human resources" and "Infrastructure," it is noticeable that a concern exists regarding information on the firms' physical and personnel structure (component 4). Component 5 ("Projection") identifies the need of information to forecast and plan the business expansion process. Components 6 and 7 - quite self-explanatory - identify the demand for information on clients, competitors, demand and location. Although at first glance one might surmise that the "Costs" category is related to "Income" or "Price," further analysis of its keywords indicate a separate identity and component. At the same time, these keywords have quite a wide meaning, not being related especially to any another category ("Costs" could easily be related to the component "Human resources," for instance). The four categories loading less than 0.5 didn't contribute significantly to the formation of any of the components.

Categories Mentioned by Functional Area

In order to provide for a better understanding of the informational needs of Southern-Brazilian companies, a new analysis was carried out, grouping the 92 categories (obtained previously) into functional areas of the company. To accomplish this new categorization, it adopted the same parameters and rules used for keyword grouping and validation.

Table 5. Functional department of the companies (Luciano, 2000)

FUNCTIONAL DEPARTMENT	DESCRIPTION
Personnel	Groups the categories related to human resources of the company (qualification, wage, etc).
Sales	Groups the categories related to positioning products or services in the market (market, demand, clients, sales volume, seasonal variation, etc).
Monitoring the Environment	Groups the categories related to the monitoring of the environment that surrounds the organization (environment, government, competitors, economic situation, etc.)
Strategy	Groups the categories related to the strategic department of the company (partnerships, image, risk, analysis, data, etc).
Finance	Groups the categories related to the financial department of the organization (costs, payback, investments, etc).
Production	Groups the categories related to the production of a good or the delivery of services (production, materials, technology, infrastructure, techniques, etc).
Product	Groups the categories related to the product or service (trust, product attributes, quality, etc).

Table 6. Functional departments of the company (number of citations)

Rkn.	Department	Freq.	%
1	Sales	1040	27.90
2	Financial	895	24.00
3	Monitoring the Environment	511	13.7
4	Strategy	426	11.4
5	Personnel	289	7.8
6	Production	289	7.8
7	Product	276	7.4
	TOTAL	3726	100

Many possibilities for functional grouping exist. This work will be based on an adaptation of the division of functional departments derived by Luciano (2000). Table 5 presents the functional areas and a short description of what is understood by each of them.

Based on this categorization, Table 6 presents the different functional areas and the number of citations corresponding to each area. Clearly, significant importance is attached to both Sales and Finance. It is interesting to observe that the companies demonstrated a great need of information on the placement of its products and services in the market, as well as the financial factors (investment, payback, etc.) involved in the process of business expansion. Next is presented the section concerning the data crossing.

Cross-Tabulation of Components with Socio-Demographic Data

We were interested to establish if the eight factors were related to socio-demographic features. Subsequently, an ANOVA was done to verify dependence among the variables.

Table 7. Relation among components, sector and time of existence

Ord.	Factors	Sector	Size	Exist. Time
1	Product	S	S	NS
2	Managerial Capabilities	NS	S	NS
3	Income	S	S	NS
4	Personnel and Physical structure	S	S	NS
5	Projection	S	NS	NS
6	Market Environment	S	NS	NS
7	Place	S	S	NS
8	Costs	NS	NS	NS

Table 8. Relationship between the sector and the components

Component/Sector	Industry		Trade		Services		General AVER	
	Aver.	N	Aver.	N	Aver.	N	Aver.	N
Product	0.26	201	0.21	169	0.13	363	0.15	2388
Income	0.14	201	0.01	169	0.13	363	0.13	2388
Pers. And Phys. Structure	0.16	201	0.19	169	0.08	363	0.19	2388
Projection	0.22	201	0.13	169	0.24	363	0.19	2388
Market Environment	0.14	201	0.14	169	0.22	363	0.16	2388
Place	0.10	201	0.32	169	0.10	363	0.13	2388

Table 9. Relationship between firm size and specific factors

Component/Sector	Micro		Small		Medium/Large		General AVER	
	Aver.	N	Aver.	N	Aver.	N	Aver	N
Product	0.18	420	0.17	238	0.24	75	0.15	2388
Income	0.19	420	0.11	238	-0.08	75	0.13	2388
Pers. And Phys. Structure	0.11	420	0.06	238	0.27	75	0.13	2388
Projection	0.07	420	0.18	238	0.28	75	0.19	2388
Market Environment	0.20	420	0.07	238	0.16	75	0.16	2388
Place	0.18	420	0.17	238	0.24	75	0.15	2388

Table 7 presents the result of this analysis. "NS" was attributed to the cross-tabulations in which the dependence was not significant and "S" to the crossings with a significant dependence ($p < 0.05$).

Of these eight components (containing 16 of the 20 categories for "Primary Needs"), only "Costs" did not demonstrate a statistically significant dependence in relation to one or more of the socio-demographic variables. Table 8 presents the results of an ANOVA for the components that presented statistically significant differences ($p < 0.05$) in relation to the company's sector. Note that the service sector had a particularly low demand for information related to physical and personnel areas. Perhaps the large number of small businesses in this sector could influence this fact. Meanwhile, the trade sector demonstrated a very low need for information related to "Income," but a high demand for information concerning "Place."

Table 9 presents the results of the ANOVA for the components that presented a statistically significant difference ($p < 0.05$) in relation to the size of the company. As expected, information on physical structure and personnel was not considered important by the micro businesses. Similarly, "Managerial Capability" — formed by the categories "Knowledge" and "Personal Attributes" — rated low for medium and large companies. This, allied to the previous analysis, demonstrates that the medium and large companies did not place an onus on information about the individual.

Conclusion

This research was begun to examine the challenge of defining the informational need of companies, specifically those in Southern-Brazil, when expanding their business. A key part of the research involved the development of an Internet-based, data-collection system as a support tool for a survey (Scornavacca, Becker & Barnes, 2003). The rigorous and systematic approach adopted in the test and validation stages allowed the researchers to identify some key problems that, when corrected, made the system less susceptible to failure during the data collection procedure.

One of the principal strengths of the developed system is the rigorous control of respondents. From the literature, this is an issue that presents one of the most challenging problems in the development of Internet surveys. Control tools helped increase response rates. On the other hand, it created several issues surrounding anonymity and confidentiality — which could deter survey participation. This provides a compelling dilemma, where the researcher is forced to choose between survey control and protecting the anonymity of respondents.

The system was both cost-efficient and allowed access to a very large and quite representative sample, providing simplicity and functionality. The performance of the data collection system, in comparison to similar studies, can be considered excellent. The percentage of respondents was at least twice that obtained in previous studies accomplished with a similar sample (which, typically, did not use a rigorous control procedure for respondents). Another interesting finding from the research is that the survey completed via the Internet reached higher hierarchical levels in the organizations than the telephone survey (i.e., the non-respondents' analysis).

The analysis described in this chapter focused on the identification and qualification of the informational demands of companies looking to expand. It is interesting to observe the convergence of the obtained answers. Some factors, such as information on the market, competition, payback and costs, are latent needs of information for most organizations - independent of their nature. However, other needs vary according to the industry, firm size and the company's age. An examination of responses categorized according to functional departments demonstrated that information concerning sales, financial and environmental aspects had the highest demand

The support received from our research partner, SEBRAE-RS, was central to reaching our targeted sample. On the other hand, the focus on a specific sample could also be considered a limitation of this research. The use of companies registered in SEBRAE-RS might have produced some bias in the identification and qualification of the informational demand. Therefore, future research based on additional sample frames would be useful in validating the results of the research. Further use of the data collection system in other situations would also help to consolidate validity.

The research discussed here has focused on some specific forms of analysis. Lexical analysis, although accepted and used in other disciplines, is not widely used in information systems research. We hope that the successful use of the techniques in this research spurs other researchers to carefully examine these tools as a credible alternative to traditional methods of analysis. The dictionaries created as part of this research are a very rich source for further investigation. On a practical level, the informational needs described in this work could be used as a guideline for the construction of a database that can contribute to the decision-making processes of Southern-Brazilian entrepreneurs contemplating business growth. The results can also help Brazilian information brokers to identify challenges and opportunities existent in that market.

References

Aaker, D.A. & Day, G.S. (1989). *Investigacion de mercados.* Mexico City: McGraw-Hill.

Albertin, A.L. (1999). *Comércio eletronico: Modelo, aspectos e contribuicoes de sua aplicacao.* Sao Paulo: Atlas.

Albertin, A.L. (2003). *Pesquisa FGV comércio eletronico no mercado brsileiro.* Sao Paulo: FGV EAESP.

Babbie, E. (1999). *Metodos de Pesquisa de survey.* Editora UFMG, Belo Horizonte.

Barnes, S.J. & Hunt, B. (2001). *E-commerce and v-Business.* Oxford: Butterworth-Heinemann.

Boyer, K.K., Olson, J.R., Calantone, R.J., & Jackson, C. (2002). Print versus electronic surveys: A comparison of two data collection methodologies. *Journal of Operations Management, 20,* 357-373.

Bradley, N. (1999). Sampling for Internet surveys: An examination of respondent selection for Internet research. *Journal of the Market Research Society, 41,* 387-395.

Churchill, G.A. (2001). *Basic marketing research.* Fort Worth,TX: Dryden Press.

Davidow, W. & Malone, M.S. (1993). *A corporacao virtual.* Sao Paulo: Pioneira.

DeGolvea, R. & Kassicieh, S.K. (2002). Brazil.com. *Thunderbird International Business Review, 44*(1), 104-117.

Drucker, P.F. (1998). *Introducao a administracao.* Sao Paulo: Editora Pioneira.

Emarketer. (2002). Latin America online: Demographics, usage & e-commerce. Retrieved June 20, 2003, from *http://www.emarketer.com/Report.aspx?latin_am*

Epstein, J., Klikerberg, W.D., Wiley, D. & McKinley, L. (2001). Insuring sample equivalence across Internet and paper-and-pencil assessments. *Computers in Human Behavior, 17,* 339-346.

Evans, P.B. & Wurster, T.S. (2000). *Blown to bits.* Boston: Harvard Business School Press.

Freitas, H. (2000) As tendencias em sistemas de informacao com base em recentes congressos. Retrieved June 15, 2003, from *http://read.adm.ufrgs.br/read13/artigo/artigo2.htm*

Freitas, H., Becker, J.L., Kladis, C.M., & Hoppen, N. (1997). *Informacaoo e decisao: Sistemas de apoio e seu impacto.* Ortiz, Porto Alegre.

Freitas, H. & Janissek, R. (2000). *Analise lexica e analise de conteudo.* Sagra Luzzatto, Porto Alegre.

Kotler, P. (1999). *Marketing para o seculo XXI.* Sao Paulo: Futura.

London, J. (2000). Faces do Brazil na web. *Exame, 34*(18), 122-128.

Luciano, E.M. (2000). *Mapeamento das variaveis essenciais ao processo decisorio nas empresas gauchas do setor industrial alimentar.* Master's thesis, PPGA-UFRGS, Porto Alegre.

Megginson, L.C., Osley, D.C. & Pietri, P. H. (1998). *Administracao – Conceitos e aplicacoes.* Sao Paulo: Editora Harbra.

Mendenhall, W. (1990). *Estadística para Administradores.* Mexico City: Iberoamerica.

Montgomery, D.B. & Weinberg, C. (1998, Winter). Toward strategic intelligence systems. *Marketing Management,* 172-180.

Mowshowitz, A. (1997). Virtual organization. *Communications of the ACM, 40*, 30-37.

Murdick, R.G. & Munson, J.C. (1988). *Sistemas de informacion administrativa.* Mexico City: Prentice-Hall Hispano Americana.

Pinsonneault, A. & Kraemer, K.L. (1993). Survey research methodology in management information systems: An assessment. *Journal of Management Information Systems, 10*, 75-105.

Porter, M. (1997). *Vantagem competitiva.* Rio de Janeiro: Campus.

Scornavacca, E., Becker, J. & Barnes, S.J. (2003). Experiences in e-survey development for IS research: Lessons from the use of automated control tools. *7th Pacific Asia Conference on Information Systems,* Adelaide, South Australia, June.

SEBRAE-RS. (2000). *Cadastro empresarial RS.* SEBRAE-RS, Porto Alegre.

Shapiro, C. & Varian, H.R. (1999). *Information rules: A strategic guide to the network economy.* Boston: Harvard Business School Press.

Simon, H.A. (1947). *Administrative behavior: A study of decision-making processes in administrative organization.* New York: MacMillan.

Simsek, Z. (1999). Sample surveys via electronic mail: A comprehensive perspective. *Revista de Administracao de Empresas, 39*, 77-83.

Tapscott, D. & Caston, A. (1993). *Paradigm shift.* New York: McGraw-Hill.

Taylor, H. (2000). Does Internet research work? *International Journal of Market Research, 41*, 51-63.

Vieira, B.L.A. (1999). *A Internet como canal de compras: Um estudo junto aos usuarios do provedor VIA-RS.* Master's thesis, PPGA-UFRGS, Porto Alegre.

Vieira, B.L.A., Viana, D.A. & Echeveste, S. (1998). Comercio eletronico via Internet: Uma abordagem exploratoria, *XXII Encontro da Associação Nacional de Programas de Pós- Graduação em Administração,* Foz do Iguaçú, September.

Endnotes

- SEBRAE-RS is the Brazilian Micro and Small Business Support Service. This is a non-profit, independent institution, supporting the development of small-sized business activity. The institution is funded by both the public and private sectors, and the country's main fostering and research entities.
- The equation resulting from the tendency line is $y = 22.602 \, Ln(x) + 1.8858$. This explains a variance equivalent $R2 = 0.9863$. Its derivative is defined by $y' = 22.602/x$. The equation of the straight line is $y = 1.087x$, $R2 = 1$, with derivative $y' = 1.087$. In order to find the tangent point, the equation $22.602/x = 1.087$ was solved, resulting in the value $x = 20.79$.

About the Authors

Sherif Kamel is an assistant professor of MIS and associate director of the Management Center at the American University in Cairo. Previously, he was director of the Regional IT Institute (1992-2001) and training manager of the Cabinet of Egypt Information and Decision Support Center (1987-1991). In 1996, he was a co-founding member of the Internet Society of Egypt.

* * *

Ala M. Abu-Samaha, BSc, MSc, PhD, has developed research interests in two major areas of the information systems discipline: information systems development methodologies and information systems/technology evaluation. Abu-Samaha has many publications in both of these areas, mainly in evaluating technical intervention in health provision, and holds a PhD in information systems from the Information Systems Research Centre (ISRC) at the Information Systems Institute, University of Salford, UK. He earned his Masters in the same area of interest from the Department of Mathematics and Computer Science, University of Salford. His master's thesis was dedicated to develop an evaluation approach to assess the success/failure of information systems/technology from a systemic multi-perspective oriented view. This framework was applied in the area of health informatics for the exchange of laboratory test results between general practitioners and hospital trusts in England. His PhD thesis was dedicated to address the issue of emerging telecommunication-based systems in terms of both development and evaluation.

Abdullah Akbar was born in 1956 and awarded his PhD in 2003 from the University of Leeds, School of Computing. He is currently a senior consultant in IS/IT at the Public Institution for Social Security, Kuwait. Presently, he is giving courses in PAAET at Kuwait University, State of Kuwait. He has published a number of research papers in prestigious International IT conferences IEEE and has also published a number of research articles in many international journals. He has also contributed to many committees regarding electronic government, strategic planning for the Ministry of Health and other businesses and financial projects.

Donald L. Amoroso is a professor and department chair of information systems at Appalachian State University (USA). Prior to his appointment at ASU, he served as an associate professor and coordinator of information and decision sciences at San Diego State University. Dr. Amoroso teaches classes in information systems, Web development, and strategy formulation with executive MBA students. He manages the hiring and evaluation of part-time faculty members, curriculum development and other academic programs. Prior to his appointment at SDSU, Dr. Amoroso worked with GE Capital as a director of enterprise solutions and with Solista/GartnerGroup as a consulting partner. He has authored 53 articles and proceedings, written five books, presented at more than 45 professional conferences and venues, and managed the Information Systems Department while at the University of Colorado, Colorado Springs. He has published in journals such as *Journal of Management Information Systems, Data Base* and *Information & Management.* Dr. Amoroso has been director of the Pacific Research Institute for Information Management and Systems (PRIISM) with the goal to disseminate information technology findings in Asian-based organizations. He has been mini-track chair for the Hawaii International Conference on Systems Sciences from 1992 through present and is track chair for the current AMCIS Conference on Social Issues in Information Systems. He was inducted into the Who's Who Worldwide in 1995 for contributions to international organizations changing the distribution of technology information within the Pacific Rim. Donald received his bachelor's degree in accounting and finance from Old Dominion University (1980) and his MBA and PhD from the University of Georgia (1984 and 1986, respectively).

Nahed Azab is an IT consultant and lecturer. She plays an active role in planning and updating the curriculum, compiling and editing the material and course work, and teaching a number of IT undergraduate and postgraduate courses at the American University in Cairo, Regional Information Technology Institute, and the Arab Academy for Sciences and Technology. In particular, Mrs. Azab focuses on e-commerce, e-marketing and MIS. She is also pursuing a PhD at the School of Computing Sciences - Middlesex University, London. Mrs. Azab obtained her MSc in business information technology from the School of Computing Sciences - Middlesex University, London (July 2002). She graduated with a BSc from the Faculty of Engineering - Ein shams University, Cairo (1984). Her 20-year career path encompassed software programming, analysis and design, computer center management, software instruction and general IT consultancy work. Mrs. Azab presented academic papers in the Conference of Information Science, Technology and Management (CISTM2004), Alexandria, Egypt, and the Information

Resources Management Association International Conference (IRMA 2003), Philadelphia, Pennsylvania, USA.

Stuart J. Barnes is an associate professor of electronic commerce at the School of Information Management, Victoria University of Wellington, New Zealand. He has been teaching and researching in the information systems field for over a decade. His academic background includes a first class degree in Economics from University College London and a PhD in business administration from Manchester Business School. His current research interests include evaluating Web site and e-commerce quality, and business applications of wireless information technologies. He has published three books and more than 50 articles in leading conferences, academic journals, professional outlets, and edited books.

Joao Luiz Becker is a professor of decision and information systems at the Federal University of Rio Grande do Sul Business School, Brazil. His academic background includes a Master of Science in Applied Mathematics (operations research) from the Institute for Pure and Applied Mathematics, Brazil, and a PhD in management science from the University of California at Los Angeles. Dr. Becker has published and presented more than a hundred articles in conferences and academic journals. His current broad research interest deals with information and decision systems, including the impact of IT in organizations, DSS, data mining, and quantitative modelling.

Allan Marcelo De Campos Costa is an IT manager in the private sector, professor at the High Institute of Economy and Business Administration of Getulio Vargas Foundation, Brazil, and professor at the Parana Catholic University. He has published scientific papers in international conferences and articles in Brazilian press. He holds a BSc in computer science from the Ponta Grossa State University - Brazil, an MBA from the Brazilian School of Public and Business Administration of Getulio Vargas Foundation - Brazil, and an MSc in information technology, management and organizational change from the Lancaster University, UK.

Dante Di Gregorio is an assistant professor in international management at the Robert O. Anderson School and Graduate School of Management at the University of New Mexico (USA). He earned a PhD in strategy at the Robert H. Smith School of Business, University of Maryland at College Park. Prior to working in academia, he served as an independent consultant and jointly founded an international business consulting firm. Dante's research spans the broad areas of strategy, international management, and entrepreneurship, and he has published in journals including *Management Science*, *Research Policy*, and *International Business Review*. In addition to the present research, some of the topics he has addressed in published and ongoing research projects include the commercialization of university technologies, the competitive dynamics of entrepreneurial action, intangible resources residing in top management teams, country risk, and the internationalization of emerging market firms.

Rajeev Dwivedi is the research scholar with the Department of Management Studies, Indian Institute of Technology Delhi, India. He is working on a PhD in the field of e-business strategy. He holds an MTech (Masters of Technology) in future studies and planning and an MS (Masters of Science) in applied physics-LASER-fiber optics. His research interest includes e-business, IT management, systems dynamics, technology management and econometrics.

Carlos Ferran is an assistant professor of management information systems at Penn State University Great Valley (USA). He received his DBA in MIS from Boston University, a Graduate Degree in MIS from Universidad Central de Venezuela, a Cum Laude Master in finance, and a Licentiate in management sciences (BS) from Universidad Metropolitana. Dr. Ferran has been a visiting professor at IESA (Venezuela), INALDE (Colombia), and IAE (Argentina). He worked in the software industry for ten years, acted as an IT/IS consultant for over ten years, and held the position of CIO for an important financial group in Venezuela.

Bardo Fraunholz is a senior lecturer in project management, enterprise modelling, and business communication systems. He has a master's in business specializing in information systems/accounting and a post graduation in legal studies specializing in IT, media and corporate law. Fraunholz spent several years in the information communication technologies sector as co-editor/board member of a publisher specializing in IT and telecommunication magazines. Currently, he is actively involved in managing/consulting with a number of projects dealing with trade and information systems. His main research interests are information systems projects, IT and law and mobile applications.

Raul Gouvea is Wells Fargo associate professor of international business at the Anderson Schools of Management, University of New Mexico (USA). He holds a PhD and an MS from the University of Illinois and a BS from Universidade Federal do Rio de Janeiro, Brazil. Gouvea has been a consultant to the World Bank. His research is in the area of international trade, international business, regional economics, technology policy and applications, and environmental issues. He has published in a variety of journals including *Journal of Latin American Studies, Revista Brasileira, Latin American Business Review, Thunderbird International Business Review, Engineering Management Journal* and, *IEEE Transactions on Engineering Management.*

Marwa M. Hafez earned her doctoral degree from the Maastricht School of Management (MsM), The Netherlands (2004). In 1997 and 1998 respectively, she received her Bachelor of Business Administration (BA) and Master of Business Administration (MBA) degrees from The American University in Cairo (AUC). Ms. Hafez is a freelance management consultant. She has conducted several training workshops and taught numerous management courses at the graduate, undergraduate, and professional levels in a multitude of academic institutions including the AUC, as well as the Cairo branches of foreign institutions such as New York Institute of Technology (NYIT) and City University (CU).

Murray E. Jennex is an assistant professor at San Diego State University (USA) and president of the Foundation for Knowledge Management (LLC). Dr. Jennex specializes in knowledge management, system analysis and design, IS security, and organizational effectiveness. He is also the editor-in-chief of the *International Journal of Knowledge Management* and co-chair of the knowledge management systems track at the Hawaii International Conference on System Sciences (HICSS). He has managed projects in applied engineering and business and information systems development and implementation. His industrial and consulting experience includes nuclear generation, electrical utilities, communications, health services, and governmental agencies. Dr. Jennex is the author of numerous publications on knowledge management, end user computing, international information systems, organizational memory systems, and software outsourcing. He holds a BA in chemistry and physics from William Jewell College, an MBA and MS in software engineering from National University, and an MS in telecommunications management and PhD in information systems from the Claremont Graduate University. Dr. Jennex is also a certified information systems security professional (CISSP) and a California registered professional mechanical engineer (PE).

Luiz Antonio Joia is an associate professor and MBA head at the Brazilian School of Public and Business Administration of Getulio Vargas Foundation and adjunct professor at Rio de Janeiro State University (Brazil). He has published two books, several chapters, and more than 50 scientific papers in international journals and conferences. He holds a BSc in civil engineering from the Militar Institute of Engineering, Brazil, and an MSc in civil engineering and DSc in engineering management from the Federal University of Rio de Janeiro, Brazil. He also holds an MSc in management studies from Oxford University, UK.

Magdi N. Kamel is an associate professor of information systems at the Naval Postgraduate School in Monterey, California (USA). He received his PhD in information systems from the Wharton School, University of Pennsylvania. His main research interest is in the analysis, design and implementation of computer-based information systems. Specifically, he is interested in data and knowledge modeling, e-commerce, enterprise resource planning, supply chain management systems, and knowledge discovery in databases and on the Web. He has lectured and consulted in these areas for many DoD and government organizations and is the author of numerous published research papers on these topics. Dr. Kamel is a recent recipient of a Fulbright grant for teaching and research in the computer and information systems area.

Purva Kansal is a lecturer at the University Business School, Panjab University, Chandigarh, India. She earned an MBA from the International Institute of Management, Himachal Pradesh University, Shimla (1998) and has a PhD in international marketing from Panjab University Chandigarh (2002). Her recent co-authored books are *Basics of Distribution Management* (Prentice Hall of India) and *Marketing Logistics* (Pearson Publishers). Her professional experience includes employment with the University Business School, National Institute of Pharmaceutical Education and Research (NIPER), helping high profile clients with process improvement and strategy formulation. Her areas of interest include supply chain management, strategic management and marketing. She has published her work extensively in refereed international and national edited books and journals.

Suleiman K. Kassicieh, Regents' Professor of Management of Technology and Black Trust Professor of Economic Development at the Anderson School of Management at the University of New Mexico (USA), is a co-founder of the Management of Technology program at the Anderson Schools. He has consulted with a number of organizations such as Los Alamos, Sandia National Laboratories and Ardesta among many others in a number of areas such as business development and strategic planning. He also works on business development with a large number of startups. Dr. Kassicieh teaches in the areas of economic development, technology strategy and equity/venture capital. Dr. Kassicieh has more than 100 published technical and management publications in such journals as *Operations Research, IEEE Transactions on Engineering Management, Journal of Technology and Engineering Management, Entrepreneurship: Theory and Practice, Operations Research and California Management Review.* He lectures nationally and internationally on the issues of economic development and technology business development. His book, *From Lab to Market: Commercialization of Public Sector Technologies,* was published in 1994 by Plenum. He is a member of IEEE, INFORMS and ACM.

Amit Kumar Kaushik has a doctorate in the area of human resource management and is currently a lecturer at the University Business School, Panjab University, Chandigarh, India. He has worked in the Indian Corporate sector and has worked with some of the leading Indian companies. He has rich experience in the area of human resource management. Besides human resources, he has a keen interest in organizational behavior and organizational development.

Abdulwahed Mohammed Khalfan is an assistant professor of MIS at the College of Business Studies, PAAET, Kuwait. He received his PhD from the University of Leeds, UK. He has published a number of research articles in many international journals and has also published a number of research papers in prestigious International IT conferences. He has been a referee in many journals and IT conferences. He is an editorial board member of *International Journal of Financial Services Management*, and member of the media committee for the electronic government project in Kuwait.

Ronald M. Lee is currently eminent scholar at the School of Business at Florida International University, Miami, Florida, USA. For the previous 10 years, he was a professor and director of the Erasmus University Research Institute for Decision Information System (EURIDIS) in Rotterdam, The Netherlands. He is a former research fellow at the Wharton School, University of Pennsylvania. Previously, he was associate professor of information systems at the University of Texas at Austin, and earlier served as a research scholar at the International Institute for Applied Systems Analysis in Vienna, Austria, and as visiting professor of management at the Universidade Nova de

Lisboa, Lisbon, Portugal. He has a PhD in decision sciences (Wharton, 1980). Lee's current research focuses on formal modeling for e-business and international trade, as well as local capacity development for emerging economies in areas of e-business, e-government, e-tourism and e-culture.

Alemayehu Molla is a lecturer in information Systems at IDPM, The University of Manchester, UK. He received his PhD in information systems from the University of Cape Town, MSc in information science, BA in business management and a Diploma in Computer Science form the Addis Ababa University, Ethiopia. His research interests include e-commerce in developing countries, e-trading, IT adoption and implementations, diffusion, use and impact of the Internet in Africa and the Middle East. His research has been published in the *Electronic Commerce Research Journal, Information & Management, Journal of Internet Banking and Commerce, Information Technologies and International Development and Journal of IT for Development.*

K. Momaya worked in the engineering consulting environment after completing his MTech at IIT Delhi, India. He did his doctoral research on international competitiveness at the University of Toronto. He has experience of conducting interdisciplinary research across continents, including research projects in Japan. At IIT Delhi, he contributes to courses in the areas of competitiveness, strategic management and technology management. Apart from teaching, he is actively involved in research projects.

Mateja Podlogar is a research and teaching assistant of business information systems, business process reengineering, e-commerce, and MIS and head of the eProcurement Laboratory, eCommerce Center, Faculty of Organizational Sciences, University of Maribor, Slovenia (http://eCom.FOV.Uni-Mb.si/Podlogar). In November 2002, she completed her PhD. The doctoral dissertation is titled, "A Model of Electronic Commerce Critical Success Factors in Procurement Process." Her current research includes: electronic commerce with emphasis on procurement and supply chain management, ERP systems, process reengineering and audit of information systems. She is involved in a Slovenia's eCommerce Project (1998-2004) involving 20 business and government organizations and she is also a member of the technical committee of the annual Bled eCommerce Conference. In 2001 she was a conference chair assistant of "The 9th European Conference on Information Systems – ECIS 2001." She has been an ISACA academic advocate since September 2004.

Andreja Pucihar is a research and teaching assistant of business information systems, business process reengineering, e-commerce, and MIS courses at the Faculty of Organizational Sciences, University of Maribor (http://eCom.FOV.Uni-Mb.si/Pucihar). She completed her PhD at the University of Maribor in September 2002 with the dissertation titled: "Entering e-Markets Success Factors." She is head of eMarkets Laboratory. Since 1995, she has been involved in the eCommerce Center and its several research and e-commerce activities, as for example the annual international Bled eConference, where she has been serving as a conference chair assistant. She participates in the Slovenia's eCommerce Project involving 20 business and government organizations. She also serves as a national contact point of the eMarketServices - European Commission sponsored project. Her current research includes: e-marketplaces, Business-to-business e-commerce, supply chain management and business process reengineering.

Nagla Rizk is an associate professor and chair of the Economics Department at the American University in Cairo. Dr. Rizk teaches and conducts research on the economics of information and communication technologies, productivity analysis in the digital economy, the development potential offered by electronic commerce, and the challenges and opportunities facing developing countries in the information age. She has published papers and given conference presentations on the above topics. For more information, see http://www.aucegypt.edu/academic/econ/Rizk.html.

Ricardo Salim is the chief software architect for Cautus Network Corporation (USA). He has more than 25 years of experience in developing and implementing enterprise software solutions for small and midsize companies. He has worked as a consultant in IT/IS to over a hundred companies and government institution in developing countries. Mr. Salim is a successful entrepreneur that has founded several successful IT/IS companies in various developing and developed countries. He holds a BS in computer science from Universidad Central de Venezuela and is currently a PhD candidate from Universidad Autónoma de Barcelona in Spain.

Eusebio Scornavacca, Jr. is a lecturer at the School of Information Management, Victoria University of Wellington, New Zealand. Before moving to Wellington, Scornavacca spent two years as a researcher at Yokohama National University, Japan. His academic background includes a bachelor's degree in business administration and a Master of Science in management of information systems from The Federal University of Rio Grande do Sul, Brazil. Eusebio has published and presented more than 30 papers in conferences and academic journals. His current research interests include mobile business, electronic commerce and e-surveys.

Aladdin Sleem is the vice president for Technical Services and Software Integration Development at Trendium, Inc. (USA), a leading software vendor in the area of network performance monitoring and service quality management. He received his BSc in systems and biomedical engineering from Cairo University, Egypt (1985). In 1996, he received his MBA in information technology management from Maastricht School of Management, The Netherlands. He also received a second Masters of Science in computer science from the University of Louisville (2000). In 2003, he earned his PhD from the University of Louisville, USA, in computer science and engineering. In addition to his academic experiences, he has a long history in software development and is a well-published author and a public speaker in the areas of electronic commerce, service quality management, network performance monitoring, operations support systems, and wireless data networks. He is a member of the IEEE Computer Society and Communications Society. His research interests include wireless data networks, distributed systems, software engineering, and computer networks.

Ramanathan Somasundaram is employed as a manager, projects at the National Institute for Smart Government, Denmark. He holds a PhD in information systems from the Department of Computer Science, Aalborg University, Denmark. He studied the diffusion of electronic public procurement in Denmark for his PhD work. He has held visiting positions at the Erasmus University Research Institute for Decision Information System (EURIDIS), Florida International University and Copenhagen Business School during his brief research career.

Kala Seetharam Sridhar is currently a fellow, National Institute of Public Finance and Policy in New Delhi, India. She earned a PhD (1998) in public policy & management from the Ohio State University, USA. Her MS (urban & regional planning) is from the University of Iowa, USA. Her research interests are in urban/regional development policies, infrastructure, local finances and developmental issues. Palgrave/Macmillan is publishing her book *Incentives for Regional Development*. She has published in journals such as *Applied Economics, Journal of Regional Analysis and Policy, Review of Regional Studies, Urban Affairs Review, Economic Development Quarterly* and *Economic & Political Weekly*.

Varadharajan Sridhar is a professor in information management at the Management Development Institute, Gurgaon, India. He received his PhD in management information systems from the University of Iowa. His current research interests include telecommunications management and policy, global electronic commerce, and global virtual teams. He has published telecommunications related articles in a number of journals including *Annals of Cases on Information Technology, European Journal of Operational Research, Journal of Heuristics, Telecommunication Systems,* and *The Journal of Regional Analysis and Policy*. He is Associate Editor of *International Journal of Business Data Communications and Networking*.

Sushil is an MTech in industrial engineering from IIT Delhi and subsequently did his PhD from the same in 1984, the title of the thesis being "Systems Modelling of Waste Management in national Planning." Currently working as a professor and chair of the Strategic Management Group in the Department of Management Studies, Dr. Sushil has a huge list of publications to his name including authoring and co-authoring nine books. He has a number of papers in referred journals and international conferences to his name. With rich consultancy experience in several organizations, Dr. Sushil specializes in areas like flexible systems management, strategic change and flexibility, creative problem solving, fundamentals of management and systems, EDP, and industrial waste processing to name a few. The founding president of Global Institute of Flexible Systems Management (1999), he is also a life member of several other national and international bodies including Systems Dynamics Society of India, Systems Society of India, and Indian Institute of Industrial Engineering.

Irénée N. Tiako is currently a lecturer of informatics in the Faculty of Economics and Management Science at University of Yaoundé II, Cameroon. Prior to current position he worked as a software engineer consultant for several companies in Yaoundé, such as PcSoft and InfoSoft. He is currently a PhD student in e-commerce at the National Advanced School of Engineering, ENSP-University of Yaoundé I, Cameroon and also affiliate with the Center for IT Research, Langston University in Oklahoma where he is working under the supervision of Prof. Dr. Pierre F. Tiako. He received a Diploma of Engineer and a Master Philosophy in information technology from the National Advanced School of Engineering, ENSP-University of Yaoundé I, Cameroon.

Pierre F. Tiako is a faculty member and director of the Center of IT Research of the Langston University School of Business in Oklahoma, USA. Prior to current position he worked as a Visiting Professor at Oklahoma State University, and as an Expert Engineer at INRIA, The French national institute for research in information technology. He also taught in several French Universities, including University of Rennes and Nancy. He is currently a member of the Institute of Electrical and Electronics Engineers, Inc. (IEEE); United States Association for Small Business and Entrepreneurship (USASBE); Association of Collegiate Business Schools and Programs (ACBSP); and the International Council for Small Business (ICSB). He received a PhD in software & information systems engineering from the French National Polytechnic Institute of Information Technology in Lorraine, France.

Chandana Unnithan is an associate lecturer in business information systems and project management. She has a master's degree by research in business computing and an MBA specializing in project management in global IT organizations. She has spent 14 years in the information communications technology sector working with IBM and TATAs of India. She is actively consulting and focussing on research relating to mobile applications in project and knowledge management processes within global information-systems-based, project-driven organizations. She has a special interest in comparative studies relating to evolution of mobile technologies, communications diffusion, development of applications and services in various economies.

Manfred Zielinski is currently an ICT administrator of research at the Rotterdam School of Management, Erasmus University Rotterdam, The Netherlands. Prior to joining the Rotterdam School of Management, he served as CTO of Avid AIM and e-business Manager at DEON (UTT Logistics). He has also served as a system administrator and Internet developer at the Erasmus University Research Institute for Decision Information System (EURIDIS). Throughout his career he has worked extensively with open source software (OSS) and is currently researching OSS for his MscBA at the Rotterdam School of Management.

Index

D

E

F

G

H

I